Simon W. Houlding

3D Geoscience Modeling

Computer Techniques for Geological Characterization

With 112 Figures

Springer-Verlag
Berlin Heidelberg New York
London Paris Tokyo
Hong Kong Barcelona
Budapest

Simon W. Houlding, MSc (Eng.), P.Eng.
8625 Saffron Place
Burnaby BC, Canada, V5A 4H9

LYNX Geosystems Inc.
400-322 Water Street
Vancouver BC, Canada, V6B 1B6

e-mail 74172,2245 @ compuserve,com

ISBN-13: 978-3-642-79014-0   e-ISBN-13: 978-3-642-79012-6
DOI: 10.1007/ 978-3-642-79012-6

CIP data applied for

Typesetting: Camera ready copy from the author/editor.

SPIN 10126638   32/3130 – 5 4 3 2 1 0 – Printed on acid-free paper

Preface

This book is a result of a career spent developing and applying computer techniques for the geosciences. The need for a geoscience modeling reference became apparent during participation in several workshops and conferences on the subject in the last three years. For organizing these, and for the lively discussions that ensued and inevitably contributed to the contents, I thank Keith Turner, Brian Kelk, George Pflug and Johnathan Raper. The total number of colleagues who contributed in various ways over the preceding years to the concepts and techniques presented is beyond count. The book is dedicated to all of them.

Compilation of the book would have been impossible without assistance from a number of colleagues who contributed directly. In particular, Ed Rychkun, Joe Ringwald, Dave Elliott, Tom Fisher and Richard Saccany reviewed parts of the text and contributed valuable comment. Mohan Srivastava reviewed and contributed to some of the geostatistical presentations. Mark Stoakes, Peter Dettlaff and Simon Wigzell assisted with computer processing of the many application examples. Anar Khanji and Randal Crombe assisted in preparation of the text and computer images. Klaus Lamers assisted with printing. The US Geological Survey, the British Columbia Ministry of Environment, Dave Elliott and others provided data for the application examples. My sincere thanks to all of them.

Credit for computer processing and visualization examples is due to Lynx Geosystems Inc. of Vancouver, Canada, for use of the 3D Geoscience Modeling System, and to Wavefront Technologies Inc. of Santa Barbara, California, for use of the Data Visualizer System.

In closing, I would not have completed this project over the course of a long year of late nights and working weekends without the encouragement and patience of Frances, Jean and David.

Simon Houlding
Vancouver, March 1994

Table of Contents

Part II. Applications in the Geosciences

Introduction

Geoscientists wear many hats, e.g. geologist, hydrogeologist, geophysicist, environmental engineer and geotechnical engineer. Whether the end objective is waste site characterization, contamination assessment, groundwater flow simulation, mineral resource evaluation, reservoir engineering or tunnel design, our common denominator is a concern with investigating and characterizing the geological subsurface.

This text is about techniques for creating computer representations of this complex environment. We call the results *models* because the term is less of a mouthful, although their derivation is not computer modeling in the generally accepted sense. To many, modeling is synonymous with numerical analysis and simulation of continuous systems represented by differential equations. There are no differential equations here; we are concerned instead with computer techniques for *geological interpretation, geostatistical prediction and graphical visualization* of inaccessible geological conditions from limited information. The over-riding concern is that our models should always be as representative as possible of the real conditions. In many cases they are used to justify decisions involving large capital expenditure.

We call this process of interpretation, prediction and visualization *geological characterization*. In the past many of us have been forced to use pencil, paper, eraser and planimeter to achieve this. Some of us have duplicated this essentially 2D approach on a computer using CAD or GIS tools. Relatively few have used computer systems such as those described within to approximate geology as a set of sections, or surfaces, or blocks, or gridcells. A singular deficiency of many of these computer techniques is a lack of suitable avenues for the geoscientist to impose interpretive control on the solution.

The geological subsurface is not a set of surfaces or sections or blocks. Neither is it a continuum except in the broadest sense of the term. Every geological unit is an irregular volume with distinguishing *characteristics*. The boundaries between units create discontinuities that are further complicated by faulting, erosion and deposition. Within this heterogeneous complexity we are concerned with *variables*, such as contaminant concentrations, mineral grades and geomechanical properties that are (generally speaking) continuously variable within the volume of a unit, but discontinuous across boundaries. At the scale of interest, the geological subsurface is thus both *discrete* in terms of geological volumes, and *pseudo-continuous* in terms of the spatial variation of its variables. The objectives of many geological characterizations are concerned with volumes defined by the occurrence of variables that exceed a specified *threshold*, orebody definition for example, or a contaminated soil volume, or a groundwater plume volume. Since the spatial distribution of a geological characteristic frequently influences the spatial variation of a variable we are forced to consider both.

To adequately represent this complex environment on a computer we must consider a semi-infinite continuum made up of discrete, irregular, discontinuous volumes which in turn control the spatial variation of variables. From a computer perspective, the geometry of irregular geological volumes is equivalent to *vector information*, and the spatial variation of variables to *raster information*. Integration of these two information types in a 3D context provides a platform for effective geological characterization on a computer. Advances in graphics workstation and software technology have made implementation of such an approach a practical proposition. Within this 3D context we can successfully integrate the necessary geological characterization tools: management of spatial information, graphical tools for geological interpretation, geostatistical tools for spatial analysis and prediction, analysis of volumes and their contents, and enhanced graphical visualization. This computerized process is termed *geoscience modeling*.

To further complicate matters, geological investigation is invariably constrained by cost, forcing us to derive models from limited information. The terms *interpretation* and *prediction* carry implicit connotations of *uncertainty*. An ability to quantify the uncertainty associated with both geological interpretation and geostatistical prediction plays an increasingly important part in the geological characterization process. It provides a logical basis for risk assessment in subsequent planning activities. In a scenario of limited information prediction techniques seldom work well on their own. They require large amounts of source data to achieve representative results, a luxury that is seldom available. The approach taken throughout this text is to recognize that efficient computerized characterization requires an integration of both *interpretation* and *prediction* techniques.

Many of the techniques covered in this text are not new in themselves. Geostatistical prediction evolved within the mining sector and has been successfully applied for several decades. Yet even within the mining sector the implementation of geostatistical uncertainty has not been fully realized. In other geoscience sectors the capabilities of geostatistics are only now beginning to be recognized. The implementation of enhanced visualization techniques within the geosciences is relatively new, although the technology has been available in other fields, such as graphics animation, for several years. What has been missing across the geosciences is an integrated, generic computer approach to geological characterization that incorporates all the required tools in a true 3D context.

In recent years the impetus behind development of such an approach has been provided by the environmental sector. On the one hand, we are concerned with remediating the subsurface before it impacts on our well-being. On the other, we are only too aware of the costs involved for both investigation and remediation, and the ensuing strain on our resources. There is an urgent requirement to minimize the cost of both site investigation and remediation by improving our characterization techniques. Much has been published in the environmental sector on the subjects of remediation and investigation but very little on characterization. Bridging this gap, and dissemination across the geosciences of an approach that achieves the objectives in all sectors, are the principal missions of this text.

A word of caution to the enthusiastic geoscience modeler! An excerpt from a paper by Philip (1991), a soil scientist, regarding the validity of computer-based models in a real-world context is appropriate.

> *From the viewpoint of natural science, and indeed from any viewpoint concerned with truth, a disquieting aspect of computer-based modeling is the gap between the model and the real-world events. There is reason to fear that the gap will not grow smaller and that worry about it will ultimately just fade away.*
>
> *Of the making of models there is no end. The components may be well-founded in natural science, may be a crude simplification, or may be no more than a black box which has taken the modeler's fancy, and, equally, the prescriptions for fitting the components together may or may not be well-based. Beyond the question of the validity of the model's machinery, there are also difficult questions of the quality both of the parametrization of the components of the model and of the data input.*
>
> *Many modelers, in their enthusiasm, seem not to be fazed by these problems but push on regardless. Many appear to be possessed by their models, to have little interest in the real-world processes, and to be oblivious to the unrealities of their parametrization and input data.*

We are well advised to bear these concerns in mind as we create complex computer models in the course of our geological characterization. The modeling techniques presented here do not constrain us to a realistic solution. Our models are founded to a large extent on interpretation and prediction from limited data, and the available data contain their own components of error and uncertainty. Besides the obvious step of cross-checking results wherever possible, there are three options available to us to guard against creating *geoscience fiction*.

The first is to keep the computerized process as interactive and accessible to the user as possible. A geological entity should be stored and visualized by the computer the same way we define it, a *what you see is what you get* approach, as opposed to a black-box solution where we feed parameters at one end and await results at the other, possibly lacking a full understanding of the internal manipulations. The second is to make extensive use of computer visualization capabilities, a realistic graphical simulation of what we have interpreted or predicted can very quickly highlight inconsistent, erroneous or unreasonable assumptions in our models. The third option is to emphasize, and to quantify wherever possible, the uncertainty associated with any of our results. This increases our awareness that any attempt at characterization is only one of a range of possibilities. The computer techniques discussed within make extensive use of all three options; used intelligently they at least ensure a modeling solution that is reasonable in a real-world context.

The text is divided into two parts. The first covers derivation and integration of suitable data structures and characterization tools in a logical, practical, three-dimensional context. It provides the necessary background to apply these tools effectively. The second provides examples of application of the computerized process to a range of geoscience problems. These will assist readers to relate the process to their own particular requirements. The text closes with a summary of new and evolving concepts and techniques aimed at more efficient computer representations. With the present rapid evolution of graphics workstation technology, some of these will likely have practical application in the near future.

The text attempts to provide a general overview of the various computer techniques applied in the past to geological characterization, together with their advantages and limitations. The body of the text focuses on a generic, integrated solution that has application across the geosciences. The text also attempts to maintain a hands-on approach throughout, with emphasis on practical application rather than theoretical detail. The objective is to provide a practical reference to the application of geological characterization computer techniques for both professionals and students in the geosciences.

Part I. Computer Techniques for Geological Characterization

1 The Geological Characterization Process

1.1 Prediction of Geological Variables

The primary objective of almost every geological characterization is concerned with predicting the *spatial variation* of one or more geological *variables*. In this context, a variable is defined as any property of the geological subsurface that exhibits spatial variability and can be measured in terms of real numeric values.

In a mineral resource evaluation of an ore deposit a typical variable is a *mineral grade*, expressed as a ratio or percentage by weight of metal to ore. Figure 1.1 is an example of mineral grade variation on a section through an ore deposit. An equivalent variable for a coal deposit is a measure of *coal quality* such as thermal unit rating or ash content. In a contaminated site assessment the variables of primary concern are soil or groundwater *contaminant concentrations*. A groundwater flow study is concerned with the distribution of *hydraulic parameters* such as conductivity; characterization of a high-level waste storage site with *geomechanical properties* such as fracture density, and so on.

All of these variables vary in some more or less continuous manner with spatial location. Ideally, we require a continuous measure of their variation throughout the region of interest. We can then perform a final evaluation, of orebody limits, or contaminated soil volumes, or groundwater flow, with precision. In the real world, however, we are governed by the economic considerations of subsurface investigation and have to settle for as many isolated, discrete measurements of the variables of interest as we can afford. We use the generic term *samples* to refer to these measured values at known locations. To represent a continuous measure of a variable we apply various prediction techniques to the samples to predict unknown values at intermediate points.

The investigative sources of this type of *sample information* are typically application dependent. They vary from mineral assays of borehole core samples from an ore deposit, to soil and groundwater contaminant concentration samples, to calculations of fracture density from mapping of exposed rock surfaces, to rock quality (RQD) values or cone penetrometer test results in a geotechnical study, to the results of well pump tests in a hydrogeological study. To make effective use of this information every sample must be located as precisely as possible in three-dimensional space. Since some of the variables we deal with, such as contaminant concentrations, may change over the time-scale of our investigation, samples may

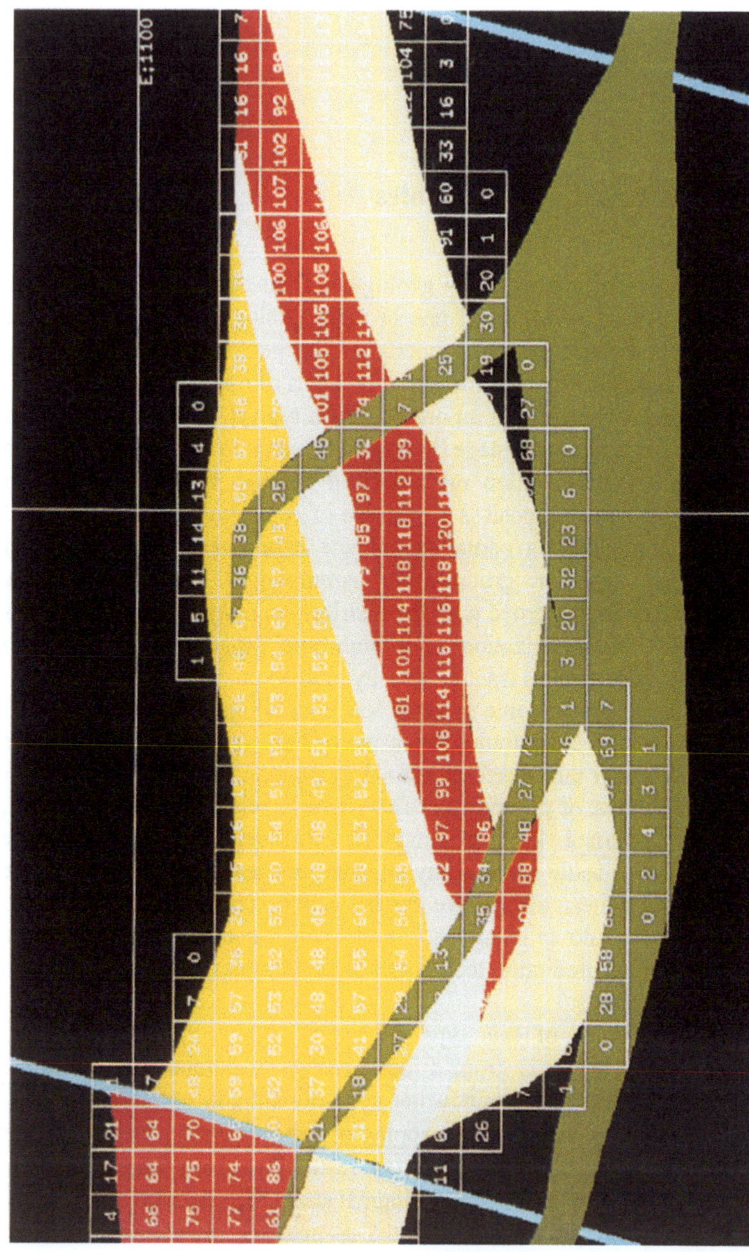

Fig. 1.1   Section through a 3D computer representation of complex, faulted, intrusive geology, with the spatial variation of a geologically influenced variable overlaid

also need locating in the time dimension. And since some sampling techniques are more reliable than others they may need to be classified by source and sample type.

1.2 Interpretation of Geological Characteristics

In a regional sense the spatial variation of an attribute is seldom determinate in mathematical terms, due largely to geological complexities that we discuss in Chapter 2. In particular, the variation is invariably influenced to some degree by stratigraphic and structural features and other discontinuities. For example, the variation of a contaminant concentration in a sand layer may be noticeably different from its variation in adjacent clay or silt layers. In a hydrogeological context the differences may be more dramatic; the conductivity values within an aquifer are likely to be several orders of magnitude greater than those in the adjacent confining layers. Similarly, the variations of mineral grades in an ore deposit are frequently influenced by lithological or mineralogical discontinuities.

These discontinuities are further compounded by the geological structure of a region. The strata themselves may be subjected to abrupt spatial discontinuity and dislocation by, for example, faulting, shearing or overturning. The combined effect of these stratigraphic and structural discontinuities is that, while the variation of a variable may be more or less continuous (and hence predictable) within individual strata, its variation in a more general sense throughout the geological continuum may be highly discontinuous. No matter what prediction technique we apply to a variable, we are unlikely to achieve an acceptable result unless we take these geological effects into account. Our prediction of spatial variation must be controlled by a prior knowledge of stratigraphic, structural and other relevant information.

This requires that our investigation of the subsurface include sufficient observations of stratigraphy, structure and other controlling factors to allow us to interpret these conditions with an acceptable degree of confidence (or its reverse, uncertainty). Interpretation essentially involves a *discretization* and *qualification* of geological space into irregular volumes with similar distinguishing *characteristics*. The controlling characteristic varies with application, the most frequently used being *lithology*, or soil and rock type. In an ore deposit evaluation we might also consider the type of *mineralization* as a controlling characteristic. In Fig. 1.1 the controlling characteristic for the variation of metal grades is the mineralogy of the four ore types that comprise the deposit. At a lower level, to estimate the cost of a large excavation, we may only be concerned with qualifying the subsurface by *excavation qualities* such as soft, firm or hard material.

There are numerous investigative sources for this type of *observation information*, including borehole core logs, well logs, geological mapping of exposed surfaces, topographical surveys, geophysical borehole logs and remote geophysical surveys. At a more subjective level, the list also includes our geological experience and intuition, and our general knowledge of the geological

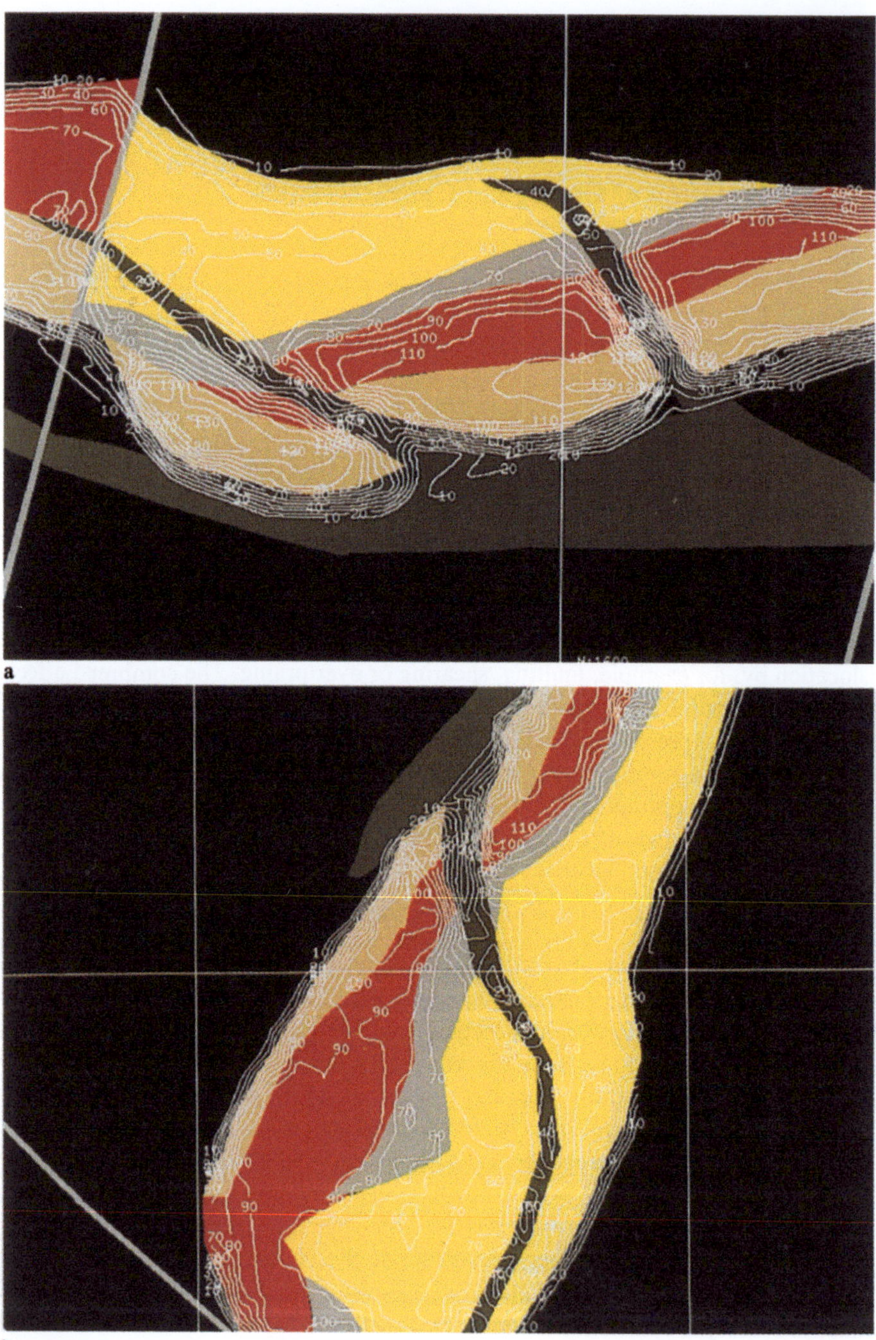

Fig. 1.2 **a** Contours (isolines) of the mineral grade variable of Fig. 1.1 displayed on a plan section of geology;   **b** the same contours displayed in vertical section

sequence and history of a region. All geological characterizations use at least one of these sources of observational information and most use several. In some instances the sample information may also be used in conjunction with characteristic observations for interpretation purposes.

Since these observations are subsequently used to interpret the geometry of discrete geological volumes, they must be located in three-dimensional space. However, the degree of precision with which this can be achieved varies considerably between the investigative sources. A topographical survey might be regarded as being reasonably precise, while the results of a geophysical survey can be regarded at best as *fuzzy* in terms of locational precision. The location of a core log observation from a deep borehole with variable trace geometry would fall somewhere between these two extremes in terms of precision. Thus, there is a need to qualify observation information according to its source so that it can be appropriately *weighted* in the (typically subjective) interpretation process.

1.3 Uncertainty and Risk

In summary then, our geological characterization must predict and interpret potentially complex geological conditions that are both discrete and pseudo-continuous in terms of their spatial variability. And the source information for this exercise is derived from a variety of investigative sources of variable quality, much of it subjective. And last, but not least, the amount of source information is invariably limited relative to the complexity of our problem. All of which means that our characterization results are subject to varying degrees of uncertainty.

Quantification of this uncertainty is a key requirement of the characterization process. Any results we produce are of limited usefulness unless we know how *good,* or representative of the real conditions, they are. Measures of uncertainty can be related to risk and used, for example, in remediation planning for a contaminated site to assess the risk of any contamination remaining after implementation of a particular remediation plan. From a mining perspective, an ore deposit with a predicted average grade of 5 g/T and a standard error of 1.5 g/T is probably a more viable, and less risky, mining proposition than another with a higher predicted average of 7 g/T but an error of 4 g/T. (The term standard error, also referred to as standard deviation, denotes a statistical measure of uncertainty.) In similar vein, an artesian water supply with a predicted potential of 10000 l/hr becomes a much less attractive proposition with the additional information that the potential could be as low as 3000 l/hr for a 90% confidence level. To be used effectively, every predicted or interpreted result should be accompanied by a measure of its uncertainty

Quantification of uncertainty can also be applied to the site investigation phase. Knowledge of the spatial distribution of uncertainty associated with prediction of a variable, based on the available samples and observations, can help us to optimize the location of subsequent, more detailed investigation. Ideally, this should be focused where it will have the greatest return in terms of reduced uncertainty. It can also help us to decide whether further investigation is justified in terms of the reduced uncertainty that might result.

To achieve a true measure of uncertainty requires that we combine the uncertainties associated with both prediction of variables and interpretation of characteristics. The former is more readily achieved since prediction is generally based on measured numerical values at known locations. In contrast, interpretation frequently includes subjective assessment of observations and its uncertainty is less easily quantified. These aspects are discussed in more detail in Chapter 10.

1.4 Spatial Analysis

Evaluation of the results of geological characterization typically involves *spatial analysis* of the variation of one or more variables. The objective is to delineate a volume whose geometrical limits are defined by the occurrence of variable values that exceed a specified *threshold* (or cutoff) value. The volume encompasses all points where the variable exceeds the threshold, and is therefore enclosed by the *isosurface* (3D contour) of the variable for the threshold value. We need to determine not only the volume contained by such an isosurface, but also the average value of the variable within and a measure of its associated uncertainty.

We generally refer to such an isosurface as a *soft boundary* since it does not coincide with any clearly definable geological discontinuity. In contrast, the boundaries that define the discrete volumes of geological units and structural discontinuities are considered *hard boundaries*. Fig. 1.2 displays contours (isolines) of a mineral grade variable on geological sections. These contours are, in effect, a series of soft boundaries; in this case, overlaid on the hard geological boundaries of mineral ore types. In figure 1.3 several of these contours have been extended into 3D *isosurfaces* of the variable, viewed in perspective.

In an environmental engineering context the threshold is typically defined by the *action level* for a soil or groundwater contaminant concentration. The resulting isosurface defines the limits of contaminated material that is potentially harmful and must be remediated. The enclosed volume, its extent and its degree of contamination generally dictate the cost of remediation. In some cases we may extend our analysis to provide a breakdown of the isosurface volume by geological type. For example, there may be significant cost differentials between remediation of contaminated sand, silt and clay materials. This requires a volumetric intersection of the isosurface with discrete geological volumes.

In a mining context the threshold is typically defined by a *cutoff grade*, below which ore cannot be economically extracted and processed. The resulting isosurface defines the limit of ore that is (theoretically) economically mineable. Its volume, average grade and geometry will ultimately determine whether the deposit represents a viable mining opportunity. In practice, a deposit may consist of several ore types, defined by their type of mineralization. As before, a breakdown

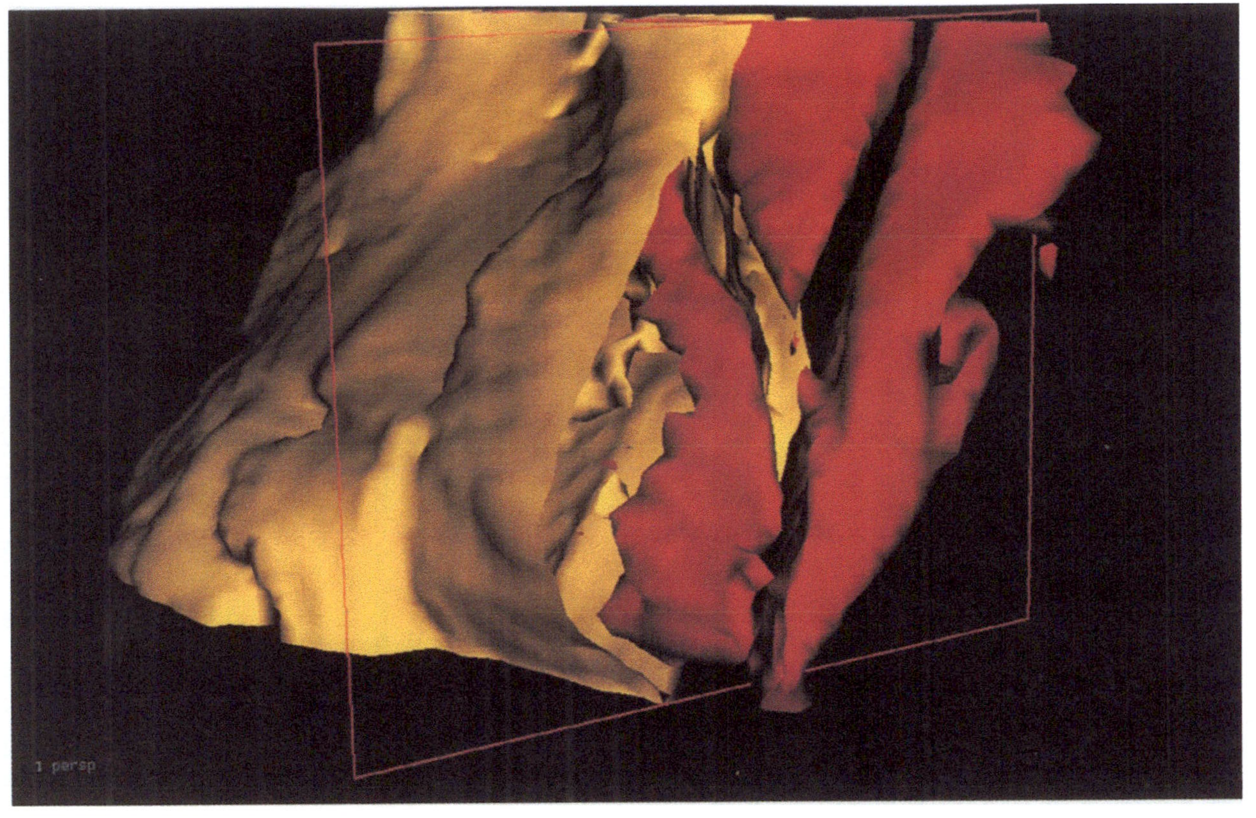

Fig. 1.3   3D view of two isosurfaces of the variable displayed in Fig. 1.1; the outer (lower value) isosurface has been partially cut away; these isosurfaces are the 3D equivalent of the 2D isolines displayed in Fig. 1.2

of the isosurface volume by ore type requires a volumetric intersection. And at a later stage, for mine planning purposes, we need to intersect these volumes with the geometries of proposed mining excavations, which requires a three-stage intersection of volumes.

In some cases we may also need to consider the intersection of two or more isosurfaces to obtain a volume defined by several thresholds for individual variables. In the computer context we use the term *volumetrics* to refer to the types of spatial analyses described above.

In a tunneling (or underground mine development) context we may be less concerned with variable isosurfaces, and focus instead on ensuring that our proposed excavation geometry does not intersect poor rock or soil conditions. We might do this either by visual reference or by analyzing the intersection of the excavation geometry with the geological units of concern.

In a hydrogeological context we are typically not concerned at this stage with spatial analysis, although an isosurface of porosity that defines the most likely migration pathways of a contaminant might be useful. We are more concerned with transferring the results of our geological characterization to a groundwater flow model for further analysis.

Visualization also figures largely in our spatial analysis of results. Whether our visualizations are on paper or a computer screen they are the primary medium for assessing, correlating and verifying spatial information, enhancing our appreciation of complex conditions, and communicating our results to others. The key to successful visualization is the availability of a wide range of graphical display tools that can be applied in appropriate combinations.

1.5   The Generic Characterization Process

The basic requirements and procedures of geological characterization can be summarized as a sequence of clearly defined steps. Whether we employ a conventional pencil and paper approach, or an integrated 3D computer modeling approach, these steps are essentially the same. The information flow is summarized schematically in Fig. 1.4 and the steps are summarized as follows.

· *Management, Correlation and Integration* of the various information sources available from site investigation. The most important, and potentially most time-consuming, aspect of this step is *spatial correlation* of the available sources of investigative information.
· *Review and Analysis* of the information sources in terms of their quality, sufficiency, scale and spatial variability. This provides an initial appreciation of the complexity and magnitude of the conditions we have to deal with. It also allows us to identify the geological characteristics that influence the spatial variation of the variables of interest.
· *Interpretation* of geological stratigraphy, structure and other relevant factors, based largely on observation of characteristics. The result is a discretization of

the subsurface continuum by controlling geological characteristics. The operative term here is *interpretation*, because we have limited information we are forced to apply our geological intuition and experience.

· *Prediction* of the spatial variation of relevant variables, based on sample information, geological interpretation and appropriate prediction techniques. The operative term in this case is *prediction*, the true conditions are seldom determinate from limited information and our predictions are always subject to some degree of uncertainty.

· *Spatial Analysis* of the interpreted and predicted information, generally in terms of volumetrics and visualization procedures, and/or transfer of the characterization results to the relevant end process, such as mine planning, groundwater flow modeling or remediation design.

A common feature of all approaches to characterization is their reliance, throughout the process, on *visualization*. At one extreme this is reflected by hand-colored geological sections on paper with sample locations and hand-drawn

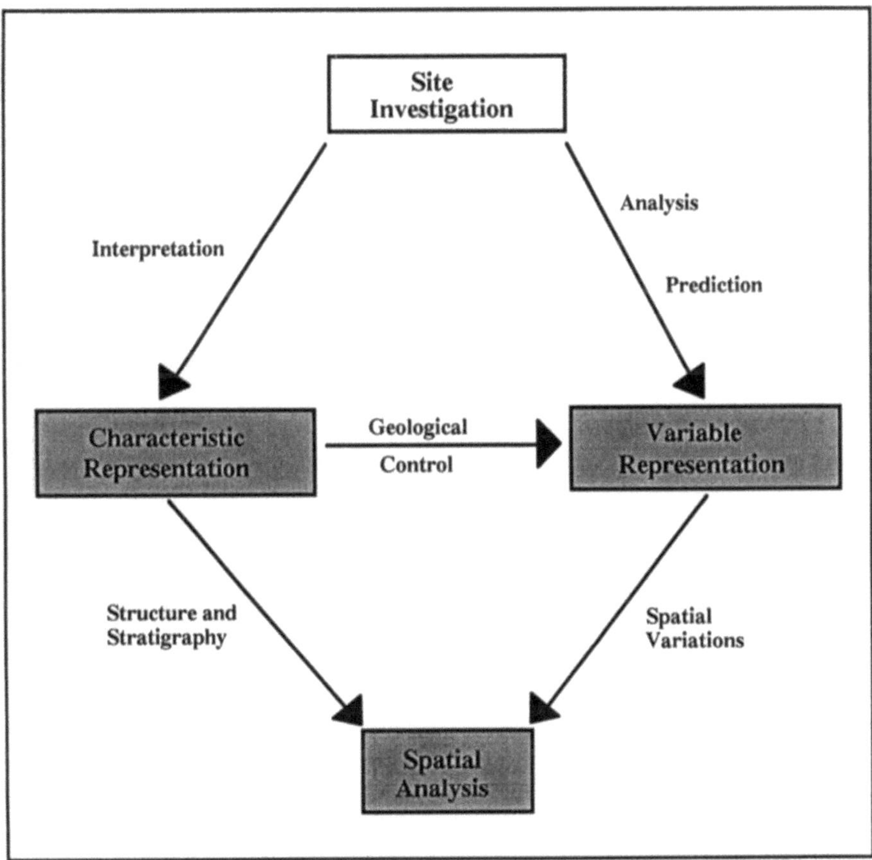

Fig. 1.4 Summary of the generic process of geological characterization

contours; at the other extreme by a 3D computer visualization of geological conditions with cut-aways, transparencies and isosurfaces.

This definition of the geological characterization process is familiar to anyone who has applied the conventional approach. Some of us may already be performing one or more of these steps with the aid of computer technology. For example, storing the information sources in a relational database or spread-sheet system; generating sample plans and contours in a geographic information system (GIS); creating sectional interpretations of geology in a CAD system; or analyzing samples with a statistical or geostatistical package.

We are concerned here with demonstrating how all of these steps can be extended and integrated on a computer, given the appropriate *integrating technology*, and how the necessary interpretive interaction with the process can be maintained throughout. To provide the necessary background it is useful to first review the conventional approaches and the evolution of various computerized approaches, together with their advantages and limitations.

1.6 Conventional Approaches to Characterization

We are all familiar with the tedious process of manually scaling and plotting the locations of samples and observations, and their values, onto a sheet of paper that represents a horizontal or vertical section through the subsurface. If the locations must first be projected from (say) a borehole trace with variable geometry, then the process becomes even more laborious.   This is the typical *management, correlation and integration* step in the conventional approach to geological characterization, performed entirely on paper. The subsequent step of *review and analysis* is conducted by visual reference to these results, possibly augmented by statistical and geostatistical analysis of the available sample values.

If we are dealing with complex geological conditions, then the next step involves a subjective *interpretation* of stratigraphic or structural discontinuities, by connecting the locations of like observations of a particular characteristic with hand-drawn lines. This may involve much erasing and redrawing as we iteratively approach a result that we intuitively feel is representative of the real conditions. The results typically define the boundaries of intersection of one or more geological units with our section. We subsequently perform a 2D geological qualification of the section by coloring or shading the interior of these boundaries appropriately to reflect their geological characteristics.

The *prediction* step involves manually contouring sample values in the plane of the section. If the geological influences are significant, then we might contour subsets of samples from different geological units independently. The prediction technique is approximately equivalent to a linear interpolation of values on the section, it ignores the third dimension entirely. If we repeat these steps on a sufficient number of sections that adequately represent the region of interest, then, in effect, we are approaching a 3D solution. In practice we seldom have sufficient information or resources to achieve this.

In the *spatial analysis* step we are reduced to planimetering the area of the contour that represents our threshold value on each section, and multiplying these results by the section spacing to approximate the volume of interest. To obtain an average variable value for this volume would require repetition of this exercise in planimetry   for each contour interval. Further, attempting to calculate the uncertainty attached to these results is generally beyond our capabilities.

An alternative conventional approach, limited in application to relatively simple layered geological conditions, is the preparation of contour maps of the depth or elevation of geological surfaces, or isopachs of thickness. This approach typically involves characterization of individual geological units or surfaces in plan, rather than a general characterization of the subsurface. Visualization is limited to contours rather than the more intuitively realistic geological sections of the sectional approach.

All the conventional approaches are tedious to implement by hand and do not lend themselves readily to revision to accommodate new information. However, within their limitations, which we discuss later, these conventional approaches to geological characterization have served us well. In the past many resource-based projects have been launched and successfully put into production on the basis of results like these. Moreover, a surprising number of characterizations are still conducted this way.

1.7 Evolution of Computerized Approaches

Dwindling mineral and oil reserves, increasing concern with environmental contamination, and steadily rising costs in all sectors have combined over the years to generate a requirement for improved precision and efficiency in the geological characterization process. Geoscientists in the energy and mining resource sectors recognized many years ago that the process was a candidate for computerization. Since then, the evolution of a computerized approach has taken three distinct and separate paths, influenced by the needs and requirements of different geoscience sectors. It is only recently, with the advent of powerful graphics workstations, that these different approaches to characterization have converged and started to proliferate in a more generic form.

In metal mining the focus of computerization has always been on predicting the variation of mineral grades, with particular application in the early stages to large porphyry (disseminated) base metal deposits. Computerization involved representation of the subsurface by a 3D set of uniform rectangular blocks, not surprisingly called *block models*, in effect a regular 3D gridded approach. Mineral grades, which represented the average value for each block, were predicted at the centroids of the blocks from sample information, using geostatistical or other prediction techniques. This was essentially analogous to a raster approach to computerization since it represented a continuous measure of the variation of a variable throughout the subsurface and a fixed spatial density of information. If necessary, a monetary value could be determined for each block based on its mineral grades and the costs of excavation and mineral processing. An optimizing algorithm could then be applied to the block model to determine the limits, in terms of block geometry, of the most profitable mining excavation, called *pit optimization*, a sophisticated form of spatial analysis. Thus, the emphasis in computerization in metal mining was weighted towards characterization by *variables* rather than characteristics.

Subsequently, this approach was extended to include geological influences by assigning a characteristic value to the centroid of each block, which could be used to apply geological control to the prediction of mineral grades. This involved significant approximations since the precision with which stratigraphy and structure could be represented was governed by the uniform size of the blocks (cf. Fig. 1.5). This highlights the principal deficiency of attempting to represent geology with a raster-like approach. The deficiency has since been remedied to some extent by the implementation of *linear octree algorithms* that allow progressive subdivision of blocks wherever greater definition is required. However, this becomes a computationally intensive and time-consuming process, even for a powerful computer.

The theory of *geostatistics* was developed and implemented within the mining sector in the 1960s as a computerized statistical approach to the spatial analysis and prediction of variables. It is based on the *theory of regionalized variables,* developed by Georges Matheron in France. It provides the *best estimate* (statistically speaking) of the unknown value of a variable at a specified location based on measured sample values at known locations. Geostatistical prediction, or *kriging* as it is generally called, can be readily applied in a 3D context, can be tailored to the spatial variability of the available sample information, and provides a measure of the uncertainty associated with each predicted value. It has been successfully applied to many mineral and coal deposits, and verified by subsequent mining operations. An anomaly in its implementation has been a general reluctance by the mining sector to make full use of the available measure of uncertainty, possibly because high grades are regarded as good news for investors, whereas high uncertainty is not! Geostatistical algorithms have been refined and extended over the years and are now generally regarded as appropriate techniques for the spatial prediction of variables in most applications.

In the energy sectors, including oil, gas and coal applications, there has been a much greater emphasis on representation of geological stratigraphy and structure. These applications developed *surface-based* approaches to computerized geological characterization, typically derived from observations of *characteristics* rather than variables. In oil and gas exploration concern has focused on identifying and representing potential reservoir formations; in coal mining, on defining the thickness and extent of coal seams and intermediate waste material. Initially, these representations involved regular,horizontal 2D *gridded surface models*. Each stratum or surface of interest in the formation is defined by elevation or depth and/or thickness values at the nodes of the grid, obtained by a variety of prediction techniques. This is the computer equivalent of the conventional approach of producing contours of structural surfaces or isopachs, except that a single grid model may contain many surfaces or strata. A gridded surface model is essentially

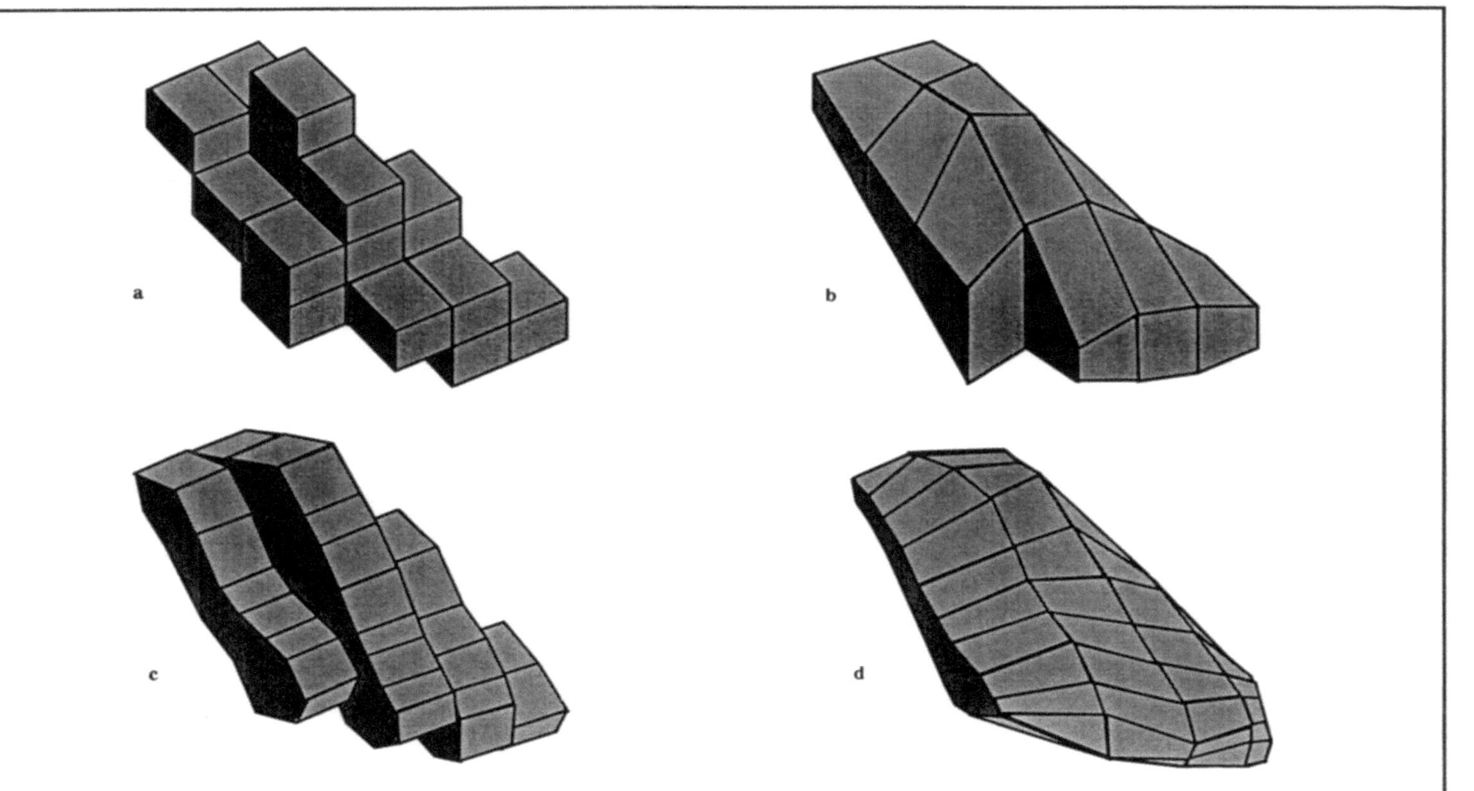

Fig. 1.5   Different model representations of an irregular geological volume: **a** regular grid (block) approach; **b** deformed grid approach; **c** sectional approach; **d** volume model approach

a raster-like approach. It assumes a continuous variation in the horizontal plane, and the degree of horizontal detail it can represent is governed by the grid-cell dimensions.

The gridded surface approach has since been refined by the implementation of 3D *deformed grid models* that are regular in plan, but irregular in the third (vertical) dimension, a combination of raster-like and vector representation (cf. Fig. 1.5). This provides an improved representation of geology, as discrete gridded volumes rather than surfaces, but creates difficulties in the integration of variable prediction techniques. More recently, similar approaches have been developed using triangulated, rather than gridded, surfaces. These provide greatly improved efficiencies in terms of flexibility and detail, and represent a true vector based approach to the problem. However the principal deficiency of most surface-based approaches is a general inability to readily accommodate structural discontinuities such as faulting, folding and overturns.

Until recently, the use of computerized spatial prediction techniques in the energy sectors has, in general, been limited to variations of the familiar distance-weighted contouring algorithms, extended to three dimensions in some cases. The application of geostatistical prediction is growing, but lags behind the mining sector.

This spate of computerization in the energy and mining resource sectors began in the late 1960s using large mainframe computers. It was justified by the capital-intensive nature of most resource-based projects. The various approaches have evolved considerably in sophistication since their inception and have recently begun to merge into more generic solutions that combine their individual advantages.

In contrast, computerization of the characterization process in the other geoscience sectors began relatively recently with the advent of graphics workstations and personal computers. Without the resource sectors' advantage of 25 odd years of experience with computerized approaches, computer applications in these sectors are still, in general, limited to a duplication of the conventional approaches. These applications typically make use of a mixed solution of readily available CAD, GIS, spreadsheet and database technologies without any serious attempt at 3D representation. As a result, much of the tediousness of manual characterization is eliminated, but the basic assumptions and their inherent deficiencies and limitations remain. The use of geostatistical techniques has been promoted and encouraged but in most cases is limited to 2D applications. Visualization techniques have advanced dramatically in recent years, principally due to developments in animation graphics that are now readily available to other industry sectors. The key to successful 3D visualization lies in ensuring that all source information and model data structures are compatible with these visualization techniques, which can generally   accommodate both vector and raster information. Thus, any modern computerized approach should provide a 3D graphics toolkit for color-mapping, light-sourcing, transparency, cut-aways, rotations, etc., which can be applied to point, line, polygon, surface, volume, grid and isosurface geometries. The advantages of this type of visualization, in terms of an enhanced appreciation and understanding of complex subsurface conditions, are obvious from the examples provided in this text.

1.8 Advantages and Limitations

All the approaches to geological characterization discussed above include a number of assumptions and varying degrees of approximation and simplification, and all have their particular advantages and limitations. As we shall see, even the conventional approach offers a certain unique advantage. In Chapter 2 we discuss some of the factors that make computerization of the process a complicated task. We arrive at the conclusion that a precise representation of every possible condition that might arise is beyond our current capabilities. However, by combining the advantages of some of the existing approaches in an integrated, computerized 3D context, and thereby eliminating most of their disadvantages, we achieve a powerful new approach to characterization that satisfies most of our current requirements. A summary of their advantages and limitations is appropriate at this stage.

The limitations of the conventional approaches are obvious. The sectional approach approximates the subsurface continuum as a set of 2D sections, and implicitly assumes that the conditions on each section extend over a finite distance in the third dimension. Correlation of information between sections is difficult and tedious, and if we require a new section at some arbitrary orientation, it must be virtually reinterpreted and repredicted. 2D spatial prediction of variables is performed on sections using projected sample locations rather than their true locations. Hand-contouring is a prediction technique that employs the equivalent of a linear interpolation algorithm. It frequently ignores geological influences, and invariably ignores directional (anisotropic) effects and spatial trends. Spatial analysis is generally limited to planimetering of contours and the resulting sectional areas are assumed constant over a finite distance. The effects of some of these approximations naturally diminish as the number of sections is increased and their spacing reduced. In reality, however, this seldom happens owing to the tedious nature of the process.

The advantage of this conventional approach is that it employs what is still the most logical and efficient technique for geological interpretation (on sections) and, hence, a natural avenue for the geoscientist to impose interpretive control on the results, a facility in which many computerized approaches are deficient. A secondary consideration is that most of us are incapable of interpreting complex conditions simultaneously in three dimensions, irrespective of whether or not we are working on a computer. We operate more efficiently if we reduce this process to several steps, beginning with the familiar approach of sectional interpretation in two dimensions.

The regular 3D grid model approach, or block model approach as it is called in the mining sector, provides an appropriate data structure for representing the spatial variation of a variable. It allows raster type information to be predicted and stored at whatever density is appropriate to the available samples and their spatial variability. However it is entirely inappropriate for representing geological structure and stratigraphy, which in reality seldom occur with block-like geometry. Also, in practice, the optimum grid-cell dimensions for representing the spatial variability of variables are seldom adequate for representing geology with acceptable precision.

The use of linear octree data structures that allow progressive subdivision of grid cells where necessary for improved representation of discontinuities is a promising approach. However, even with the power of modern graphics workstations, this approach is too computation-intensive when applied to many geological conditions. For example, a faulted, inclined, thin-seam orebody may be only centimeters thick but several kilometers in extent, requiring an inordinate number of subdivided cells for acceptable representation. In addition, the use of variable cell dimensions conflicts with several aspects of geostatistical prediction theory.

The regular 3D grid model therefore provides an acceptable approach for representing a continuous measure of the spatial variation of a variable. Provided that it is first, used in conjunction with appropriate prediction techniques and second, that it is suitably integrated with a more acceptable approach for representing geological structure and stratigraphy.

Geostatistics is established as a spatial prediction technique in the mining sector and has growing acceptance in the energy and environmental sectors. As we shall see, it lends itself to integration with geological control and can account for spatial anisotropy and underlying trends. It is also *justifiable* as a prediction technique since the prediction parameters are based on the spatial variability of the samples themselves. This factor is particularly important in the characterization of contaminated sites where all results may have to be legally defensible. Last, but not least, it is the only prediction technique in general use that is capable of providing a measure of uncertainty for a predicted value. Geostatistics has certain disadvantages in some situations, to which other techniques may be more applicable, but in general these are outweighed by the advantages. We discuss geostatistics in detail in Chapters 5 and 8.

The gridded surface approaches used by the energy sectors to represent geological structure and stratigraphy work well for relatively simple formations. They essentially require that geology be representable by continuous surfaces, whether these are subsequently translated into volumes or not. This creates difficulties, which are not insurmountable, and approximations in. accounting for fault discontinuities or discontinuous (pinched-out) geological units. However, the difficulties of representing overfolded or thrusted or steeply dipping (near-vertical) structures are virtually insurmountable. In particular, surface-based grid model approaches lack an efficient, direct avenue for the geoscientist to apply interpretive control; the internal computer representation of geology is divorced from the interpretive process. A more efficient approach is required to deal with geological complexity. We elaborate on these requirements in Chapter 3.

1.9 Errors and Approximations

Because so many of the existing approaches utilize grid modeling to a greater or lesser degree, a discussion of the approximations inherent to representing geology by rectangular discretization is called for. A series of 2D tests on the discretization of relatively simple shapes is sufficient to demonstrate the significance of these approximations. Extension of discretization into the third dimension merely magnifies the potential for error. The objective of these tests is to demonstrate the potential for error, they are not intended to produce definitive results. The findings are applicable in some degree to all forms of grid modeling, including regular 3D grids, variable (octree) grids and vertically deformed grids.

The tests are performed with the simple polygonal shapes reported in Fig. 1.6a. We can imagine these to represent the areas of intersection of geological units with an interpretation plane. The tests involve discretization of the areas into a range of rectangle sizes; the relative rectangle sizes for discretization are also reported in Fig. 1.6a. We are interested in two aspects of each discretization: how closely it represents the total area enclosed by the polygon and how accurately the polygon boundary is represented. A grid cell is included in the discretization if its centroid falls within the boundary and excluded if not. This is common practice in grid modeling. Figure 1.6b illustrates a discretization of one of the shapes considered. The area error is determined as the absolute value of the difference between the discretized area and the polygonal area. The boundary error is determined as the average of the erroneously included and excluded areas resulting from discretization.

The results of the tests are reported numerically in Table 1.1 and graphically as upper-bound error envelopes in Fig. 1.7. To reduce all results to a dimensionless form the errors are expressed as percentages of the real polygon area, and the grid-cell size R as a function of the complexity and size of the respective polygon shape.

$$f(R) = RP/(4A)$$

$$\text{where,} \quad P = \text{polygon perimeter;}$$
$$A = \text{polygon area.}$$

For those readers with an inquiring frame of mind, this dimensionless expression results in an f(R) value of unity for a square polygon discretized by a single cell of the same dimension. Inspection of Table 1.1 shows that the area error varies significantly, primarily because in some instances the erroneously included and excluded areas balance each other, resulting in a small total error. The boundary error results are generally more consistent. They are also more meaningful in a geological characterization context since we are typically required to represent not isolated geological shapes, but numerous shapes which share common boundaries. The results also show that errors greater than 5%, which is probably an acceptable threshold, are likely to occur for f(R) values greater than 0.10, and we have yet to consider the third dimension! To relate these results to reality consider the example of a geological characterization that is concerned with representing units of the

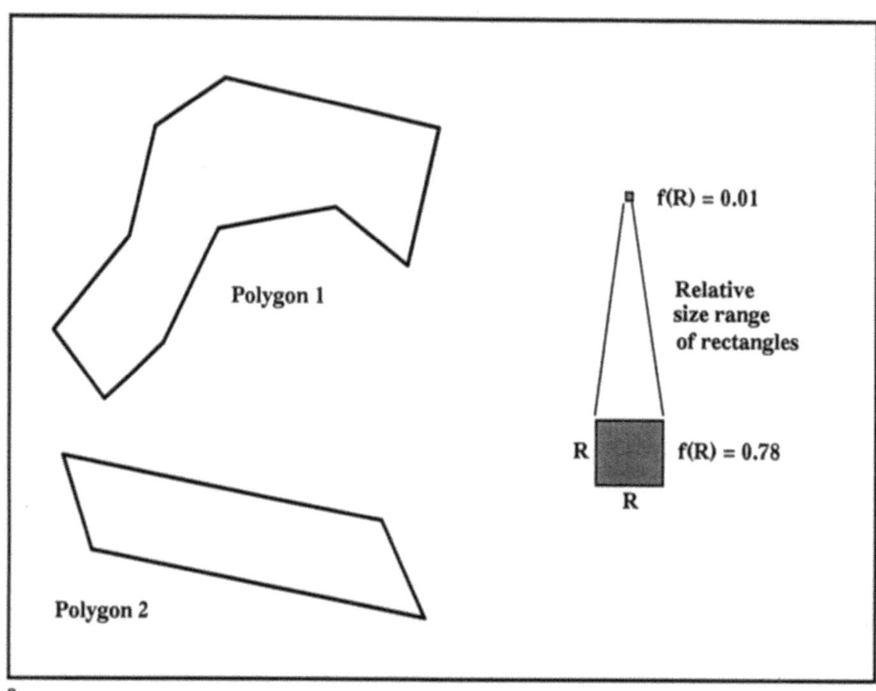

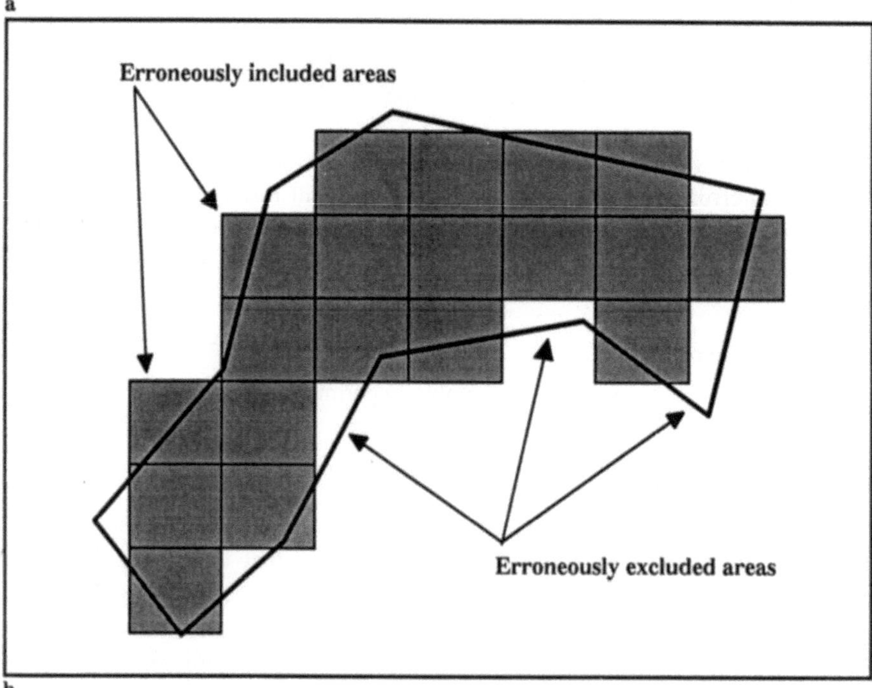

Fig. 1.6   **a** Polygon shapes and discretization size range used in tests; **b** polygon area/boundary approximation example

Discretization size function f(R)	Area error (%)	Boundary error (%)
Polygon 1		
0.53	4	19
0.27	4	13
0.13	2	7
0.06	1	3
0.04	1	2
0.03	< 1	1
Polygon 2		
0.78	43	32
0.39	29	16
0.20	11	8
0.08	1	3
0.04	1	1
0.02	< 1	< 1

Table 1.1   Results of polygon discretization tests

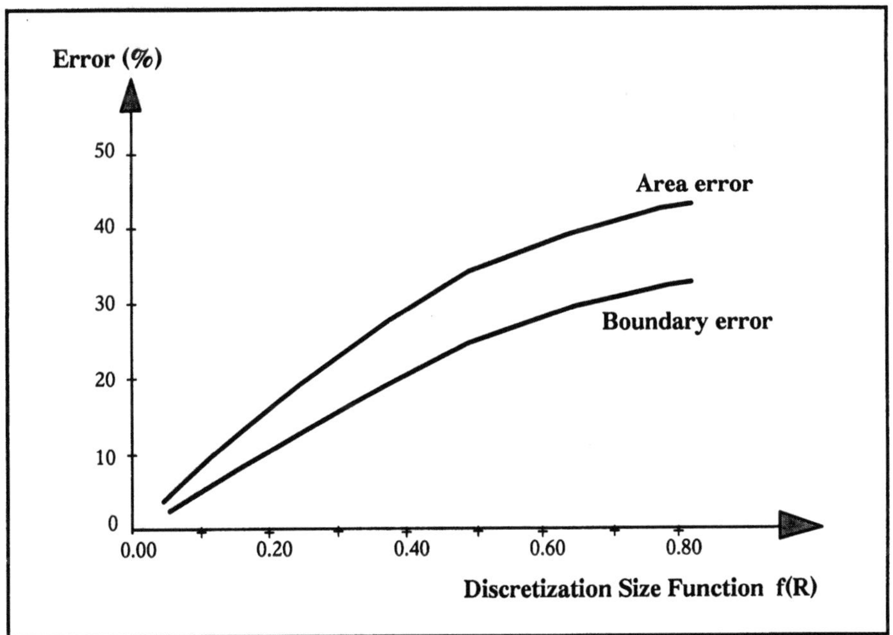

Fig. 1.7   Potential error curves derived from polygon discretization tests results

order of 5 m thickness over a region of 500 x 500 x 200 m, which is not untypical. To ensure 95% accuracy we require a cell size of 0.5 m or less, resulting in a total of 400 million cells. The average computer workstation can handle 1 or 2 million cells comfortably and 10 million if pushed! The total would obviously be reduced if we were to apply a variable (octree) grid approach, the amount of reduction being inversely proportional to the complexity and level of detail in our characterization. In many cases, the total requirement will still be unmanageably large. These results provide considerable support for the thesis that, in most cases, grid modeling approaches are inadequate for representing discrete geological volumes.

A questionable refinement of grid modeling that has been applied in some instances involves fitting smooth surfaces to discretized shapes for visualization purposes. Not only is this a contradiction of the *what you see is what you get* approach, but the geoscientist also loses control of the result. Because we are typically dealing with hard, definable stratigraphic and structural boundaries, it is infinitely preferable that they are represented directly as such within the computer. This approach brings the geoscientist much closer to the modeling process.

In summary, many of the components of a computerized approach to characterization have evolved from intensive development and application in the energy and mineral resource sectors. The principal deficiencies have been a lack of suitable approaches for representing complex geology, and a lack of integration of the necessary components in a true 3D context. The solution lies in retaining those components of the existing approaches that excel at the tasks they were originally intended for, such as regular grid modeling for representing variables, geostatistical prediction techniques and enhanced computer visualization, and integrating them with a new vector-based approach for representing irregular geological volumes.

2 Complicating Factors for Computerization

2.1 Geological Complexity

As geoscientists we are all too familiar with the potential complexities of the geological subsurface. The objective of this chapter is not to elaborate on this subject, but rather to review the complexities in the context of their impact on computerization of the geological characterization process.

We have already established that the process requires initial *interpretation of geological characteristics* as discrete irregular volumes, followed by *prediction of geological variables* in terms of their spatial variation within these volumes. For effective computerization we have also established that we require a *vector representation* of discrete geological volumes that is integrated with *raster representations* of the spatial variations of variables. The considerable volume of information which results, and all the source information resulting from investigation, must somehow be entered, correlated, integrated, interpreted, predicted, stored, revised, visualized and evaluated in the computer in appropriate data structures. This process is influenced by a host of complicating factors that, in the conventional approach and earlier computerized approaches, force us to make significant simplifying assumptions. The principal complicating factors are mostly related to geological complexity. They are summarized below and subsequently discussed in more detail.

- *Geometrical Complications* arise because of the potential complexity of geological stratigraphy and structure. Computerization requires a *data structure* that can efficiently accommodate the geometry and characteristics of discrete, highly irregular, 3D geological volumes.
- *Interpretive Control* over the computer representation of geological structure and stratigraphy requires that the geometry and characteristics of geological volumes must be definable in an *interactive context* that is complementary to the iterative, subjective process of interpretation.
- *Spatial Variation Complications* arise because of differences in the spatial variation of variables caused by geological influences, anisotropic effects, underlying trends and, in some cases, the migratory behavior of the variables themselves. These differences require tools for *spatial variability analysis*, and *prediction techniques* that can be tailored to each situation.

- *Information Sources* for geological characterization derive from a bewildering array of investigative methods and techniques. Each has unique data relationships between geometric, sample and observation information. This variety must be catered for by provision of suitably generic *data structures*.
- *Geological Variety* is infinite, from simple layered stratigraphy, requiring little or no interpretation, to highly complex folded and faulted conditions. The computerized process must cater for both extremes efficiently by provision of appropriate *modeling alternatives*.
- *The Time Dimension* adds a fourth level of complexity to the entire process that cannot be adequately resolved at this stage. We are reduced to representing temporal change in our characterization as a series of *snapshots in time*.
- *Quantification of Uncertainty* is rapidly becoming a mandatory requirement of geological characterization, particularly with regard to environmental concerns. Effective computerization requires the facility to *quantify, evaluate and visualize uncertainty* in a 3D context.

In the conventional approaches we simply cannot accommodate these complications and make simplifying assumptions instead. The errors introduced by these assumptions are compounded by a general inability to quantify their effect. We have no means of measuring how *good* our interpretations and predictions are.

The earlier computerized approaches accommodate some of the complications to varying degrees, depending on the geological conditions to which they are applied. Most work reasonably well for the conditions and objectives for which they are designed, but none have generic application across either geological variety or the varying objectives of the different geoscience sectors.

2.2 Geometrical Complications

Geological conditions are shaped by a host of different processes that have occurred in an infinite variety of combinations throughout geological history. Most of the regional scale geological conditions that we deal with consist of strata that have been initially formed with relatively simple geometry by depositional processes such as sedimentation or volcanic outflows. Complications have been introduced by intermediate or subsequent erosional activity, volcanic intrusions, or metamorphic processes. The most significant complications, from the point of view of computer representation, have been created by cycles of tectonic activity that have caused buckling, folding, shearing, faulting, displacement and in some cases overturning of the original strata. In extreme cases the result can be likened to a *3D puzzle* of highly irregular, closely-fitting shapes. To interpret the geometry of these shapes requires a considerable volume of information, and an efficient avenue for communicating this information to the computer. Figure 2.1 illustrates

Fig. 2.1   3D view of a computer representation of an inclined mineral ore deposit with intruding basalt dikes (olive green) and multiple fault discontinuities; fault surfaces are shown semi-transparent

the results of 3D interpretation of an inclined, faulted mineral deposit with intrusive dikes of barren material that has some of these elements of complexity.

On a more local scale, typical of contaminated site characterization, we might be concerned solely with sedimentary soil geology, e.g., alternating layers of silt and sand interspersed with clay lenses and complicated by old erosion channels now filled with coarse sand and gravel. Figure 2.2 shows a vertical section through an interpretation of this type of sedimentary soil geology. Even this relatively simple geological scenario requires a significant volume of geometrical information to achieve an acceptable 3D representation.

Interpretation of conditions like these is practically impossible for the computer to perform on its own, given the limited investigative information that is typically available. This means that much of the information must be provided by our own interpretation of the available information; this aspect is discussed in the next section. The impact of these geometrical complications is a requirement for a *data structure* that satisfies the following criteria.

- It must efficiently accommodate large quantities of geometrical and characteristic information relating to geological volumes. A data object in this context is an irregular volume, distinguished by geometry and characteristic values.
- The process of defining these volumes must be compatible with the requirements of interactive geological interpretation, which is the source of much of the information.
- In particular, the data structure must be capable of integration with the spatial prediction of variables, in order to apply the necessary geological control.

Finally, the potential geometrical complexity is such that it can only be represented efficiently in a true 3D context. All geometrical information must be referenced to an orthogonal 3D coordinate system, which defines location as well as shape.

2.3 The Need for Interpretational Control

Interpretation of complex geological conditions from limited information is a complicated task in itself, whether we perform it manually on paper or with the assistance of a computer. In the conventional approach we have considerably simplified this essentially 3D task by limiting it to a set of 2D sections. This has the advantage of reducing the complications to manageable proportions on each section.

We must also consider that the computer screen is a 2D medium. Although we can achieve effective visual simulation of three dimensions by means of perspective transformations, efficient interaction with such a simulation to define 3D locations is not so readily achievable. Attempts to define off-screen locations are confused by the ambiguity of the perspective transformation. On the other

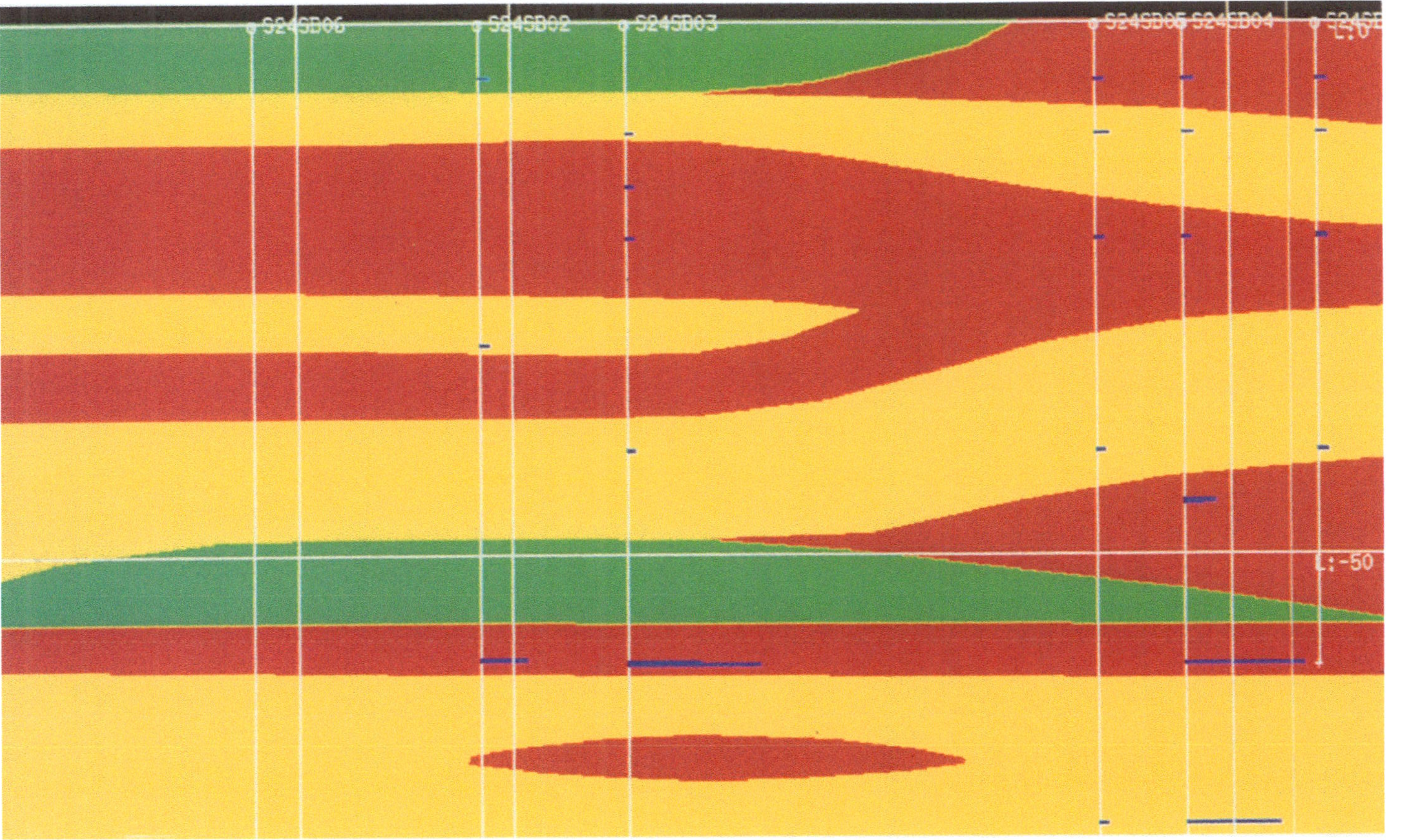

Fig. 2.2 Viewplane section through a 3D model of sedimentary soil geology with contaminant sample information from borehole investigation overlaid in histogram format

hand, interactive definition of geometry on a 2D visualization of a geological section is much more efficient and precise.

It is advantageous both to the geoscientist performing the interpretation and to computerization of the process itself if this 2D approach to interpretation can be maintained. Assuming, of course, that we subsequently have the means to extend and correlate these initial 2D interpretations in the third dimension to achieve a complete 3D interpretation. By reducing interpretation to a two-stage process we both simplify the process and maintain an approach that is familiar to all geoscientists. However, this dictates that the data structure for discrete geological volumes must be compatible with an initial definition of geometry derived from a sectional approach to interpretation and, subsequently, allow modification of this geometry to accommodate interpretation in the third dimension.

The increases in interpretive effort, information input and complexity required to progress from a series of conventional 2D interpretations to a full 3D interpretation should not be underestimated. The progression invariably highlights inconsistencies between the initial sectional interpretations and requires considerable reinterpretation. This aspect has little to do with the functionality of the computerized approach. It results from the discovery that 2D interpretations that appear reasonable enough when viewed in isolation frequently turn out to be unreasonable when we attempt to extend, correlate and integrate them into a geologically consistent 3D configuration. It is no easy task to define the geometry of potentially highly irregular shapes in 3D space and at the same time satisfy the constraints of geological consistency and integrity.

This fact has been brought home to the author on numerous occasions, including an instance during implementation of computer systems at a mine site in Australia. The engineers responsible for mine planning complained that they were experiencing extreme difficulty designing excavations that could accommodate the geologist's interpretations of the ore deposit on a series of hand-drawn sections. The inconsistencies became immediately apparent as soon as the sections were digitized into the computer and viewed in a 3D progression. Considerable re-interpretation was initiated, much to the relief of the engineers! In summary, a complex 3D geological interpretation on a computer generally results in a more geologically consistent result but may require substantial interpretive input. Interactive tools that make this task as efficient as possible are a prime requirement in any computerized approach.

2.4 Spatial Variation Complexity

The geological processes that have combined over many millions of years to produce the geological environment are so many and so varied in their degree of application that the resulting conditions might in some cases be described as chaotic. We have already discussed the effects of major stratigraphic and structural discontinuities. In the context and scale of geological units these might be termed *macro effects*, and the resulting discrete geological volumes require vector representation for computerization.

Superimposed on these macro effects we have to consider what might be termed *micro effects* in the geological context, since they influence spatial variations within the geological units. Minor structural discontinuities result in fissures, fracture sets and shear zones with their own spatial variation. The processes of deposition and intrusion result in irregularities such as bedding planes, facies, lenses and veins within geological units. Metamorphic processes caused by extreme pressures and temperatures have superimposed additional variational effects. The materials themselves are subject to spatial variability of properties such as degree of consolidation, density, porosity, cohesion, strength, elasticity and mineralogy. Weathering due to atmospheric exposure, and leaching or calcification due to groundwater activity, may have occurred. The net result is that the spatial variation of a variable is likely to be indeterminate, even within the volume of a geological unit, since it is influenced to varying degrees by many of these effects.

However, the spatial variation is seldom completely random; it has after all resulted from a number of individually determinate processes. This is the underlying supposition of Matheron's study in 1970 of what he termed *regionalized variables*. His geostatistical theory assumes that all variables exhibit a statistically measurable degree of continuity over an identifiable region, which may of course be quite small. We discuss these aspects in detail in Chapter 5.

It goes without saying that the variation of physical properties such as hardness, porosity, void ratio and fracture density may vary abruptly between adjacent geological units; sometimes by several orders of magnitude. The differences in porosity between an aquifer or an oil reservoir and the adjacent confining layers are extreme examples. Within individual units these properties typically exhibit less dramatic, although in most cases still indeterminate, spatial variation. If such properties are the variables of our geological characterization, then it is logical that prediction of their spatial variation should consider individual geological units separately. This is the only way we can accommodate abrupt changes in scale across geological boundaries.

If our variables are contaminant concentrations then, even though they have only recently been introduced to the subsurface, we must consider that their spatial variations are typically influenced to a significant degree by these variations of physical properties. Other influences include the properties of the contaminants themselves, their mode of introduction to the subsurface, surface precipitation rates and groundwater behavior. The combined effects of these influences on contaminant migration behavior is such that their variation is subject not only to abrupt differences in scale but also to differences in their *rate of change with distance* (or spatial variability). Within a geological unit we may also have to consider anisotropy in their spatial variation, for example, contaminant concentrations may vary more rapidly in the vertical direction than horizontally. Or there may be an underlying spatial trend, due to directional groundwater flow (for example) that must be considered separately from the superimposed variation. Variables such as mineral grades may also have been affected by the variation of physical properties. In many cases, however, minerals have been deposited or intruded simultaneously with the creation of geological units, and subsequently subjected, in varying degrees, to a range of metamorphic processes, each with its own unique spatial variation.

In summary, every variable which we consider is likely to have a unique spatial variation, and this variation is likely to differ from one geological unit to the next, may be directionally dependent, and possibly consists of several superimposed variations. With this level of complexity no single prediction algorithm is going to provide an acceptable, realistic representation in all cases. We require a technique that can be tailored to the observed conditions for each variable within each geological unit. And this, in turn, requires a method of determining the *apparent spatial variability* of the measured sample values in order to select an appropriate prediction technique and parameters for each case.

These requirements provided the original impetus for the development of analytical geostatistics. Semi-variogram analysis, the principal analytical tool of geostatistics, provides a statistical measure of the spatial variability of randomly located samples. Provided such a measure is obtainable from the available samples then geostatistical prediction will, in general, provide the best representation of the true spatial variation. There exists the additional requirement that geostatistical prediction must include the capability of recognizing and representing the abrupt differences in spatial variation that occur at geological boundaries.

2.5 Information Source Variety

Investigation of the geological environment includes a wide range of investigative techniques, all of which create sources of information that must be accommodated in different ways. The most common include drilling observations and samples, point samples, borehole geophysics, surface and remote geophysics, geological mapping and sectional interpretations, topographical surveys and site plans. All of this information must be correlated and integrated in a 3D context before we can use it effectively.

This creates complications in computerization of the characterization process in terms of *data structures*. Each of the information sources has its own implied geometry that locates the associated observations and samples in 3D space. A point sample obviously requires only three coordinates, or two coordinates and an elevation, to precisely locate it. However, any information that results from drill-core or borehole analysis is located by its association with the geometry of a line or trace. The trace geometry is described in terms of collar coordinates and downhole survey measurements of distance, azimuth and inclination (or dip), and all information is associated with downhole distance intervals along the trace.

Information relating to surfaces resulting from a topographical survey or a reduced geophysical survey is associated with a plane in space, in many cases the horizontal plane at zero elevation. On the other hand, a geological interpretation may be associated with a section plane which has a more arbitrary orientation in space. These examples represent information that is essentially 2D, or 2.5D if it includes offsets normal to the plane, and is located in 3D space by the orientation of the associated reference plane. A geological map or a site plan represents 2D information that really does not have a fixed location in the third dimension, unless it is mapped onto a surface. Within this planar framework we must accommodate a range of data entities that describe information locations and extents such as points, lines, polygons, contours and traverses.

For some applications it is preferable to store all observation and sample information in a relational database. This provides maximum flexibility in terms of analytical options. However, it also creates several major complications. The current level of development in relational database theory does not yet support 2D spatial information efficiently, let alone 3D information. This means either that the geometrical aspects must be managed externally to the database and interfaced to it, or that all information must be stored in a nonrelational database tailored to the requirements of geological characterization. In addition, there is a strong requirement to maintain the integrity of data entities, such as a borehole log with observations and samples, or a topographical survey with elevation points or contours. We need to be able to retrieve, analyze, visualize, manipulate and edit these as complete entities. Yet, at the same time, we also need to be able to access any individual data item, such as a sample value, stored within these data entities. These aspects are discussed in detail in Chapter 4.

2.6 Geological Variety

We have already stressed the potential complexity of the geological environment. To achieve acceptable computer representation of these conditions requires that the computerized process should incorporate interactive graphics tools that allow us to apply geological intuition and experience to the result. However, in many cases of geological characterization, the stratigraphy and structure of the region are relatively simple. The available observations may be adequate for more automated computer representation that does not require our interpretive intervention.

For the computerized process to be both generic and efficient in its application to a wide variety of conditions, it must cater to both extremes of geological complexity. This dictates that the data structure used for representation of discrete geological volumes must be compatible with both interactive interpretation and more automated computer generation. Wherever geology is simple, we should beable to represent it simply and efficiently, with the computer doing most of the work, and we should be able to combine these results with interactive interpretation wherever complexity demands it.

2.7 Time Dimension

Time adds a fourth dimension to the geological characterization process. Fortunately, the timeslices with which we are typically concerned are infinitesimal in the geological timescale and we are safe in assuming static conditions with regard to stratigraphy and structure. In most cases, the spatial variation of a variable can also be considered static for characterization purposes. The exceptions include environmental engineering studies of subsurface contamination, particularly those in which groundwater flow is a primary influence. The spatial variation of contaminant concentrations is quite likely to change during the period of investigation, which might be several years.

To accommodate a 4D spatial and temporal variation of a variable would exponentially increase the volume of information to be managed, requiring a continuous measure of all variables with time as well as in orthogonal spatial directions. It would also require a fourth dimension extension of prediction techniques (which is theoretically possible with geostatistical theory).

In practice, even in environmental engineering, we seldom monitor contaminant concentrations on a continuous basis. We are more likely to obtain sets of measurements at finite intervals. At this stage in the evolution of the computerized process it is therefore more logical that our characterizations should be treated as 3D *snapshots in time* at discrete time intervals, reflecting different stages of dynamic conditions.

2.8 Measurement of Uncertainty

Uncertainty effectively adds yet another dimension to the characterization process. Ideally, every item of observed, sampled, interpreted and predicted information should have a measure of uncertainty associated with it. To satisfy our requirement for a logical basis for risk assessment we should be able to analyze, manipulate and visualize uncertainty and its spatial variation in a similar fashion to variables.

Unfortunately this ideal is only attainable in certain areas at this stage. In many instances, formal procedures or algorithms for quantifying uncertainty have not yet evolved. For example, any measure of uncertainty that we might attach to observed or sampled information would be largely subjective without considerable research into the precision and margin for error involved in the various investigative techniques. Sides (1992) reports the results of a comparison of ore deposit limits interpreted from borehole observations with surveyed observations following mine excavation. The results indicate a potential for error of the order of 1 m in any direction for each 100 m of drilling depth. However, these results are most likely dependent, to a large degree, on geological complexity. The uncertainty associated with geological interpretation would be similarly subjective and based on our intuitive assessment rather than calculation. As discussed in Chapter 10, such subjective measures of uncertainty can be accommodated by the computerized process, provided we are prepared to attach a numerical value to our intuitive assessment of uncertainty.

The one area in which we can readily obtain a calculated, statistical measure of uncertainty is in the spatial prediction of variables. By using appropriate geostatistical prediction techniques we can obtain not only a predicted (most likely) value at a point but also a measure of its uncertainty. The resulting spatial variation of uncertainty can generally be analyzed, manipulated and visualized for risk assessment purposes. This ability to provide a statistical measure of uncertainty is unique to geostatistical prediction, an advantage that is not available from other prediction techniques. These aspects are discussed further in Chapter 10.

3 Features of an Integrated 3D Computer Approach

3.1 Overview

We have reviewed the requirements of geological characterization, the factors that complicate computerization of the process, and the various computer techniques that have evolved from applications in the different geoscience sectors. The principal data sources, data flow, data structures and operations involved in the computerized process are illustrated schematically in Fig. 3.1. This chapter is devoted to summarizing the principal features of a computerized approach that meets the identified requirements and addresses the complications.

Much of the technology is not new. The approach utilizes the advantages of many of the techniques that have been developed and applied in the past. These include conventional statistical analysis; geostatistical analysis of spatial variability; geostatistical prediction of variables; the use of 3D grid (or raster) models as data structures for representing the spatial variation of variables; use of the triangulated surface approach for creating models of relatively simple geological conditions; and enhanced visualization techniques that have evolved from graphics animation.

What is new is that these existing characterization tools have been integrated and extended by a technology which provides a unique geometrical characterization of discrete geological volumes. Its uniqueness derives from its ability to represent either simple or complex irregular shapes with equal efficiency and precision; to accommodate the interactive geological interpretation process; to apply geological control to the variable prediction process; and to provide a variety of precise volumetric determinations. We call this technology *volume modeling*. It has been developed specifically to meet the requirements of geological characterization and to overcome the limitations and deficiencies of earlier techniques.

This chapter introduces the integrated approach in sufficient detail to provide an overview of the computerized process. Some of the features may still be relatively unfamiliar to those with limited experience in geoscience modeling, however the remaining chapters of Part I provide detailed discussions of each of the principal features summarized here.

3.2 Spatial Data Management

We have already stated that the information on which characterization is based comes in a variety of types and formats from different investigative sources. Assembly and correlation of this information into a usable form are frequently the most time-consuming steps in the entire characterization process. This is a fact of life that we must deal with whether we are using a computer or not. The simple but tedious process of transforming this information into computer data has the advantage that it generally forces us to standardize and correlate the information to some extent. Having all borehole logs in a standard data format is an obvious advantage, even if the vehicle for this is a simple spreadsheet or database system. Similarly, entering site plans, topographical surveys and geophysical surveys into a GIS or CAD system generally ensures that they are spatially correlated to at least a limited degree.

However, for effective geological characterization, we have to recognize that we are dealing with 3D conditions that are potentially highly variable and discontinuous in all dimensions. We will not be able to satisfy our end objective of determining irregular volumes and their contents with acceptable precision unless we integrate all of the information within a common 3D spatial framework. This means that every item of information, including sample, observation, volume and variable data, must be precisely located in orthogonal coordinate space. If we can achieve this then all of the available information can be accessed, visualized and analyzed selectively in the correct spatial reference. These requirements are not satisfied by conventional GIS or CAD technologies.

We must also maintain the necessary data associations. We need to be able to access a borehole log both as a collective unit of information (in a familiar log format) and as a set of individual samples and observations in their correct locations. Maintenance of the implicit geometrical relationships of a borehole log is particularly important whenever the availability of the data is spread over time. For example, the core-log observations might be available weeks, or even months, ahead of the sample information, yet all information must still be spatially correlated.

Similarly, the integrity and data associations of a contour map, survey or geological map must be maintained. Each of these data entities contains geometrical *features* (points, lines, polygons, etc.) with associated characteristic or variable information. As before, these data items must be accessible individually or as a collective data entity. These requirements of data association, implicit geometry and spatial relationship are not satisfied by conventional database technology.

If we carefully analyze the contents and associations of all possible sources of investigative information, then we find that the majority can be accommodated by two simple data structures. The first assumes that items of information are located by reference to a line in space; if we know the three-dimensional geometry of the line then we can derive the location of an information item from its distance along the line. This satisfies our requirements for any information which is

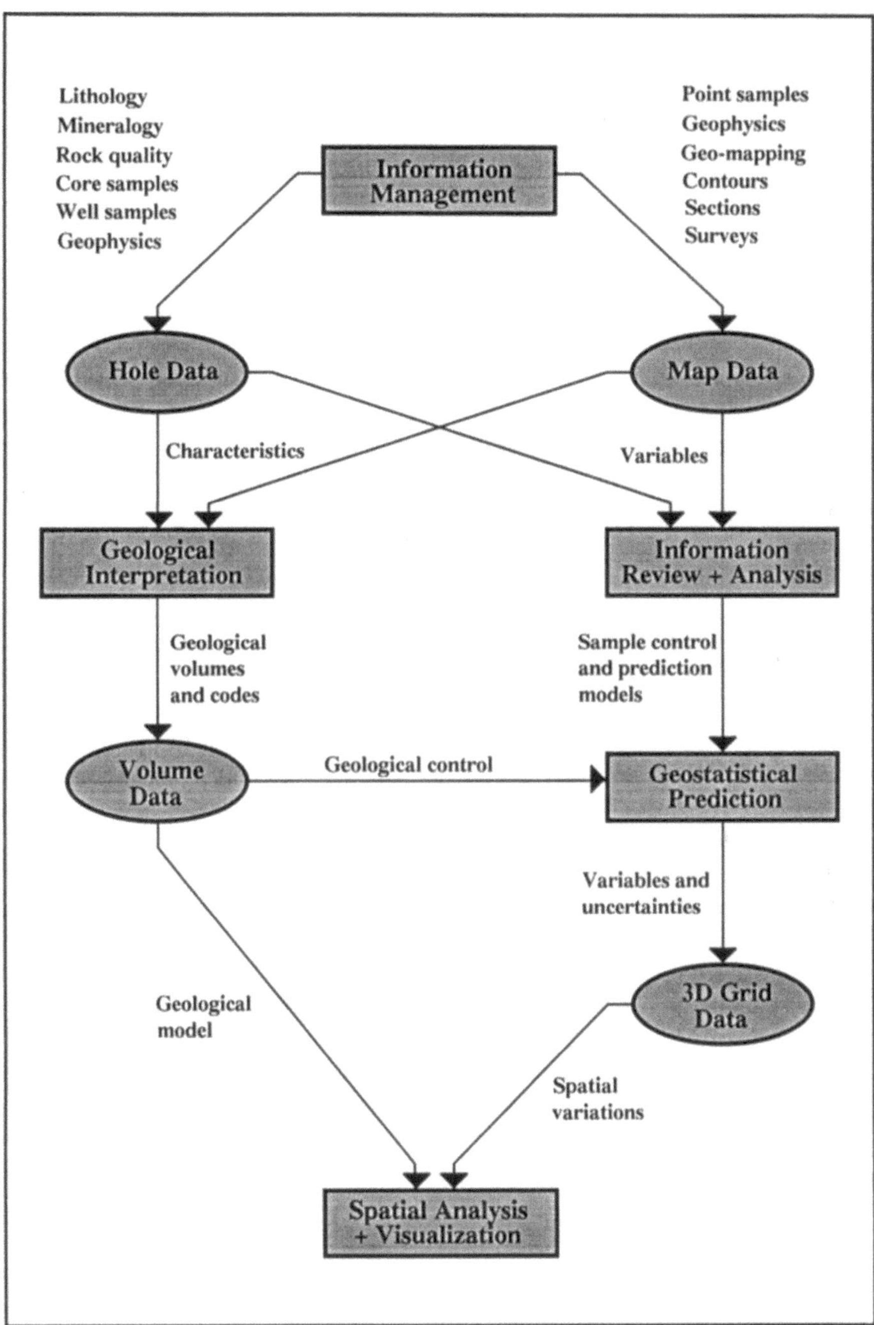

Fig. 3.1   Geological characterization - the principal data sources, data flow, data structures and operations of the computerized approach

recorded in a borehole log-like format. The borehole trace geometry can be readily derived from its collar coordinates and survey information by using a suitable computer algorithm. We can then store all such information as an entity with its implied geometrical relationships, retrieve it in a familiar log format for review or editing, and visualize it in its entirety, yet still access individual observations and samples as and when we need to for spatial analysis or prediction.

For simplicity, we use the collective term *hole data* (cf. Fig. 3.1) to refer to investigative information which fits this data structure. It includes all information derived from borehole observations and samples, well logs, geophysical logs, testpits and geotechnical line mapping. The basic components of the data structure include *identifiers* such as name, category and time subset; *locators* such as collar or starting coordinates; *geometry* described by downhole distance, azimuth and inclination readings; and the *primary information content* consisting of the observations and sample values themselves.

The second data structure assumes that information is located by geometrical reference to a plane at some arbitrary spatial location and orientation. If we know the location and orientation of the plane, then we can derive the true three-dimensional location of any information item by using an appropriate coordinate transformation. This accommodates our requirements for any map-like information, including contours, geological sections and mapping, site plans, geophysical surveys and point samples and observations, all of which can be entered and stored with reference to an appropriate planar orientation.

We use the collective term *map data* (cf. Fig. 3.1) to refer to all information which fits this data structure. As before, the data structure consists of basic components. The *identifiers* include the map name and description; the *locators* consist of the coordinates of the reference plane origin and its azimuth and inclination; the *geometry* is described by appropriate features such as points, lines and polygons and their offsets from the plane; and the *primary information content* consists of observations and sample values associated with these features.

Using these simple data structures, we are able to enter our source information in familiar formats by keying-in borehole logs, or digitizing contour maps, or importing them from external sources, and we can maintain the necessary data associations. Yet all of the information is integrated within a common 3D framework and is readily accessible to the characterization process. By using a familiar, *real-world* coordinate system of northings, eastings, elevations, azimuths and inclinations our entire geological characterization becomes more usable and accessible.

The hole data structure and map data structure described above meet our data management requirements for all source information derived from investigation. To complete our characterization we also require appropriate data structures in which to store information relating to geological volumes and variables, the models which result from our characterization of the source information.

The requirements for representation of irregular, discrete geological volumes are met by a *volume data structure* (cf. Fig. 3.1) which consists of three polygonal boundaries on parallel, offset planes and links which join the boundaries in the third dimension. By generalizing this data structure so that the boundaries can be defined at any spatial orientation, and attaching a geological characteristic to the enclosed volume, we achieve a unique geometric characterization of geology. Any irregular shape or volume can be defined to whatever level of detail is appropriate by a set of these *volume modeling components*, as we shall call them, defined at appropriate spatial orientations. The *identifiers* of the data structure include a geological unit identifier such as clay, silt, sandstone, limestone, mineralization, or whatever is appropriate to our characterization, a component name and a description. The *locators* include the coordinates and rotations of the plane from which the component is initially defined. The *geometry* consists of the three boundaries, their offsets (or thicknesses) from the plane and their intermediate links. The *primary information content* is a code that we attach to the enclosed volume to represent its characteristic value.

The volume data structure supports a vector-based approach to representation of hard geological boundaries. As we shall see, it is also compatible with all other characterization requirements. It allows us to perform interactive interpretation, initially on 2D geological sections, and subsequently in the third dimension. It also lends itself to rapid generation of simple geological models from surface information, using a simple triangulated form of component geometry. We can readily obtain geological sections at any orientation through the resulting geological model. We can spatially integrate the geological model with the 3D grid data structure described below to achieve the necessary degree of control over the variable prediction process, and we can perform a variety of precise volumetric analyses. The data structure also lends itself to graphical editing to accommodate refinements to an interpretation generated by new investigative information. The volume data structure is the key component that allows integration of the computerized process. It is discussed in detail in Chapters 4 and 6.

The requirements for representation of the spatial variation of a variable are met by a regular *3D grid data structure* (cf. Fig. 3.1) that encompasses the region of our characterization. Variable values are predicted and stored at the centroids of the grid cells. Thus, a continuous measure of the variable is represented by a finite set of values at discrete, regularly spaced points in space. The *identifiers* of the data structure include the grid name and description, and the name of the variable. The *locators* include the coordinates of the grid origin and its rotations. The grid *geometry* is implicitly defined by the regular grid-cell dimensions and the numbers of grid cells in orthogonal directions. The *primary information content* is comprised of a predicted variable value at the centroid of each grid cell, together with a measure of its uncertainty. The 3D grid data structure provides us with an appropriate vehicle for representation of raster-like information. It is compatible with all variable prediction techniques and represents a continuous spatial measure of a variable at whatever density of information is appropriate.

With the four data structures described above we can manage all information relating to our characterization effectively. The data structures allow us to integrate different types of information in 3D space, to retrieve and maintain information in an appropriate manner, and to visualize and analyze it effectively. Each of the data structures is discussed in more detail in Chapter 4.

3.3 Time Data Management

We have already established in Chapter 2 that the majority of characterizations for which time is significant can be catered to by a sequence of data and models representing discrete points on the time axis. In some cases, this is catered to merely by naming our data and models according to an appropriate time convention.

However, many of these dynamic situations involve groundwater observations and samples taken from monitoring wells and stored as hole data. Well identities, locations and geometry do not change with time, but sample and observation values do. In these cases, it is convenient to be able to create *hole data subsets*, each of which contains the observation and sample information pertaining to a specific time interval.

3.4 Visualization Tools

At this stage in our characterization we are ready to review and evaluate the available information in terms of its quality, sufficiency, scale, range and spatial variability. We also obtain an initial appreciation of the complexity and scale of the conditions which we must subsequently interpret and predict. This section is concerned with *visualization tools* that can be applied to visual review and analysis.

To discuss computer visualization effectively we need to introduce the concepts of the *viewplane* and the *viewpoint*. The viewplane is equivalent to the familiar concept of a geological section. In the conventional approach we first choose an *origin* (starting point) for the section in terms of northing and easting coordinates and an elevation, followed by an *orientation* in terms of azimuth (plan rotation about the vertical axis) and inclination (vertical rotation about the horizontal axis in the plane of the section). We then select a suitable *scale* based on the physical extent of the information we wish to visualize and the size of our visualization medium, in this case a sheet of paper. For reference purposes, we usually superimpose a *coordinate grid* at appropriate intervals. Finally, we plot a projection of the information to be visualized, typically including all information within a certain distance either side of the section, which means that our section has a defined *thickness*. If the thickness is zero, then we visualize only that information which actually intersects the section.

This familiar procedure can be readily and efficiently duplicated on the computer. The only real difference is that the paper medium is exchanged for the computer screen.

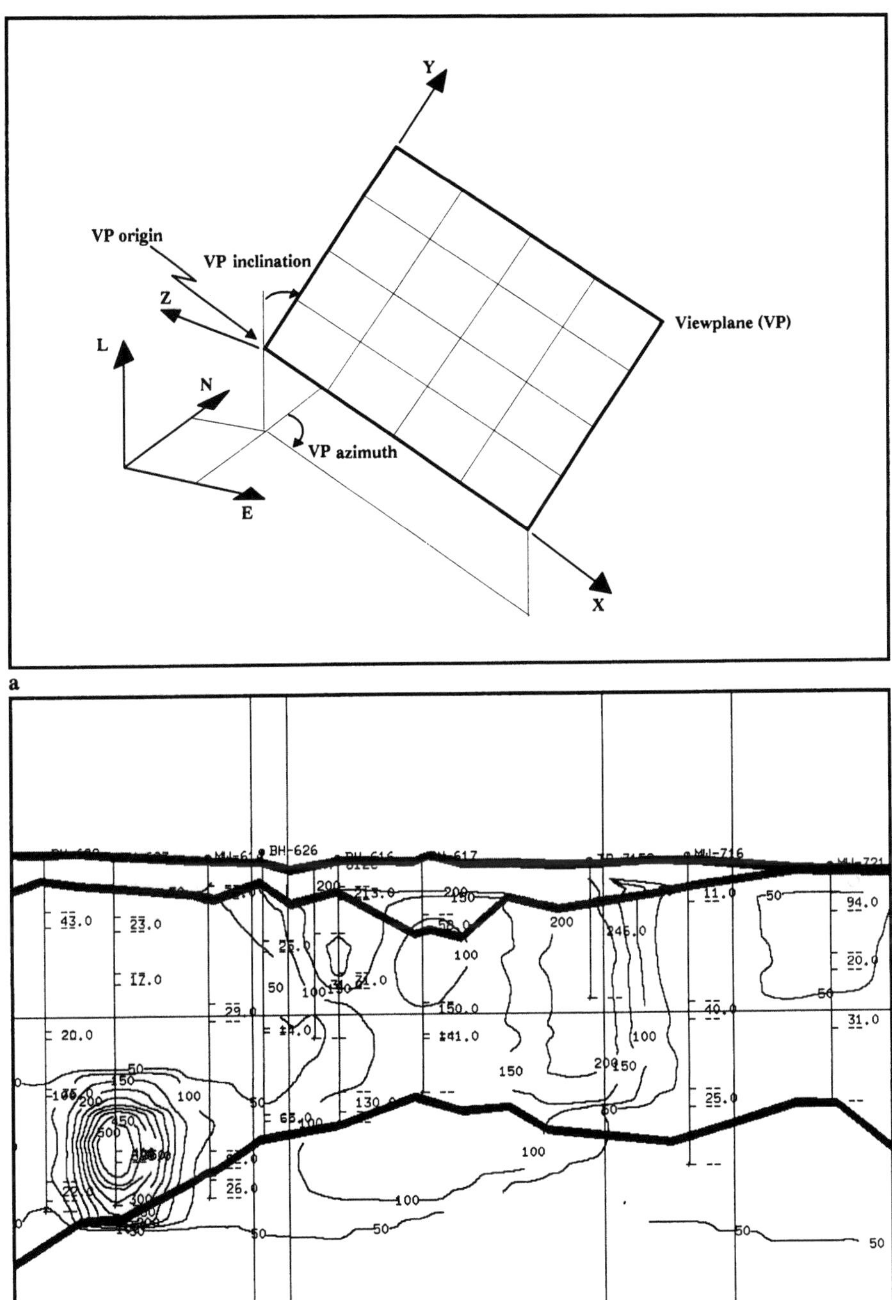

Fig. 3.2   **a** Definition of viewplane conventions, global NEL coordinates and local XYZ coordinates; **b** contour display of sample values of a variable on a vertical viewplane section, overlaid on strata horizons, borehole traces and sample values from site investigation

By assigning real world coordinates, rotations and scales to the screen it becomes a viewplane on which we can visualize selected information; the computer equivalent of a conventional geological section on paper (cf. Fig. 3.2a). Because the computer retrieves, sorts, projects and visualizes information so much faster (and with much greater precision!) than we can, this approach introduces tremendous range and flexibility into the visualization process.   With simple interactive procedures we can reorient the viewplane, change the scale, zoom in on a selected feature, respecify the information content, or change the graphical display format. Moreover the new visualization is available in seconds (instead of days!). An appreciation of the spatial variability and complexity of the available information is rapidly obtained by *walking through* the region of characterization with appropriate viewplane definitions and display formats.

The success of the viewplane approach to visualization is obviously dependent on a suitable range of display formats with which to visualize both a wide variety of investigative information and, ultimately, our models of geology and variables. Besides display of observations and samples against the geometry of their associated features, either as values, histograms or continuous curves, the visualization tools should also optimize the use of color-coding, texturing and contouring, all of which are appropriate in different situations. Figures 3.2b and 3.3 provide examples of some of these visualization options.

We maintain this 2D viewplane approach to visualization, firstly, because it is immediately and intuitively familiar, even to geoscientists who have never used a computer and, secondly, because it is the most efficient medium for precise interaction. In contrast, 3D visualization (described below) provides a much more realistic image but is relatively inefficient for precise interaction. Both visualization techniques have their place in the characterization process.

In the conventional approach the closest equivalent to a 3D visualization is probably a perspective view of a fence section interpretation of geology. To achieve a perspective transformation we define a *viewpoint* in terms of real coordinates, scale the information to be visualized according to its distance from the viewpoint and project the results onto a 2D medium. Because this is such a time-consuming process on paper, we seldom apply it in the conventional approach. Even on a computer the algorithms for perspective transformation and visibility ordering require considerable processing effort for large amounts of information. Unless we have a more powerful than average workstation, 3D visualization tends to lose interactivity as the amount of information increases. It also introduces ambiguity to interaction with the computer screen due to perspective distortion of all dimensions. In particular, distances normal to the screen are undefined in the context of user:screen interaction. Despite these deficiencies it is an extremely effective visualization procedure. Its value to the characterization process lies in its ability to provide a medium for enhanced appreciation of complex conditions, for rapid verification of the validity of 3D interpretations, and for presentation to nontechnical audiences. This last consideration is of particular importance in the environmental sector where the results of a characterization are frequently exposed to litigation procedures or public review.

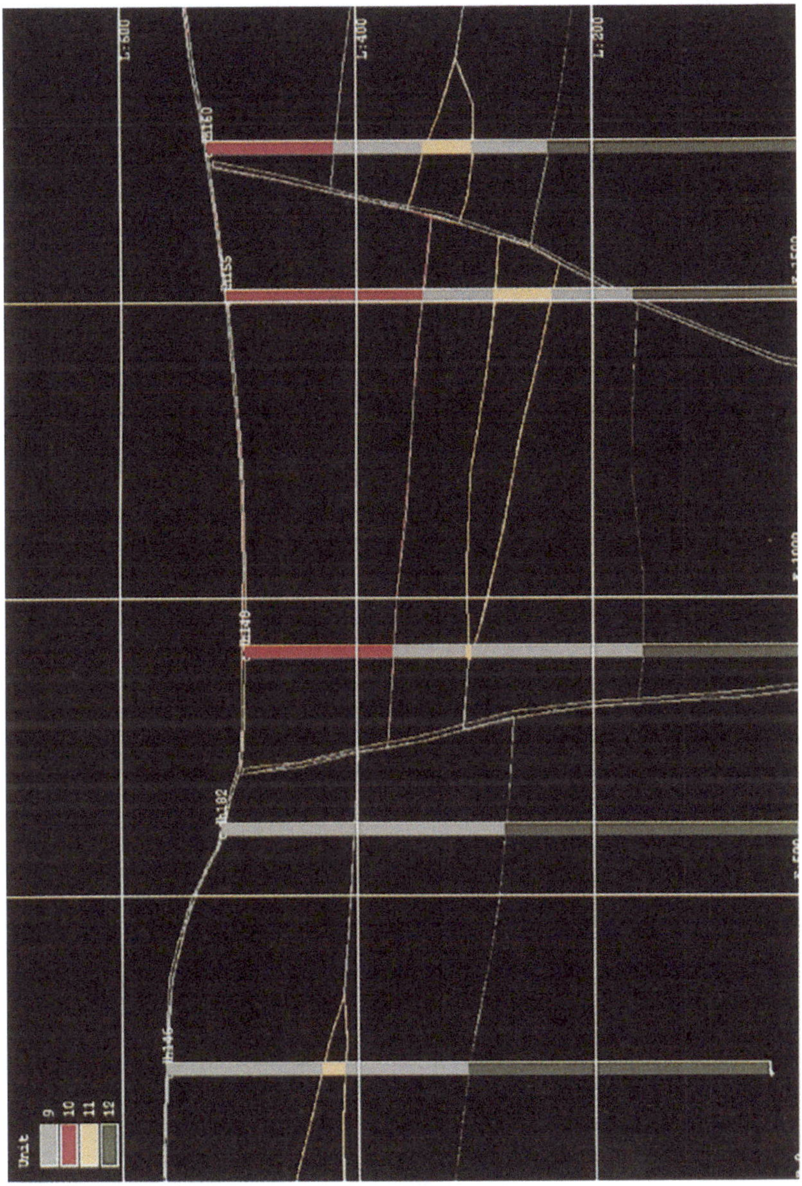

Fig. 3.3 Viewplane section for geological interpretation, with topography, fault surfaces and borehole lithology intersections displayed as background information for interpretation of geological boundaries

Effective 3D visualization requires additional tools to those described for 2D viewplane visualization. These include light-sourcing to emphasize geometrical relief, 2D and 3D cut-aways and selective transparency to expose hidden information, and the ability to interactively manipulate the view. Appropriate rotations and translations allow us to view the information from different perspectives, effectively *walking around* the information by changing our viewpoint. This form of *enhanced visualization* allows us to view any information, model or isosurface in three dimensions, e.g. a cut-away model of soil strata superimposed on a visualization of contaminated soil volumes with a site plan superimposed for spatial reference, or a model of geological structure and stratigraphy cut away to expose an isosurface of ore-grade material. Examples of 3D visualization in a variety of display formats are scattered throughout this text.

Despite the fact that many of us, out of necessity, have become adept at constructing mental images of complex subsurface conditions and regard enhanced visualization as nothing more than *pretty pictures*, there are significant advantages to be gained. It improves our understanding and appreciation of complex, inaccessible conditions, and enhances our ability to interpret and verify them. This contention is supported by the numerous examples of computer visualization throughout the text. It also provides a useful presentation aid for explaining our results to investors, regulatory authorities and clients who are unfamiliar with geological complexity.

In summary, by combining 2D viewplane visualization with 3D perspective visualization we achieve the best of both worlds in the computerized approach. The former provides a familiar working medium for efficient and precise interaction, the latter provides enhanced realism and an efficient presentation medium. By providing both techniques with appropriate visualization tools that can be applied to any of the data structures described earlier, we satisfy all requirements of the computerized approach with respect to visualization.

3.5 Statistical Analysis

Statistical analysis should be standard practice in any geological characterization. It provides us with an appreciation of the scale, distribution and correlation of sample and observation values. It also assists us in identifying samples of suspect quality. The standard tools of statistical analysis play an invaluable part in the characterization process, however, they should be familiar to all of us and do not require detailed discussion, a summary is sufficient.

The ability to *sort and differentiate* sample data and to *select subpopulations* by value, location, time period or characteristic value is a prerequisite. A prime objective of the data review and evaluation step is to determine whether or not there are significant differences in the distribution of sample values between various subpopulations. For example, at a contaminated site: are the contaminant concentrations in sand material significantly different to those in silt material?; are the differences in concentration between two time periods significant?

Alternatively, in a mineral deposit: do the mineral grades from one zone of mineralization exhibit a different distribution of values to those from another zone? (possibly indicating several phases of mineralization). To answer questions like these we require basic statistical tools such as the ability to calculate *means, minima and maxima, median values and standard deviations* for all subpopulations of interest. These allow us to quantify whatever differences exist and assist in identifying the subpopulations and distinguishing factors which are relevant to our characterization.

More sophisticated statistical tools can tell us a lot more about our sample data. A *correlation analysis* of the sample values for two variables indicates the degree of spatial correlation between them. For example, if high concentrations of one contaminant generally coincide with high values of another, and vice versa, then their spatial variability is similar; they probably originate from the same source and exhibit similar migratory behavior. If not, then they probably come from different sources. In a coal deposit, if there is a high degree of correlation between the negative qualities of ash content and sulfur content, then our characterization of the deposit (and subsequent planning and extraction) is considerably simplified. For rapid visual assessment, the results of a correlation analysis should be presented graphically, as an orthogonal plot of one variable against the other, as well as numerically.

Other primary tools of statistical analysis include *frequency histogram analysis* and *probability analysis*. Both provide us with an appreciation of the distribution of values within a subpopulation. A frequency histogram analysis is a graphical plot of the frequency of occurrence of sample values within specified value intervals. It allows us to assess whether the values approximate a statistically normal distribution, or a log-normal or exponential distribution, which may be an important factor in subsequent analysis of their spatial variability. For example, soil contamination frequently exhibits a log-normal distribution of values. We are unlikely to achieve an acceptable representation of its spatial variation unless we are aware of this behavior and take it into account. To be effective, the analysis should also provide a measure of how well the distribution of values approximates a statistically normal (bell-shaped) distribution. A probability analysis essentially provides similar information in a different format. The results are presented as a graphical plot of the probability of exceeding a value against the value itself. A statistically normal distribution of values results in the familiar s-shaped probability plot.

To complete our statistical analysis toolkit we require the ability to apply *data transformations* to our sample values. If we suspect a log-normal distribution of values, then we can confirm this by applying a logarithmic transformation and then repeating the frequency histogram analysis; the results should now approximate a normal distribution. Other useful data transformations include indicator and rank order transforms. An indicator transform converts sample values to zeros and ones depending on whether they are less than or greater than a specified threshold value, which might be a mining cut-off grade, or an environmental action level. A rank order transform reorders the sample values in terms of increasing value. All of these transformations are relevant in certain situations. In many cases the results of statistical analysis become meaningful only after application of the appropriate data transformation.

Besides providing us with a review and appreciation of the statistical relevance of our sample data, these tools also have application to verification (or otherwise) of data quality. In particular, they can assist in identifying samples of suspect quality. It is common practice during investigation to gather two sets of samples from the same locations and have them analyzed at different laboratories. Statistical analysis allows us to determine whether there are significant differences between the two subpopulations, possibly due to faulty analytical equipment. Alternatively, samples may be mixed or incorrectly labeled at the investigation stage, or the values may be incorrectly entered into the computer. Many such occurrences can be visually detected as *spikes* or *outliers* in the results of a correlation analysis or a frequency histogram analysis. On occasion the distribution of sample values may be distorted by the too frequent occurrence of values set to the *non detect limit*. The unrepresentative samples may have to be removed from the subpopulation before meaningful results are obtained.

Examples of the use of statistical tools are presented in Chapter 5 and in the application examples in Part II of the text.

3.6 Spatial Variability Analysis

The principal tool of spatial variability analysis is the *geostatistical semi-variogram*, and the principal objective is to quantify the spatial variability of the available sample values. These results provide a logical basis for subsequent prediction of the spatial variation of variables. Geostatistical analysis is probably less familiar than conventional statistics to many readers and a more detailed discussion is called for. However, discussion focuses on how geostatistical analysis is integrated into the characterization process, rather than a justification of its theory and advantages which are dealt with in more detail in Chapter 5. It is sufficient at this stage to state that it is the accepted method for obtaining a statistical measure of the spatial variability of information.

In the conventional approach to characterization, we typically limit the spatial analysis step to contouring of sample values on plans and sections. In many cases, a 2D linear spatial relationship between samples values is assumed, which is invariably incorrect, and the fact that many of the samples are not located in the analytical plane is ignored. With 3D geostatistical analysis we can improve considerably on these unnecessary approximations.

The semi-variogram analyzes all possible pair combinations in the subpopulation of samples considered. For each pair of samples the square of the difference in value and the distance apart are determined. Because even a small number of samples may have thousands of pair combinations, the results are *lumped* into distance intervals and averaged for each interval. The results are a measure of the *spatial variability* of the samples and are typically presented as a graphical plot of difference squared against distance apart. It should be emphasized that the semi-variogram is concerned with *differences in value* and *distances* rather than with actual values and locations. What we are looking for is a meaningful relationship between difference and distance; if this exists, then it is immediately apparent as a trend in the graphical results. They typically show that the difference in value increases with distance within a certain range, which is what we would intuitively expect. What this means is that the *rate at which difference changes with distance* is similar, and measurable, for all samples within this moving *range of influence* of each other. This applies throughout the sample subpopulation, even though the spatial distribution of the values themselves may appear random or indeterminate.

Although many of the current applications of semi-variogram analysis, particularly within the environmental engineering sector, are limited to 2D, the analysis is readily extended into the third dimension. It can also be directionally controlled, which allows us to selectively analyze sample pairs that have a particular spatial orientation. For example, we can determine whether a degree of anisotropy is present in the spatial variability of the samples by comparing the semi-variogram results for pairs with a horizontal orientation to those for vertically oriented pairs. By applying it to appropriate subpopulations of the available samples, we can also determine whether the differences in spatial variability between samples from materials with different characteristics are significant. Both of these conditions occur frequently in practice. If we are aware of their existence then we can take them into account in the prediction step.

These are not the only complications that occur in practice. As in statistical analysis, we may have to apply a *data transformation* to our sample values before we achieve meaningful semi-variogram results. If the sample values are log-normally distributed, then we need to apply a logarithmic transformation before performing a semi-variogram analysis. Alternatively, the semi-variogram results may be distorted by an underlying trend that is spatially determinate. For example, groundwater contamination samples may exhibit a dominant trend in a certain direction due to groundwater flow, on which the inherent variability is superimposed. We can recognize such conditions by applying appropriate *trend analysis* techniques to the samples. If the condition exists, then the determinate and indeterminate components of the spatial variability are isolated and treated independently in both the analytical step and the prediction step.

The complications discussed above often occur in combination, which means that we may have to work our way through an exhaustive list of analytical options with regard to geological and directional influences, data transformations and trend analyses before we achieve a meaningful result. This requires a comprehensive spatial analysis toolkit. On occasion, even after exhausting all the possibilities, we may fail to distinguish any meaningful spatial relationship in the available samples; the number of samples may be too small or, less frequently, their spatial variability is too great, i.e. approaching truly random conditions. However, this is still a useful result. In the former case, we have a logical basis for taking more samples, which should improve the semi-variogram results. If it does not, then we know that further sampling would be wasted and we must do the best we can with what we have.

The greatest advantage of semi-variogram analysis is that we can use the results to derive a semi-variogram relationship that can be applied to predicting values between the samples. We call such a relationship a *prediction model*. This model is obviously only as good as the degree to which it represents the spatial variability reflected by the semi-variogram results. To satisfy all conditions of spatial variability we require a range of *model types* and a method of fitting them to the results, which is usually done interactively. We also need a means of verifying these models before we use them in the prediction step. The best way of doing this is by *cross-validation* against the sample values themselves. This involves successively eliminating each sample from the population and predicting a value in its place from the remaining samples. By comparing each predicted value against the corresponding measured value we obtain a measure of the *goodness* of the prediction model. We discuss prediction models in more detail in Chapter 5, and there are numerous supporting examples of semi-variogram application throughout the text.

Despite the apparent complications, geostatistics provides a logical approach to spatial analysis, firstly, to quantify spatial variability and, secondly, to select prediction techniques and models that are appropriate to the available information. Prediction of the spatial variation of each variable, within each geological unit, can then be individually tailored to the variability of the available samples.

3.7 Interactive Interpretation

Interpretation is required wherever a significant degree of geological complexity is present, since we seldom have enough information to define the conditions directly. In the conventional approach we interpret geological structure and stratigraphy on plans and sections, using plotted observations of characteristics and profiles of surfaces such as topography and faults for visual reference, e.g., by correlating borehole intersections of a geological unit and interpreting the profile of the unit between intersections or by interpreting stratigraphy from a seismic profile.

In this way, we simplify interpretation to manageable proportions by initially limiting it to two dimensions. Most of us are incapable of the mental gymnastics required for simultaneous interpretation in three dimensions anyway! After repeating this process on a number of sections we might attempt to correlate the sectional interpretations on a transverse section which intersects them. If we are dealing with complex geological conditions, then this obviously becomes an iterative process, with much editing and refinement before we achieve an acceptable result. Despite its iterative, tedious nature this is still an effective method of interpretation.

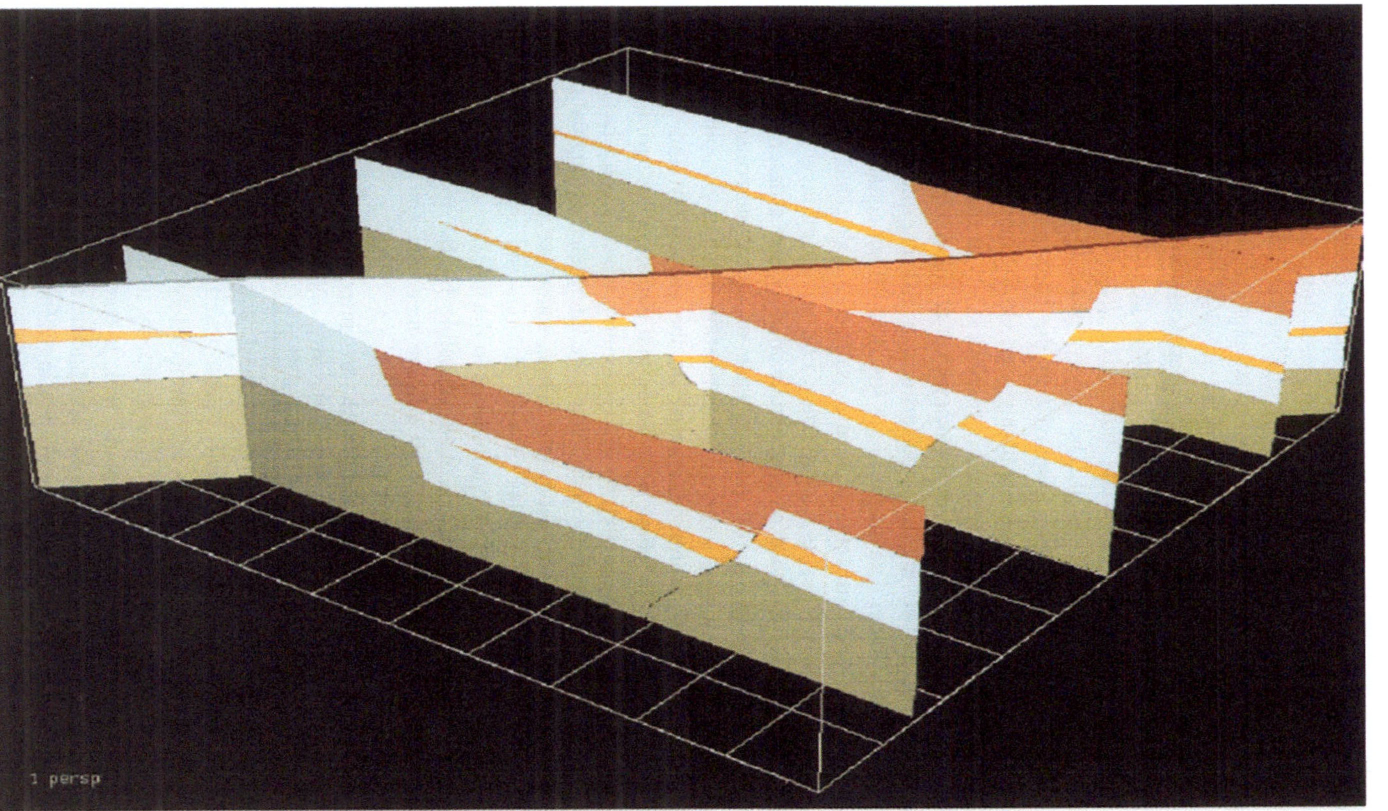

Fig. 3.4   3D perspective view of a geological fence section model compiled from vertical viewplane interpretations using interactive volume modeling techniques

Even on the computer, interpretation progress is governed by our capacity for mental assessment of the available information rather than the rate at which we can draw or define boundaries. Computer algorithms are notoriously inefficient at direct interpretation of geology, except in rather simple cases.

The computerized approach effectively duplicates this conventional, user-driven interpretation process by making use of the vector-based *volume modeling technique* discussed earlier. The *viewplane* (computer screen) is substituted for the paper section and whatever information is relevant to our interpretation is displayed in the background in appropriate graphical formats. The *volume data structure* is substituted for the hand-drawn geological profiles and its boundaries are defined by means of appropriate interactive *drawing tools*, the computer equivalent of the conventional pencil. The concept of *background* and *foreground* requires explanation. Information displayed in the background is for reference purposes only and cannot be modified in any way. On the other hand, information defined or retrieved to the foreground is accessible to any of the available drawing and editing tools. Figure 3.3 provides an example of an interpretation viewplane.

Since at this stage we are only performing a 2D interpretation, all three boundaries of the volume data structure are (by default) set equal and assigned a nominal thickness. The data structure is also assigned a code which designates its characteristic and determines its display color. For ease of retrieval in subsequent operations it is finally assigned an appropriate identity and a description. In the context of volume modeling, we refer to each of these interpreted entities as *components*, and our finished sectional interpretation includes a number of them, one for each geological profile interpreted on the section. At this stage each component is a flat, irregular shape with a nominal thickness, located in the plane of the geological section simulated by the viewplane, which in turn has whatever spatial orientation we chose at the start. Thus, the interpreted geological profile is precisely located in 3D space. At a later stage, when we have completed our interpretation on a sufficient number of such sections and are satisfied with the result, we extend and modify these components in the third dimension to achieve a complete 3D interpretation. This is made possible by the unique geometrical characterization of volumes provided by volume modeling and the volume data structure.

This interactive computerized approach to interpretation has significant advantages over the conventional approach. Any information defined in the foreground is immediately added to the background display and, therefore, can be used for reference purposes in the interpretation of other geological profiles. Since the viewplane can be easily and rapidly manipulated to a new orientation, the validity of any interpreted information can immediately be verified with respect to adjacent or orthogonal interpretations. The geometry of the volume data structure allows the intersection of the viewplane with a component to be rapidly and precisely determined and displayed in the appropriate color. This is the equivalent of the conventional approach of manually scaling information off several sections and replotting it on a new intersecting section, but much less laborious and error-prone. The data structure, and the drawing tools provided, also allow rapid editing and refinement of geological profiles, again, infinitely more efficient and less tedious then modifying a paper section! At any stage in the process we can also move to a 3D perspective visualization to obtain an overview and a visual correlation of the results.

With these interpretation tools we can rapidly achieve an enhanced appreciation of complex conditions and, using a familiar yet efficient approach, construct a detailed interpretation of geology on a set of 2D sections. The result is a 3D fence model of geology that represents the *boundaries and areas of intersection* of discrete geological volumes with a set of section planes. Figure 3.4 provides an example. The boundaries of intersection form the basis of our final interpretation of the *geological volumes* themselves.

3.8 Modeling Discrete Geological Volumes

It is in the second stage of interpretation, of extending and correlating the initial interpretations in the third dimension, that the conventional and computer-assisted approaches diverge. This effectively requires that we extend the 2D interpreted *boundaries of intersection* of each geological unit into *irregular 3D volume representations*. In the conventional approach we typically make the implicit assumption that the area (and profile) of each unit is constant from mid-way between a pair of sections to mid-way between the next pair. For complex geology this is a gross approximation, although the errors are obviously reduced as the spacing of the sectional interpretations decreases.

With the appropriate volume modeling techniques and the volume data structure, we can eliminate the approximations of both the conventional approach and earlier computer techniques. The data structure is designed to accommodate the required extension and correlation (interpretation) in the third dimension. The derivation of these irregular volumes is therefore based on the initial sectional interpretations. In this way we satisfy a principal requirement of the computerized approach, in that the final representation of geology includes our interpretive input. We can also accommodate to a large degree all of the three-dimensional complexities introduced by geological discontinuities such as faulting, folding, bifurcation and pinch-outs, since the technique also allows us to exercise interpretive control in the third dimension. Finally, the resulting volumes retain the code classifications assigned earlier, thus satisfying our requirement for *qualification* of subsurface space by geological characteristics.

The key to successful application of this approach is provision of an appropriate set of *interpretation tools* for deriving volumes from areas. For each volume modeling component, the first step is to extend the thickness normal to the plane containing the initial interpreted profile. The result is an extruded volume with a constant but irregular profile, the equivalent of the conventional assumption that profiles are constant between sections. However, each extended profile is now modified by relocating the viewplane to the offset profile plane, using the same

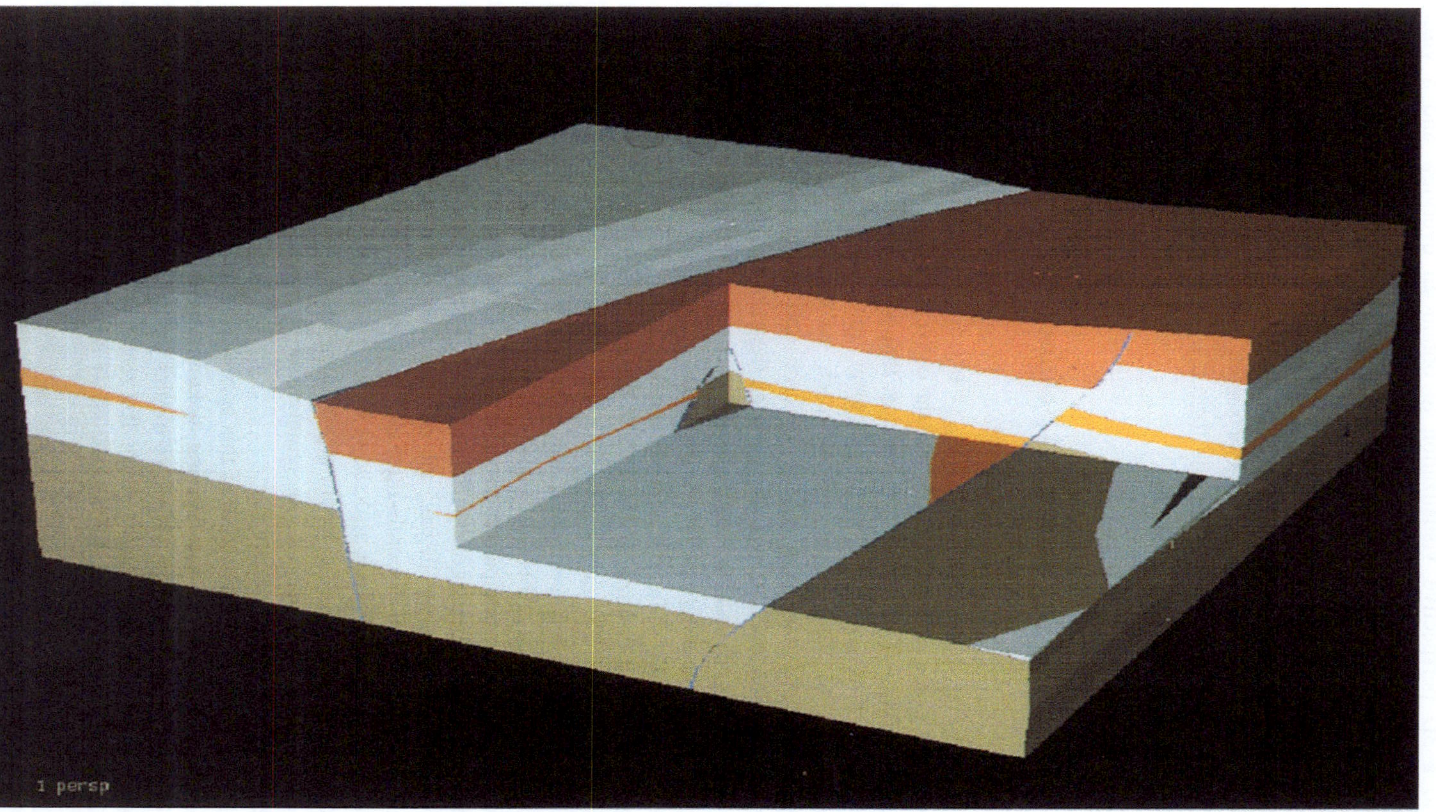

Fig. 3.5   Perspective view of a geological interpretation model developed from the fence sections shown in Fig. 3.4 using interactive volume modeling techniques

interactive drawing tools that we applied to the initial interpretation. In effect, the extended profiles are reinterpreted to match the background information (topography, fault surfaces, observation data, sample data and interpreted geological volumes) on the offset viewplane. The resulting component geometry consists of three irregular interpretation profiles joined by links, and can accommodate extremely complex shapes. In practice, a complete geological unit is represented by a set of contiguous components, whose extended profiles generally match each other on common planes. This is not always the case, however, as the geometry may be interpreted to accommodate pinch-outs, bifurcations and faults.

In many cases geological units show some degree of continuity in the third dimension and interactive interpretation is unnecessary. Where appropriate, *profile matching algorithms* are applied to pairs of components to achieve an interpolated profile between them.   On the other hand, if we are dealing with extreme geological complexity, then it may be necessary to define new components with appropriate viewplane orientations to fill in undefined voids in our initial interpretation, for example, near the intersection of inclined fault surfaces. Since each profile of a component accommodates a large amount of detail (several hundred points' worth), and each geological unit is represented by a number of components (as many as necessary), the degree of detail interpreted is limited only by our own perseverance and time constraints. Since the volume modeling technique is vector-based (as opposed to a raster or grid technique), it allows us to achieve precise representations of the *hard*, irregular boundaries which define geological structure and stratigraphy. It also allows us to readily update and refine our interpretation to accommodate new investigative information.

The final computer representation, which we call a *geological model*, is equivalent to a three-dimensional map of geology. The 2D geological areas of a conventional map are replaced by irregular 3D geological volumes that can be visualized on any intersecting plane or cut-away through the model. Figure 3.5 provides an example of a finished geological interpretation, derived from the initial sectional interpretations shown in Fig. 3.4. The theory of volume modeling and its application to the geological characterization process are discussed in detail in Chapters 4 and 6.

3.9 Modeling from Geological Surfaces

Characterization does not always require interactive interpretation of geology. If we are dealing with simple layered formations, then the available information may be perfectly adequate for a more direct definition of geological volumes. In these cases we use a *surface-based approach* to create geological models. We proceed by first generating triangulated surfaces that represent geological horizons or contacts, and then filling the volumes between them with volume modeling components that have a simple, triangulated geometry. In this way we rapidly create 3D geological models of relatively simple conditions without any need to resort to interactive interpretation.

This approach is made possible by the ability of the volume data structure to assume a wide variety of geometrical shapes, in this case, a triangulated prism configuration with a linear variation of thickness. We can extend this approach, making it applicable to more complex conditions, by provision of a range of *surface manipulation tools*. These tools include the ability to clip a surface to a polygonal boundary, to intersect one surface with another and truncate it, to manipulate surface information with logical and arithmetic expressions, to map one surface onto another, and to allow surfaces to be generated from any spatial orientation. These tools allow us to model moderately complex conditions, e.g., representation of discontinuous, pinched-out geological units, or representation of faulted conditions by modeling a geological unit in one fault block at a time. The surface-based approach is also applied to advantage during interpretation by representing surfaces such as topography and faults by assigning them a nominal uniform thickness. The approach and its application are discussed in detail in Chapter 7.

3.10 Modeling Geological Variables

Our objective in this step is to predict the *spatial variation* of variables of interest, based on available sample values, appropriate prediction techniques and the spatial constraints of our geological model. As discussed earlier, the equivalent in the conventional approach would consist of contouring sample values on plans and sections using a simple interpolation algorithm. As a spatial prediction technique this is an unjustifiable simplification of conditions and does little to satisfy the requirements of three-dimensionality or the identified complications. In the computerized approach we extend all aspects of the spatial prediction process into three dimensions. We also use prediction techniques and models that are tailored to the spatial variability of the samples.

The appropriate technique and prediction model for each variable within each geological unit have generally been determined earlier by spatial analysis of the samples using geostatistical semi-variogram analysis. In most cases *geostatistical kriging* is the prediction technique of choice, and an appropriate semi-variogram relationship is used as the prediction model. If an acceptable semi-variogram result is unobtainable, then we might resort to an alternative prediction technique, such as an inverse distance-weighting algorithm. In those cases where an appropriate technique cannot be established, which means that the spatial variability is not measurable, almost any technique is as good as the next. We focus on geostatistical kriging since it is appropriate in the majority of cases, and the only technique which provides us with a measure of the uncertainty attached to the result. Even in those cases where we cannot measure the spatial variability, it provides at least as good a result as other techniques. In short, it is the most universally applicable of the prediction techniques at our disposal. We discuss the technique in more detail in the next section.

The data structure for the predicted values is the *3D grid data structure* discussed earlier. The objective is to predict values at the centroids of all grid cells, thus providing a continuous measure of a variable throughout the region of our characterization. The process involves centering a *search volume* on a grid-cell centroid and then searching the database (both hole data and map data) for the subset of samples which lie within this search volume. The predicted value at the centroid is based on this sample subset, the distances of the samples from the centroid, and the selected prediction model. This process is common to all prediction techniques; they all employ some form of *distance-weighting algorithm* so that samples close to the centroid have greater influence than those at greater distances. If anisotropy is a factor in our prediction, then the distance-weighting and the search volume are modified accordingly. In these cases the search volume is an ellipsoid rather than a sphere. If no samples are found within the search volume, then the resulting predicted value is *undefined*, a result which must be accommodated by the data structure.

There are obviously other factors besides the choice of prediction technique which affect the result, such as the search volume dimensions and the grid-cell dimensions. With geostatistical kriging the search volume dimensions are often set to the established ranges of influence of the semi-variogram relationship. However, if the sample density is highly variable, then we might increase these dimensions to avoid predicting too many undefined values in regions of low sample density. We may also wish to place restrictions on the spatial relationship of the samples to the grid-cell centroid, e.g., only predicting a value if at least three of the search volume quadrants are occupied by samples.

The grid-cell dimensions control the resulting density of predicted information. We obviously prefer that the density is compatible with the required degree of precision in our characterization. However, the use of grid cells that are very small relative to the average sample density and/or the established ranges of influence results in much wasted effort in the prediction process and a false impression of precision. Ideally, the grid-cell dimensions should be between two to five times less than the ranges of influence and the average sample spacing in the corresponding directions. This means that the grid cells should be able to assume any rectangular configuration. If the sample variability is high in a particular direction, then the resulting density of predicted information in that direction should also be high. For certain conditions, for example, in the case of an inclined ore deposit, it is also an advantage to be able to orient the 3D grid model appropriately by means of azimuth and inclination rotations.

The most important requirement, as emphasized earlier in several contexts, is the ability to *control the prediction process* with the spatial constraints of the *geological model*. The prediction process must be provided with information regarding the geological volumes intersected by each grid cell. We achieve this by volumetrically intersecting each grid cell with the geological model prior to prediction. The resulting volumes of intersection are stored with their characteristic

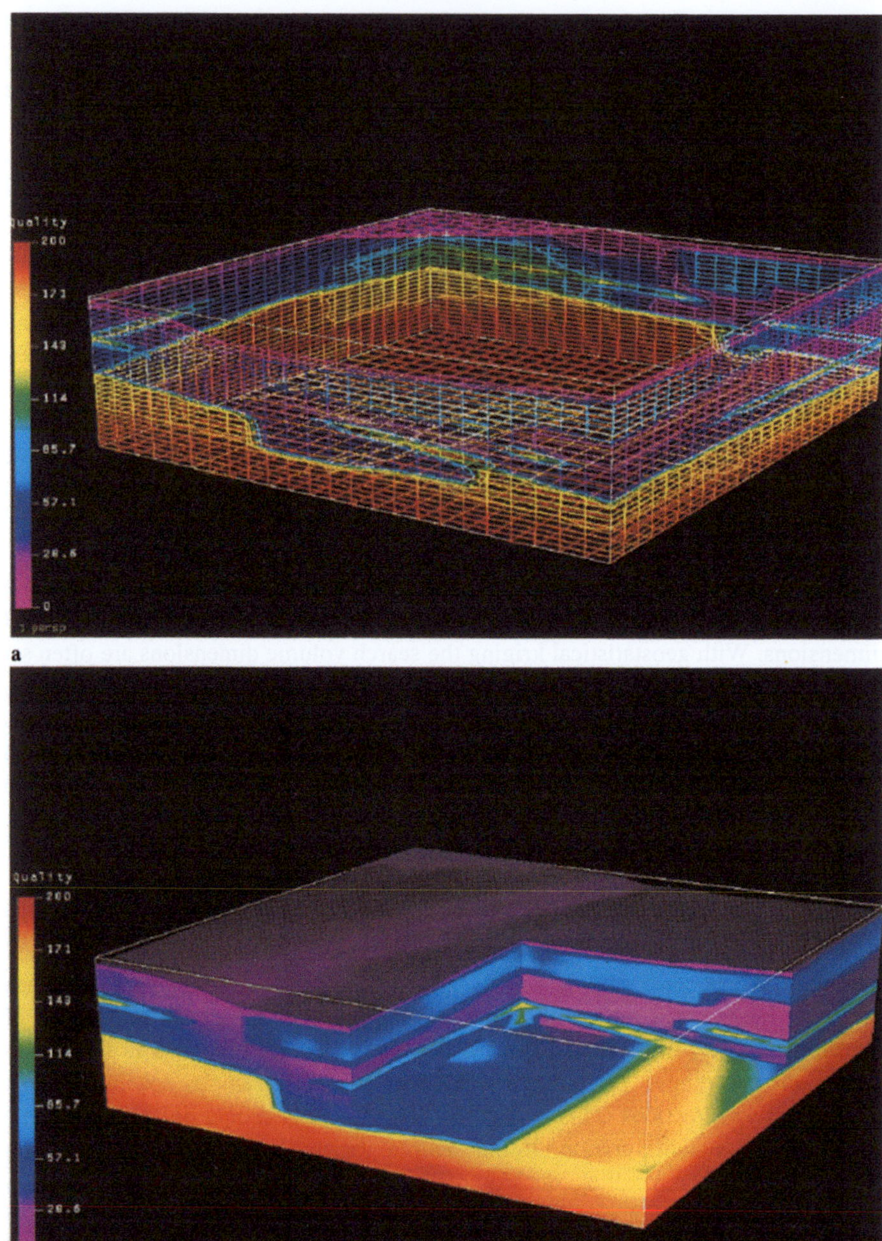

Fig. 3.6   a 3D grid model geometry, color-coded to the predicted spatial variation of a rock quality variable; **b** a solid visualization of the same grid model with a 3D cut-away to expose the internal variation of the variable

values in the data structure of the grid model. Then, for example, if we are predicting variable values for sand material, a value is predicted for a particular grid cell only if it contains a sand intersection volume. The value is based on a subpopulation of samples from sand material. Of course, that grid cell might also intersect clay and silt volumes, in which case corresponding values are subsequently predicted for these materials as well, based on different sample subpopulations and appropriate prediction techniques in each case. If necessary, following completion of the prediction process, these values can be combined by a volume-weighted compositing algorithm to obtain an average value of the variable for the grid cell.

In this way we achieve an appropriate degree of geological control over the variable prediction process. The vector-based geological model of characteristics is spatially and analytically integrated with the raster-based grid model of variables, thus retaining the advantages of both modeling approaches and eliminating the identified deficiencies. Figure 3.6 provides an example of a 3D grid model representation of spatial variation. In this case the variable is a measure of rock quality that has been predicted under the geological control of the volume model representation reported in Fig. 3.5.

3.11 Geostatistical Prediction and Uncertainty

We focus on *geostatistical kriging* as the prediction technique in this discussion since, given the necessary constraints, it is applicable to most conditions. Some of us may already be familiar with the technique by exposure to geostatistical packages that are freely available in the public domain. These typically suffer from two major deficiencies in that (in most cases) they are limited to two dimensions and make little or no provision for geological control. A deficiency of kriging (and other spatial prediction techniques) is that it tends to *smooth* the variation of an attribute across discontinuities. As we have already discussed, this deficiency can only be eliminated by *constraining* the process within the volume of a geological unit.

Kriging has recognized advantages over other prediction techniques. These advantages exist whether it is applied in 2D or 3D and irrespective of geological control. They can be summarized as follows.

· *Smoothing*: Kriging smoothes, or regresses, predicted values based on the proportion of total sample variability accounted for by *random noise*. The noisier the sample set, the less the individual samples represent their immediate vicinity and the more they are smoothed, and the greater the associated uncertainty.
· *Declustering*: The kriging weight assigned to a sample is lowered to the degree that its information is duplicated by nearby, highly correlated samples with little variability. This helps mitigate the effects of variable sample density caused, for example, by the tendency to oversample *hot spots* during investigation.

· *Anisotropy*: When samples are more highly correlated (less variable) in a particular direction, then the kriging weights will be greater for samples in that direction.
· *Precision*: Given a semi-variogram relationship that is representative of the region to be predicted, kriging predicts the most precise values possible from the available samples. In practice, this is only approximated since the semi-variogram is itself derived from the available samples and may not be truly representative of the region.
· *Uncertainty*: Kriging predicts not only a value but also a measure of the uncertainty associated with the value. This uncertainty is not just a fortunate bonus, it is an integral part of the prediction result. The predicted value loses much of its meaning if it is considered in isolation. The uncertainty is a measure of the *goodness* or *reliability* of the predicted value and a measure of its *possible variability*; results that other prediction techniques are unable to provide.

Kriging algorithms assume normal distributions of values, which is why we sometimes have to apply *data transforms* to our sample values prior to prediction, in order to achieve a distribution which approaches normality. In these cases the predicted values also include a transformation. We can apply *back-transforms* following prediction to obtain real values, although this is not always necessary or desirable. The two most commonly used data transforms are *logarithmic* and *indicator* transforms, as discussed earlier. Indicator transforms are particularly useful in environmental engineering, and represent an instance when back-transformation is unnecessary. They involve the substitution of zeros and ones for the real sample values, depending on whether the latter are less than or greater than a specified threshold value. If the threshold value is set equal to the regulatory action level for the variable (in this case a contaminant concentration), then the resulting predicted values represent the spatial variation of the probability of exceeding the level. This is a particularly useful result for effective remediation planning and risk assessment.

Besides data transforms, there are other variations on kriging that we may require. A selection of semi-variogram relationships, or *prediction models*, is one. These are mathematical relationships which define spatial variability in a kriging context, the most commonly used is a spherical model, followed by linear and exponential models (and other more esoteric varieties). Which model to use is determined by the observed trend in the semi-variogram analysis results; each of the models has a distinctive geometry. If anisotropy is a factor, then the *ranges of influence* in orthogonal directions, which form part of the model definition, are set appropriately.

Another variation on kriging involves a distinction between *point kriging* and *volume kriging*. Point kriging, like all other prediction techniques, provides a predicted value at the grid-cell centroid, whereas volume kriging determines the average value throughout the volume of the grid cell and assigns it to the grid-cell centroid. The differences are generally marginal, however, point kriging is more appropriate when the predicted values are destined to be used for interpolating isosurfaces or contours and their enclosed volumes. Volume kriging, on the other hand, is more appropriate if the predicted values are used for determining the contents of predefined volumes, such as those enclosed by excavation geometry.

If an underlying trend has been determined, then this component is predicted independently by *trend surface prediction* using polynomial regression. The trend component and the kriging component are then combined to obtain the final predicted values.

Finally, we need to consider the *measure of uncertainty*. Kriging algorithms compute this measure for every predicted value of a variable, and store it with the predicted value in the 3D grid data structure. It is usually computed as the statistical *standard error* (or standard deviation) of the possible distribution of values at a point. Statistically speaking, the predicted value at a point is the most likely value but, with 68% confidence, the value could lie anywhere within a standard error either side of the predicted value, and with 96% confidence it could lie anywhere within two standard errors. If the samples in the vicinity of the point are highly variable and do not correlate well, then the computed measure of uncertainty is large relative to the predicted value, and vice versa. Thus, at completion of the prediction process, we have achieved a continuous measure of the spatial variation of both the variable and its geostatistical uncertainty.

Uncertainty impacts on our characterization in two areas, namely sampling control and risk assessment. Since we know its spatial variation, we can identify regions of high uncertainty by inspection. This knowledge, together with other factors discussed in Chapter 10, allows us to optimize the location of any additional investigative sampling. New samples that are located in regions of high uncertainty have, in general, the greatest impact on reducing the overall uncertainty of our characterization. This is a much more logical and justifiable approach to sampling control than blindly applying infill sampling on a gridded basis, or concentrated sampling around known hot spots. It can help considerably in minimizing the costs of investigation.

Since the measures of uncertainty are stored alongside the predicted values the two can be manipulated to provide the spatial variation of *best case* or *worst case scenarios*. For example, in a contaminated site characterization, we might add a multiple of the uncertainty values to the predicted values to obtain the worst case scenario with a certain statistical confidence. Alternatively, in a mining context we might subtract a multiple of the uncertainty from the predicted mineral grades to obtain the worst case mining scenario. This capability has obvious and significant advantages in the risk assessment and planning steps, discussed in more detail in Chapter 10. However, a final word of caution, in some cases the uncertainty may be so large, possibly because we cannot afford sufficient sampling, that we prefer to ignore it!

This completes our summary of geostatistical prediction and uncertainty. As in other steps in the characterization process, the prime requirement is for a range of options, or *prediction toolkit*, that enables us to deal with all possible conditions. All of the spatial prediction techniques presented above are discussed in more detail in Chapter 8.

3.12 Spatial Analysis of Volumes

At this stage we have interpreted geological structure and stratigraphy, and predicted spatial variations of the variables of interest. We also have an appreciation of the uncertainties involved in our prediction, and whether or not we need to initiate further investigation before we evaluate the results, in which case an update cycle of the previous steps is required. This step is concerned with *spatial analysis* of the interpreted and predicted results.

What we do in this step depends on the objectives of our geological characterization. However, a large degree of commonality in function between the requirements of the different geoscience sectors is still evident. In most cases evaluation involves the determination of volumes and analysis of their contents. In the conventional approach we might planimeter contours or profiles on sections, multiply these areas by the section spacing and sum the results to obtain an approximation of true volumes.

Volume modeling technology and its integration with variable prediction provide us with a range of options for performing more precise volumetrics analyses. This includes determination of the volume of any irregular shape, or the volume of its intersection with another irregular shape, or the volume of an isosurface (three-dimensional contour) of a variable value, or the intersection of an isosurface with an irregular shape, and the ability to analyze the contents of any volume or intersection in terms of contained variable values. Isosurfaces, or soft boundaries, are determined by inspection and interpolation of the spatial variation of a variable stored in the 3D grid data structure, whereas the hard boundaries of discrete geological volumes are obtained directly from the volume model representation in terms of the volume data structure; or from volume model representations of excavation geometry, which we cover briefly in Chapter 9. Visualization of results also figures largely in the final characterization step of spatial analysis.

If we are characterizing a mineral deposit, then our primary objective is to determine the 3D limits of *ore grade* material, and to analyze the contained volume in terms of its mineral grades. In this case the ore limits are defined by the isosurface for a so-called *cutoff* grade, or threshold. The volume of this isosurface is the *probable*, or most likely, ore volume. In view of the fact that a large investment of capital is dependent on the outcome of our characterization, we might repeat this analysis, with manipulation of the uncertainty values, to obtain best case and worst case ore volumes for a specified confidence level.

If we are characterizing a contaminated site, then the process is essentially similar. We are concerned with *contaminant concentrations* and *regulatory action levels*, rather than mineral grades and cutoffs. We are also more concerned with the risk of leaving behind unacceptable contamination, rather than (as in the mining context) not finding what we have predicted. In this case we require isosurfaces which define contaminated soil volumes, or groundwater plume volumes. As before, the uncertainty values can be applied to the determination of worst case contaminated volumes in order to minimize the risk of leaving behind any unremediated material. Figure 3.7 provides an example of the spatial

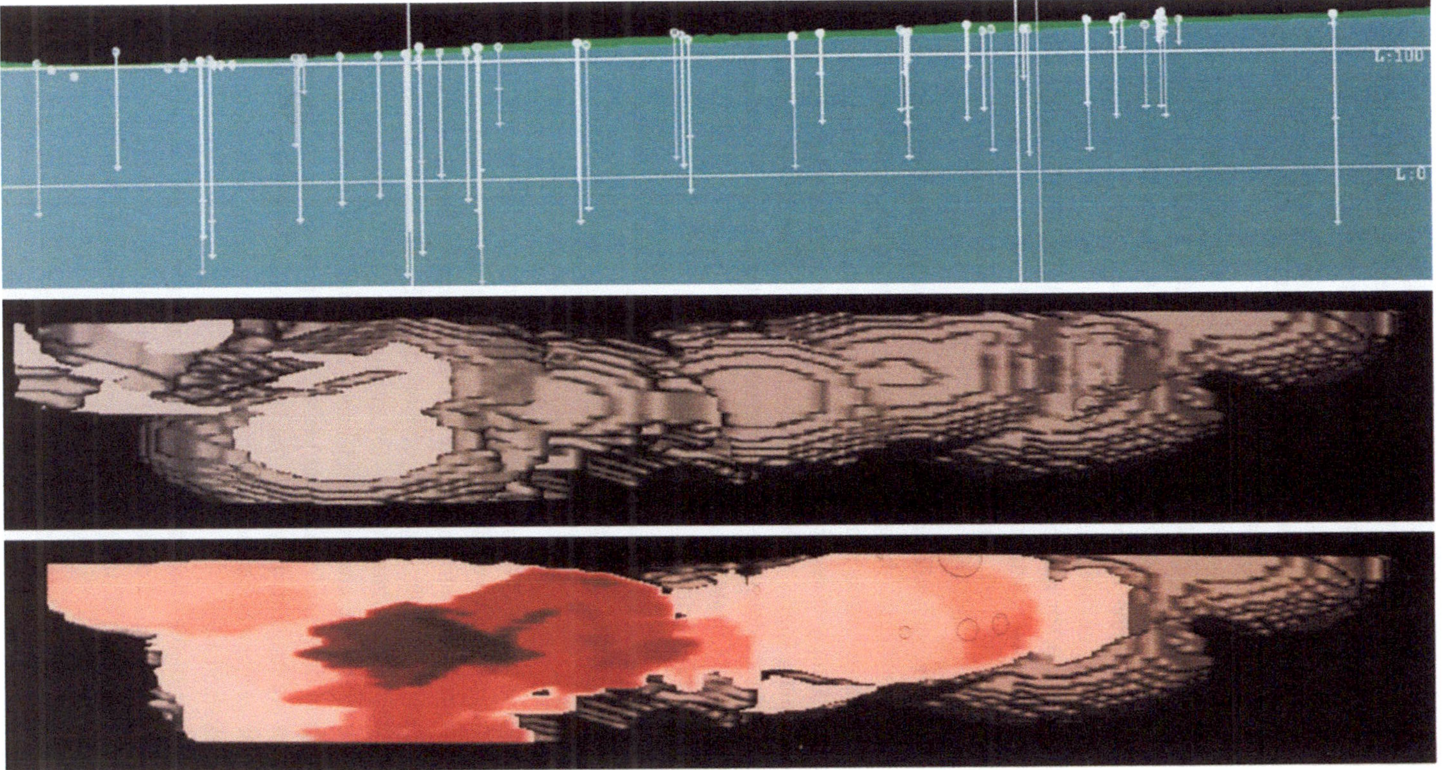

Fig. 3.7   **above** Section through contaminated groundwater site showing monitoring wells and saturated/unsaturated zones; **centre** isosurface defining VOC contaminant plume; **below** vertical cut-section through plume showing predicted spatial variation of VOC contamination

analysis of a contaminated groundwater plume volume, derived from a predicted spatial variation of contamination based on monitoring well samples.

As hydrogeologists, we are frequently more concerned with variables such as seepage conductance and porosity, the location of impermeable barriers such as dikes and highly permeable regions such as shear zones, and with identifying possible contaminant migration pathways. The latter are obtained by analyzing isosurfaces of seepage conductance.

In tunnel and excavation design we are concerned with locating materials with poor excavation qualities, identified by a variable which measures rock quality or fracture density (for example) or by lithological type (such as a sand layer), and the location and orientation of shear zones and fault surfaces. In this context our spatial analysis may be more visual than volumetric.

Evaluation of an oil or gas reservoir requires its volume, configuration, porosity and other factors which determine its potential yield and the siting of extraction wells. Evaluation of an aquifer as a potential water supply has similar requirements.

All of the above, and other applications of geological characterization such as geothermal site evaluation, hazardous waste site characterization and landslide risk assessment, require spatial analysis in terms of volumetrics and visualization to some degree. In many cases, the final results must be transferable to the relevant end-process technology, such as mine planning, remediation planning or groundwater flow modeling, in an appropriate computer format. Spatial analysis is covered in detail in Chapter 9.

3.13 Advantages of Process Integration

Earlier approaches to computerization contain many of the features presented here, but in fragmented form. This results in continual shuffling and manipulation of data from one format to another and back again, which, besides being tedious, seriously jeopardizes the integrity of our data. Every time we manipulate information from one data format to another we run the risk of losing or transposing information; and the maintenance associated with numerous copies of the same data in different formats can present a data processing nightmare! These inefficiencies are magnified by both the large volumes of data that we typically deal with for geological characterization and by the current lack of any widely available data standards.

The efficiency of the characterization process and the integrity of our data are enhanced by providing all of the necessary characterization tools in an integrated 3D context and storing all information in compatible data structures.

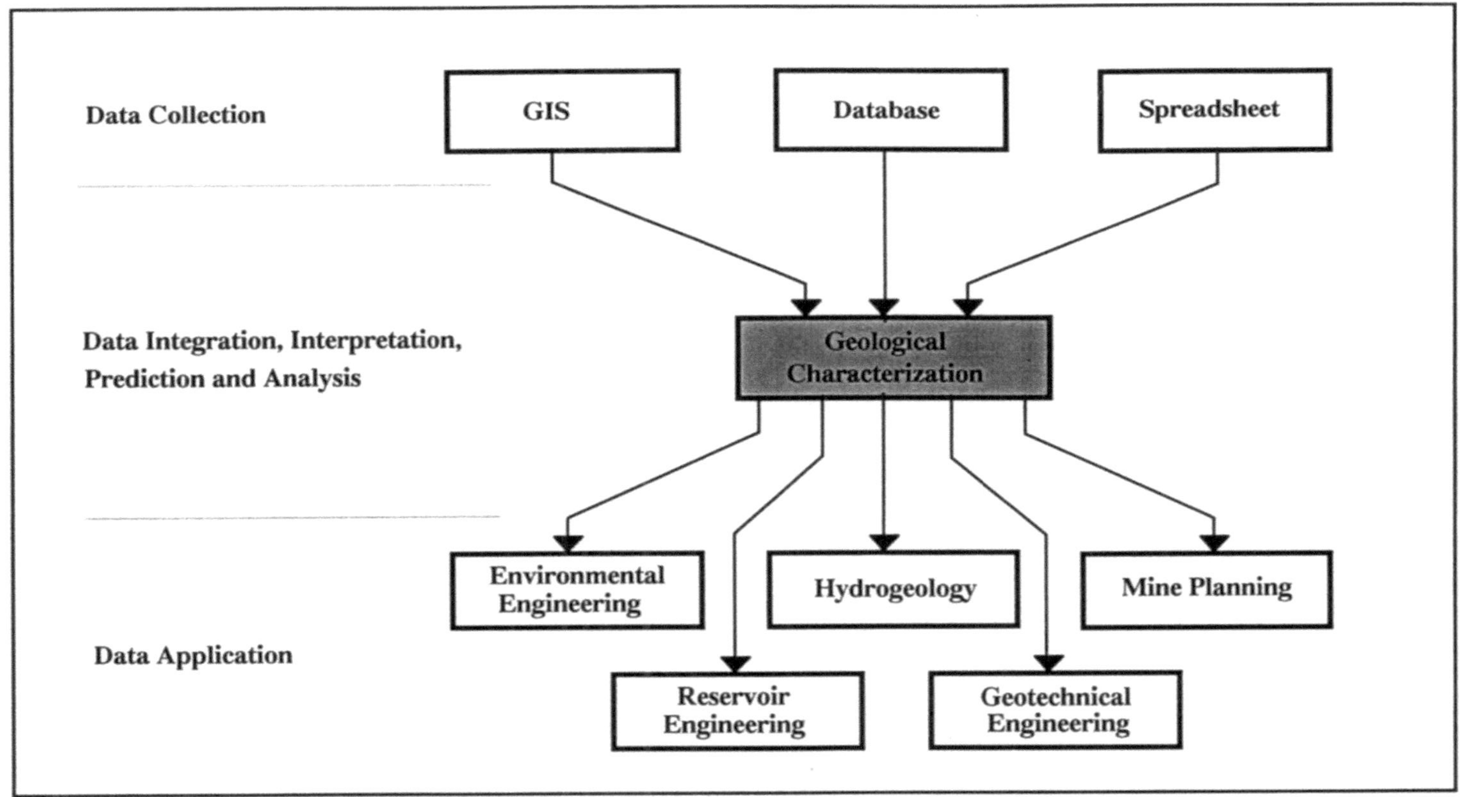

Fig. 3.8   Geological characterization - its role in the geosciences and requirements for data interfacing with external computer technologies

3.14 Technology Interfacing

We have briefly discussed the requirement for interfacing the results of characterization with other, external technologies (cf. Fig. 3.8). Any of the source information stored in hole data or map data structures, the characterization information stored in volume data or 3D grid data structures, and various analytical reports must be exportable in standard ASCII files to ensure transferability. Because spatial data standards have not yet evolved for some of these data structures, the best approach is to use *self-defining file formats*. Each data file then becomes a self-contained entity since the format, or structure, of the file is described by header records at the beginning. In addition, there are certain low-level de facto data standards that must be catered to, including HPGL graphics files, DXF drawing files and PostScript image files.

There is an additional requirement for interfacing with preprocess technologies as well. In many instances some or all of the source information for characterization is already in a computer format. For example, information may have been collected during investigation in a spreadsheet, database or GIS system, or derived from a geophysical reduction system. As discussed above, the best way to meet these import requirements is by making provision for self-defining file formats.

3.15 Process Summary

This completes our preliminary discussion of computerized geological characterization. The computerized approach parallels and extends the conventional approach, yet retains many of the concepts and methods which are familiar to us. The principal features can be summarized as follows.

- *3D spatial management* and integration of all characterization information.
- *Visual, statistical and geostatistical analysis* of information and its spatial variability.
- *3D geological interpretation* and modeling.
- *Geostatistical prediction* of the spatial variation of variables and their uncertainty.
- *Spatial analysis* of the results involving volumetrics, visualization and provision for interfacing with other technologies.

We have also established that the key, underlying component of the technology is volume modeling, since it provides the ability to interpret and model geology, to control the prediction of variables, to perform volumetrics analysis and to visualize models effectively. In short, volume modeling is crucial to most of the steps in the computerized geological characterization process. The remaining chapters in Part I are concerned with detailed descriptions of the principal tools of computerized characterization in terms of their underlying concepts and considerations, theoretical details, and how we apply them in practice.

4 Spatial Data Types and Structures

4.1 Rationale for Spatial Data Management

The initial step in any geological characterization involves *management, correlation and integration* of the variety of data types that represent our source information. Efficient organization and management of large quantities of data of various types and structures are critical to successful computerization of the geological characterization process. On a typical characterization project we must deal with anywhere from 20 to several thousand borehole logs, each of which may contain several hundred intervals, and each interval may have anywhere from 10 to a 100 observations and measurements associated with it. We must also deal with numerous maps in digital form, each of which may contain tens (or hundreds) of thousands of points. All of this information has limited use to us unless we can integrate it in a spatial context, and selectively retrieve and display it in appropriate fashion.

These considerations require effective management of *data location* and *data identity*. To achieve this we have to develop a suitable rationale for categorizing and organizing the data. These problems are identical to those faced by conventional geographical information systems, except that GIS technology is limited to two spatial dimensions, whereas we are faced with three.

4.2 Data Categories and Their Organization

There are many ways we can categorize the wide variety of investigative information that provides the basis for any characterization. Strictly from a *data perspective*, we must deal with *observations*, which have alpha (character string) or integer values, and *measurements*, which have real, numeric values. Observations are invariably concerned with information relating to characteristics, such as lithology, mineralogy, geological member or sequence identity, and geological structure. Most of these derive from visual observation, either of borehole cores, surface outcrops or excavations. On the other hand, measurements are generally, but not exclusively, concerned with variables such as contaminant concentrations (from laboratory samples), mineral grades (from assays), reservoir porosities and saturations (from a variety of borehole geophysical recordings), and geomechanical properties (from in-situ tests or samples). Measurements may also be the source of characteristic information, for example, lithology may be indirectly derived from either borehole or surface geophysical recordings. Soil strata information may be derived from cone penetrometer tests, or from ground penetrating radar.

From an *application perspective*, we have already discussed in detail the categorization of information into *characteristics*, which form the basis of our interpretations, and *variables*, which provide the source for predicting spatial variations. In general, characteristics are observable geological qualities that we apply to a qualification of subsurface space; variables are measurable qualities that we apply to a quantification of their spatial variation within the subsurface.

We can also categorize information from an *investigative source perspective*. However, since there are potentially several score of these, each with its own peculiar recording formats and conventions, catering to them individually would be prohibitively messy and confusing in a computerized context.

The common feature of every item of information, no matter what data type or application or source it belongs to, is that it is associated with a location and an extent (point, line, area, volume or grid) in 3D space. How its location and extent are defined is, in all cases, a function of an implicit data geometry. Thus, in the computerized approach, data categorization from a *data geometry perspective* is the most efficient and relevant means of organizing all of our data. This is the basis for the four data structures introduced in Chapter 3. Thus, all observations and measurements are represented either as *hole data*, associated with points and intervals along a line or trace with a 3D configuration, or as *map data*, associated with points, lines and polygons referenced to a plane with a 3D orientation. The spatial extent of each characteristic is represented as *volume data*, in which characteristic values are associated with irregular volume geometries. The spatial variation of each variable is represented as *3D grid data*, in which a continuous measure of the spatial variation is simulated by values at the grid-cell centroids.

These implicit data geometries and their data associations prevent us from making efficient use of relational database technology. This will remain the case until spatial locations, spatial objects (points, lines, polygons, volumes and 3D grids), and (equally important) spatial queries can be accommodated efficiently in a relational context.

Data structure	Data identity	Data location and geometry	Characteristic data sources	Variable data sources
Hole data	Hole data subset Subset description Hole name Hole category Hole region	Collar location Survey records Distance intervals	Lithology intersections Mineral intersections Rock qualities Borehole geophysics	Core samples Well samples Test-pit samples
Map data	Map name Map description	Map origin Map orientation Feature geometry	Geological mapping Geological sections Site plans and surveys Surface contours Surface geophysics	Point samples Variable contours
Volume data	Volume type Volume name Volume description Component name Component description	Component origin Component orientation Component geometry	Geological codes	
3D grid data	Grid name Grid description	Grid origin Grid orientation Grid dimensions Cell dimensions	Geological intersections and codes	Predicted variables and uncertainties

Table 4.1   Geological characterization - the principal data structures and definitions required by the computerized process

4.3 Spatial Integration of Data Structures

The geometries of the four data structures introduced in Chapter 3 are the spatial objects that we must deal with. Besides an implicit geometry, each of these data structures has several other *attributes* of importance. These include an *identity*, consisting of one or more components, a *location* in terms of global coordinates and rotations, and associated *observation* and *measurement* values. These are summarized in Table 4.1. Identity, location and geometry are discussed for each of the data structures in the next four sections, followed by a discussion of how we deal with observations and measurements, and their identities. This organizational framework provides a platform for the entire characterization process.

At the lowest organizational level we need what we shall call a *global coordinate convention* within which we can define the location of any and all data objects. For convenience, we use a familiar NEL orthogonal system of northing, easting, elevation, azimuth (plan rotation) and inclination (vertical rotation). Although we are concerned with 3D space we can afford to neglect the third component of rotation, since we seldom stand on our heads to view geological information! In contrast, the internal geometry of any data object is typically defined in terms of what we shall call a *local coordinate convention* of orthogonal XYZ coordinates. These two coordinate conventions are defined for the various data structures in Figs. 4.1 through 4.4. They provide the necessary framework for spatial integration of all data structures.

In order to manage information effectively we require a means of readily transforming the location of any item of information from one coordinate system to another, as discussed earlier in Chapter 3 with reference to viewplane transformations (cf. Fig. 3.2a). The global coordinate reference for any point stored in the local coordinates of a particular data structure can be readily obtained by a coordinate transformation of the form :

$$P_i = P_0 + T \bullet P'_i$$

or, in expanded matrix notation,

$$
\begin{bmatrix} N_i \\ E_i \\ L_i \end{bmatrix} = \begin{bmatrix} N_0 \\ E_0 \\ L_0 \end{bmatrix} + \begin{bmatrix} \cos\Theta & \sin\Theta.\sin\Phi & -\sin\Theta.\cos\Phi \\ \sin\Theta & -\cos\Theta.\sin\Phi & \cos\Theta.\cos\Phi \\ 0 & \cos\Phi & \sin\Phi \end{bmatrix} \bullet \begin{bmatrix} X_i \\ Y_i \\ Z_i \end{bmatrix}
$$

$$\text{where} \quad \Theta = \text{global azimuth rotation}$$
$$\Phi = \text{global inclination rotation.}$$

Although perfectly adequate for our purposes, this is a mathematically inelegant transformation; it results from attempting to satisfy geological convention rather than computational aesthetics. A more elegant relationship would be obtained if

YXZ coordinates were equivalent to NEL coordinates, respectively, for $\Theta = \Phi = 0$. The inverse of the above relationship provides a reverse transformation from global to local coordinates, thus

$$\mathbf{P'}_i = \mathbf{T}^{-1} \bullet \left(\mathbf{P}_i - \mathbf{P}_0\right).$$

These are the standard coordinate transformations for converting global to local coordinates and vice versa. Similar transformations are used throughout for most data structure and viewplane operations.

Another common feature of all data structures is alpha-numeric naming, a component of the identity of any data structure to which wildcard retrieval functions can be applied. By employing naming conventions that are appropriate to our characterization, we can use familiar wildcard functions to rapidly and efficiently retrieve appropriate subpopulations of our data. This is of great advantage when we are dealing with large numbers of a particular data structure.

4.4 Hole Data Structure

In most characterizations the bulk of our investigative information is stored as hole data. However, it may derive from a number of different sources, such as boreholes, monitoring wells, exploration trenches or face mapping, and several sampling techniques may have been used for a particular source. For convenient differentiation of hole data by source and/or technique, we employ a *category* component in the identity, in addition to the name component. Because some characterizations cover a large area we also employ a *region* component that allows us to differentiate hole data by property, land parcel or mineral claim. Examples of these identity components can be seen in the hole data report shown in Fig. 4.1a.

Contaminated site characterizations frequently involve a number of observations or samples from the same data source at different points in time. The identity, location and geometry of the data structure remain constant, but the observation or sample values may change. To cope with this complication we employ yet another component of identity, in the form of hole data *subsets* and appropriate *descriptions*. This arrangement allows us to store a particular data source in a number of subsets, each representing a different time interval and containing relevant observation and sample values.

The location of a hole data structure is simply defined by northing, easting and elevation (NEL) coordinates that locate the *origin* (or start point) of its local geometry; the collar in the case of a borehole or monitoring well (cf. Fig. 4.1).

Definition of the geometry of the data structure is more complex. In the general case it is represented by a smooth line (or trace) with direction and curvature

```
PROJECT:TEST2.89                                              USER:swh
SUBSET: #0                                            RAW DRILLHOLE DATA

DRILLHOLE REPORT FOR SUBSET #0, HOLE:80-2, CATEGORY:80

HOLE #          NORTH        EAST      ELVN       LGTH       Sf1       Sf2 RG'N CG
--------    ----------   ----------  --------   --------   --------   -------- ---- --
80-2         22788.00    38006.00   1553.00     264.30                          80

  #           DIST       AZIM         DIP       #          DIST       AZIM       DIP
----        ----------   ----------  ----------  ----    ----------  ----------  ----------
0001          0.00      180.00       45.00     0002       5.20      189.00     45.00
0003         77.40      188.00       39.00     0004     168.80      189.00     32.00
0005        260.30      190.00       29.00

  #          DIST Lith     Sg     Cu     Zn      Ag     Au  Cu_eq Unit
----       -------- ------  -----  ------  ------  -------  -----  ------  --------
0001          3.40 zone1    2.00
0003         41.30 zone2    2.90
0004        177.60 zone2    2.90
0006        177.70 zone3    2.90
0008        196.00 zone5    2.80
0009        199.00 zone7    2.80
0014        212.20 zone7    2.88   0.29   0.43    7.52   0.12   0.56
0015        215.20 zone8    3.42   2.63   1.61   18.38   0.62   3.57 ORE
0016        216.90 zone8    4.24   2.68   2.72   19.44   0.51   4.16 ORE
0017        218.10 zone9    2.70   0.59   0.08    4.10   0.00   0.67 ORE
0018        219.50 zone9    2.70   0.14   0.06    1.70   0.00   0.18 ORE
```

a

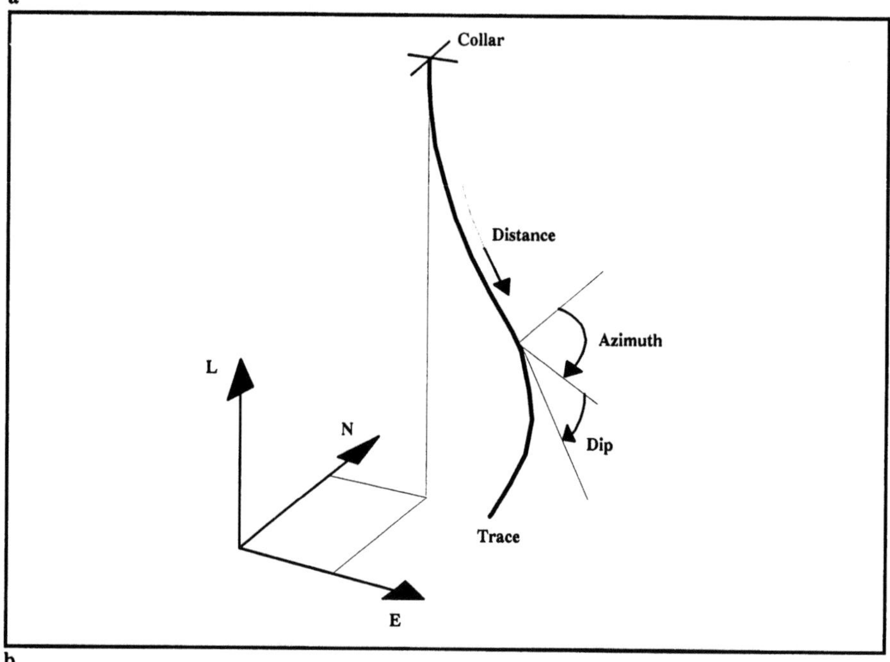

b

Fig. 4.1   **a** Hole data report including identity, location, downhole survey information, and downhole observation and sample information;   **b** geometrical definition of the hole data structure

varying throughout its length. In practice the direction is measured in terms of azimuth and dip (cf. Fig. 4.1b) at a finite number of discrete points (called survey points) along the length of the data source. Examples of survey records consisting of distance (from the origin), azimuth and dip readings are shown in Fig. 4.1a. To obtain an acceptable geometrical representation of the data source (boreholes generally do not have kinks in them!), the directional change between two points is spread evenly in small increments over the distance between them.

Thus, the generalized geometry can be represented in terms of global NEL coordinates by a simple incremental function of the form

$$N_i = N_{i-1} + \Delta S_i \cdot \cos(\Phi_{i-1} + \Delta\Phi_i) \cdot \cos(\Theta_{i-1} + \Delta\Theta_i)$$
$$E_i = E_{i-1} + \Delta S_i \cdot \cos(\Phi_{i-1} + \Delta\Phi_i) \cdot \sin(\Theta_{i-1} + \Delta\Theta_i)$$
$$L_i = L_{i-1} - \Delta S_i \cdot \sin(\Phi_{i-1} + \Delta\Phi_i)$$

$$\text{where} \quad \Delta S_i = \text{distance increment}$$
$$\Delta\Phi_i = \text{dip increment}$$
$$\Delta\Theta_i = \text{azimuth increment}$$
$$i = 0 \quad \text{represents the local origin of the geometry.}$$

This generalized representation of geometry allows any line configuration and orientation. The default condition (no survey records) is a straight, vertical line geometry. In the extreme case we may have to deal with several hundred survey records and/or negative dip values, in which case the geometry is oriented upwards (a frequent result of underground investigation of mineral deposits).

The intervals with which observation and sample values are associated are in turn defined by distance from the origin (cf. Fig. 4.1a). Their global NEL locations and extents are obtained by *mapping* the distances onto the line geometry.

Thus, for any form of spatial analysis or visualization, we must first apply the geometrical transformation described above. To project hole data information onto a viewplane for visualization purposes we must also apply an inverse transformation of the form described in Section 4.3, thereby transforming all data locations to the local coordinates of the viewplane. On the other hand, for convenient editing and reporting purposes we need to access a hole data source in the flat text format illustrated in Fig. 4.1a, similar to the familiar borehole log format.

4.5 Map Data Structure

The number of map data structures to be dealt with on a project is typically between 10 and a 100, although they may be individually quite large. They

```
PROJECT:OPEN PIT MINE DESIGN                         Sun Jan 23 12:40:14 1994
                                                          USER: SWH
MAP ID:RD                        DESCRIPTION:Haul Road Figure - MAS 09-06-93
ORIGIN:5097000,308000,147,90,90

Group         y            x            z          Elvn
-----  ------------ ------------ ------------ ------------
crest       15.00       -25.06        12.50        12.50
            15.00         0.11        12.50        12.50
            15.00       100.10        12.50        12.50
             4.75       100.33        12.50        12.50
             4.75       149.98        12.50        12.50
toe         10.11       -25.06         0.00         0.00
             9.88        -0.12         0.00         0.00
             0.09         0.11         0.00         0.00
             0.09       100.10         0.00         0.00
             0.09       149.98         0.00         0.00
line         9.88         0.34
            14.77       100.10
             4.98       100.33
             0.09         0.11
             9.88         0.34
line         6.62        33.21
             7.31        49.75
             8.25        47.66
             7.31        49.75
             6.38        47.66
```

a

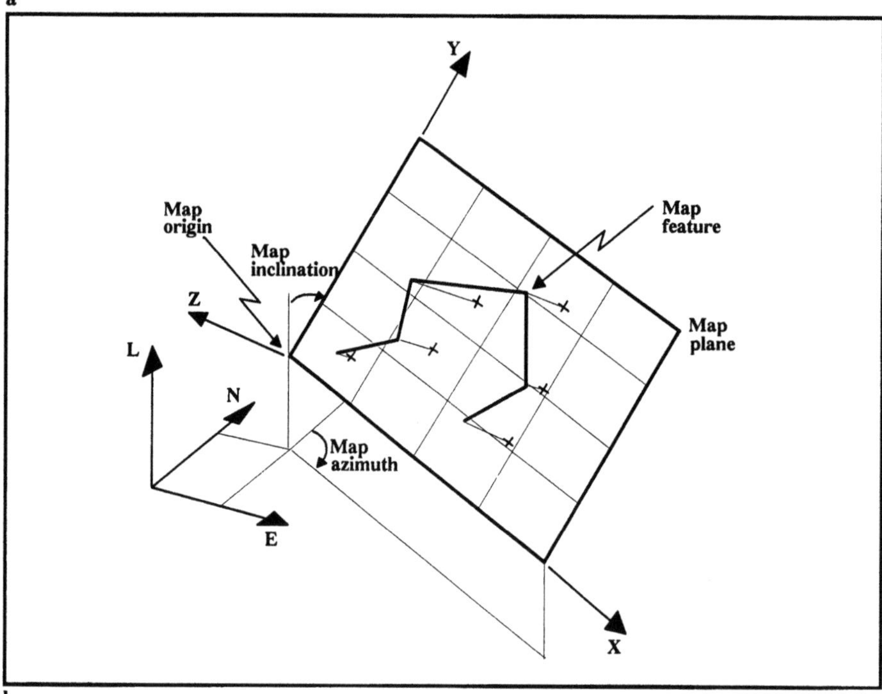

b

Fig. 4.2 **a** Map data report including identity, location, feature type and local geometry information; **b** geometrical definition of the map data structure

can generally be readily differentiated by employing an appropriate *name and description* convention only; category and region components are unnecessary in this context.

The *location* of a map data source is defined by the global NEL coordinates of its origin, and the azimuth and inclination rotations of its plane. These parameters also define the local XYZ coordinate system (cf. Fig. 4.2b). The global reference for any point stored in the data structure can then be readily obtained by the standard coordinate transformation described earlier in Section 4.3.

In a map data context these transformations allow us to define (or import) information conveniently and efficiently in a local coordinate system by digitizing a contour map, for example, or the geological boundaries of a cross section, and yet the information can be immediately and precisely located in 3D space. We only have to ensure that our viewplane coincides with the defined location and orientation of the map data structure. The geometry of map data can then be defined in terms of *features* such as points, lines and polygons in the local coordinate space.

Each point within a map feature is located by X and Y coordinates in the plane of the map and, where appropriate, a Z coordinate normal to the map plane; a Z value of zero indicates that the point lies in the map plane (cf. Fig. 4.2b). One or more observation and sample values can then be associated with each feature. A variety of feature types is required to deal with different data sources. If we are dealing with discrete sample points, then their locations are represented by *point features* with XYZ coordinates, and each point may have any number of associated observation and sample values. In the case of line contours (or isolines), each is represented by a *contour feature* containing a multitude of points with constant Z value, and (typically) a single associated variable value, assigned to the length of the feature. On the other hand, a survey traverse is represented by a *traverse feature* with different Z values and multiple characteristic or variable values at each point in the feature. The same feature is also useful for representing geophysical traverses. Geological mapping is represented by closed *polygon features* with an associated characteristic value attached to the enclosed area. Site plans and engineering design information are represented by simple *line features* with a zero Z value that lie in the map plane.

All of these map data features can be readily located in global coordinate space by transformations of the form discussed in Section 4.3 and subsequently projected onto any viewplane for visualization purposes by the equivalent inverse transformation.

Since we are frequently dealing with large numbers of features in a single map data structure, they are generally selected and retrieved by simply pointing at them interactively in a graphical display; there is typically no necessity to assign any further identity to them. In certain cases, however, it may be more convenient to access map data in a flat, text file format (cf. Fig. 4.2a) for simultaneous editing of a large number of features, for example.

4.6 Volume Data Structure

The primary objective of the volume data structure is to provide us with a means of qualifying subsurface space by assigning characteristic values to discrete, irregular geological volumes. However, it must also satisfy two important secondary objectives, which we have already identified in an earlier discussion. The first is concerned with providing an avenue for interactively imposing our interpretational control on the characterization process. The second concerns the application of spatial (geological) control to the geostatistical prediction of variable values. The volume data structure satisfies all of these objectives by employing the unique geometrical characterization of irregular volumes defined in Fig. 4.3a.

A complete geological characterization may require several geological volumes (or models), each consisting of tens of thousands of these volume elements, called components for convenience. Further, a single geological unit within a model may be represented by several thousand components. We therefore require a convenient and efficient means of identifying component subsets. To achieve this we employ a hierarchical naming structure (identity) consisting of volume type, volume (model) name and description, and component name and description. The *volume type* determines whether we are dealing with geological or engineering (excavation) volumes. The *volume name* allows us to employ several volume models within a characterization, either to represent different regions, or to represent the different spatial distributions of several characteristics, e.g. lithology and mineralization. The *component name* is split into two subunits, each of which employs wildcard functionality and accepts customized naming conventions. With this arrangement we can readily retrieve all components in a volume, or all components representing a particular geological unit or subunit, or all components with a particular characteristic value, or just a single component.

Unlike the map data structure where all of the features contained are referenced to a single map plane, every component within a volume data structure is referenced to its own component plane, with its own location, orientation and local XYZ coordinates. This arrangement provides the high degree of flexibility we require for characterizing complex conditions. Each component of a volume can then be defined and referenced to the plane which is most appropriate in terms of geological complexity and/or continuity and available information. Thus, in similar fashion to the map data structure, each component is first located in space by means of global NEL coordinates and rotations that define the origin and orientation of the component plane (cf. Fig. 4.3a). The geometry of the component is then defined in terms of its local XYZ coordinates.

The geometry consists of three *closed polygonal boundaries*, a mid-plane (MP) boundary coincident with the component plane; and two parallel boundaries extended normally on either side of the component plane, referred to as the fore-plane (FP) and back-plane (BP) boundary, respectively (cf. Fig. 4.3a). These boundaries are connected by straight line *links* that join the corresponding points in each of them. The Z coordinate values of each of the extended boundaries are

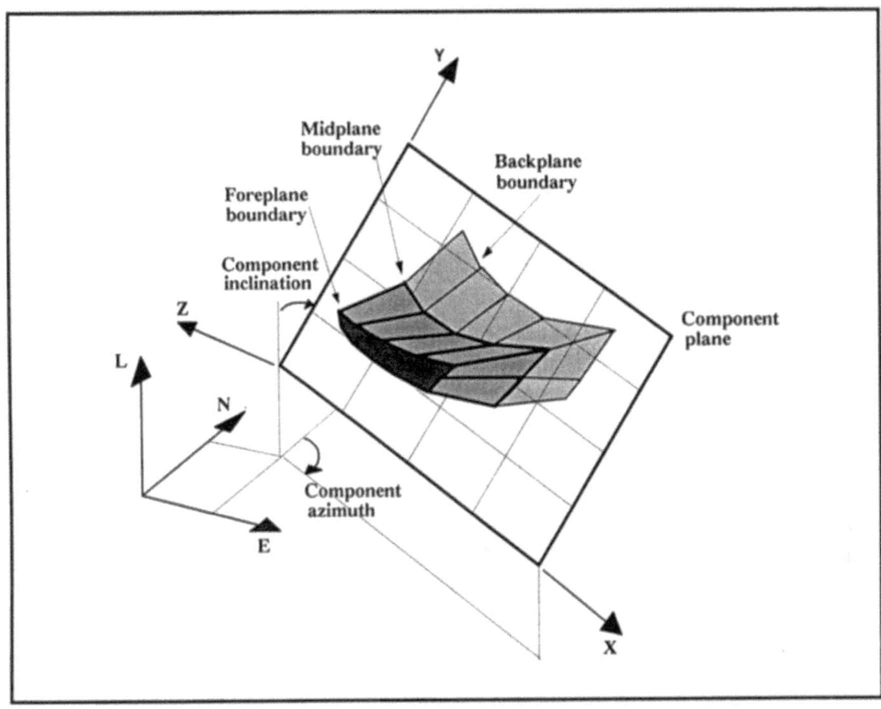

a

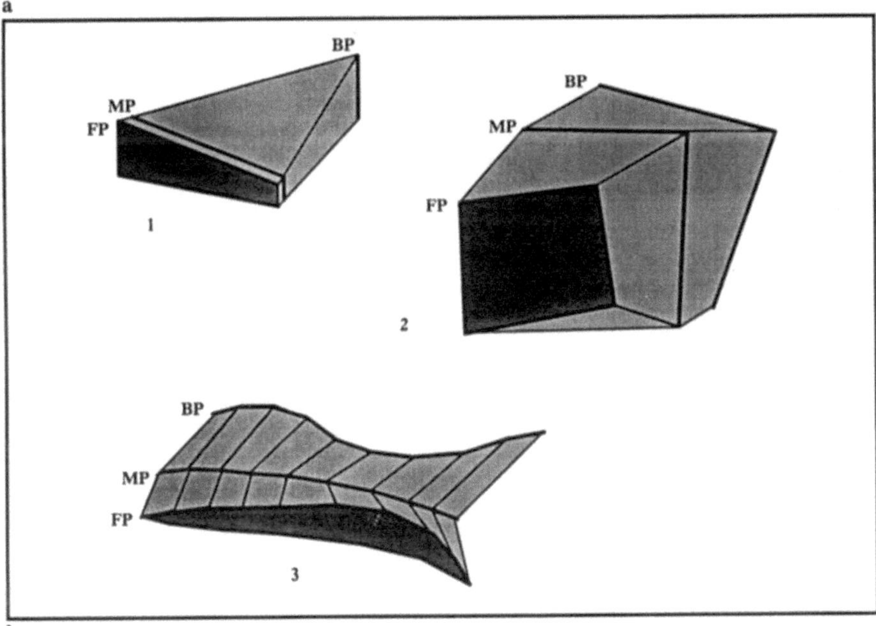

b

Fig. 4.3 **a** Geometrical definition of the volume data structure; **b** examples of volume data geometry: **1** simple triangulated component; **2** fault block component; **3** general component

constant, determined by specified *fore- and back-thickness* values. By default these thicknesses are initially set to a nominal value and the extended boundaries are equal and immediately adjacent to the mid-plane boundary. Each thickness and boundary can subsequently be modified independently to represent a wide range of geometrical shapes (cf. Fig. 4.3b). Component configurations vary from a simple triangulated shape, which can be rapidly generated between triangulated surfaces, to complex, highly irregular shapes that must be defined by interactive interpretation. The only constraints are that the extended boundaries should be on opposite sides of the component plane; an individual boundary must be convex, i.e. not cross over itself; and links between boundaries should not cross in the third dimension through excessive relative twisting of adjacent boundaries.

Each boundary consists of a set of points, located by XYZ coordinates, and straight line segments joining the points. The amount of detail that can be incorporated in a boundary is dependent on the number of points, which in turn is (generally) dependent on our own diligence and the requirements of our characterization. Any irregular shape can be represented by as many of these components as are necessary to achieve the required degree of volume detail. The enclosed volume can be qualified by assigning an appropriate characteristic value to each component.

Just why we should choose to represent irregular volumes in this way will become more apparent when we discuss application of volume modeling to the characterization process in Chapter 6. In summary,

- The ability to define the component plane at any appropriate location and orientation is of great advantage in the geological interpretation process.
- The initial (nominal thickness) state of a component is readily applicable to interactive interpretation of geological boundaries on 2D sections and plans; and the ability to extend and modify the boundaries in the third dimension satisfies our requirements for 3D interpretation.
- The geometrical characterization of a component lends itself to efficient computation of boundaries of intersection with planes at any orientation, which satisfies our requirement for visualizing sections through geological volumes.
- The geometrical characterization also lends itself to efficient computation of volumes, and volumes of intersection between components; this provides the basis for integration between geological modeling and geostatistical prediction, as discussed in Section 4.7.

The global coordinate references for any point contained in a volume data structure is readily obtained by a standard transformation of the form described in Section 4.3. Its local coordinate reference for some other plane (typically a viewplane) can then be obtained by the inverse transformation.

4.7 3D Grid Data Structure

The 3D grid data structure provides an efficient medium for representing the spatial variation of our characterization variables. It simulates a spatially continuous measure of a variable by assigning values which are representative of a finite, small volume to fixed points at regular intervals. The fixed points are the centroids of each grid cell and the assigned values are the predicted values at the centroids. We call this a raster approach because every point in the data structure represents a finite equal volume.Since we typically employ only one 3D grid structure for a characterization, and seldom more than two, a simple *name* and *description* are adequate for identity purposes.

The 3D grid is *located* by global NEL coordinates and rotations that apply to its origin (cf. Fig. 4.4b). Note that it is occasionally useful to be able to incline the grid for characterization of mineral deposits and hydrogeological conditions.

The *geometry* of a 3D grid is defined within the local XYZ coordinate framework in terms of cell dimensions, cell indices and numbers of cells in each direction (cf. Fig. 4.4a). The points of interest within the grid are the cell centroids, since all predicted variable values are assigned to these points. The local coordinates of the centroid of the cell (i,j,k) can be readily obtained from an expression of the form

$$\mathbf{P'}_{ijk} = \begin{bmatrix} \mathbf{X}_{ijk} \\ \mathbf{Y}_{ijk} \\ \mathbf{Z}_{ijk} \end{bmatrix} = \begin{bmatrix} \Delta X \cdot (i-1) + \Delta X / 2 \\ \Delta Y \cdot (j-1) + \Delta Y / 2 \\ \Delta Z \cdot (k-1) + \Delta Z / 2 \end{bmatrix}$$

and the global coordinates are derived from a standard transformation of the form (cf. Section 4.3)

$$\mathbf{P}_{ijk} = \mathbf{P}_0 + \mathbf{T} \cdot \mathbf{P'}_{ijk}$$

For visualization purposes these are in turn transformed to the local coordinates of a viewplane using the inverse form of the expression.

In order to apply spatial (geological) control to the variable prediction process we must integrate the 3D grid data structure with the volume data structure. Essentially, we need to know the *characteristic values* and *volumes* of all geological materials intersected by each cell in the grid. We achieve this by performing a volumetric intersection between each cell and a volume model that represents the spatial distribution of the characteristic. The volumetric analysis techniques used for this are discussed in Chapter 6. In the context of the 3D grid data structure, this requires that the data structure accommodate a list of volumes and characteristic values for each cell. When we come to predicting the variable values associated with a particular characteristic value, we can then interrogate this

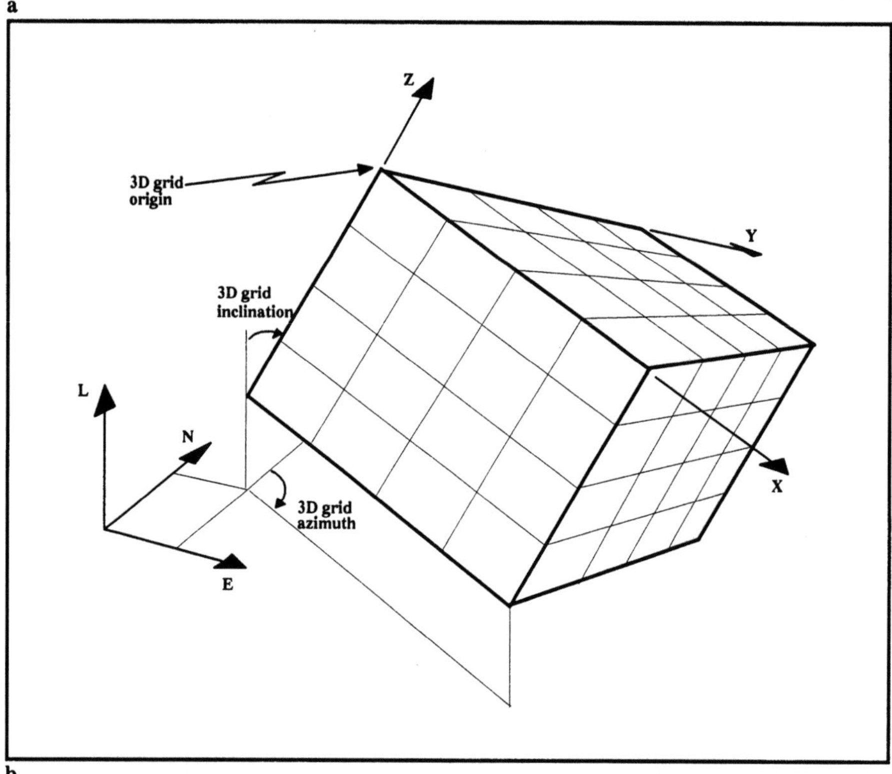

a

b

Fig. 4.4   a 3d grid data structure definition and report; **b** geometrical definition of the 3D grid data structure

list to establish whether that characteristic value exists within the cell. Only if it does will a variable value be predicted.

This requires that the data structure should also accommodate a list of values for each cell, one for each characteristic value, for every variable that we may wish to predict. And because prediction uncertainty plays a large part in the characterization process, every variable value is (generally) accompanied by a measure of its uncertainty. This adds up to a lot of values! If we are dealing with ten characteristic values and the same number of variables, then the total amounts to a possible 220 values per cell and, in the typical case, we could be dealing with a 3D grid containing a million or more cells.

4.8 Data Definitions and Identities

We have established a methodology for identifying and locating data structures and, through the unique, implicit geometry of each data structure, a methodology for locating the associated characteristic and variable values in 3D space. These values are derived from observations and samples from site investigation (the source data), from our own interpretation of the spatial distribution of geological characteristics, and from subsequent prediction of the spatial variation of variables.

Every characterization project has its own list of relevant characteristics and variables. At the start of a project this list must be defined, in terms of appropriate identities, types, reporting formats, etc., to provide a data framework that accommodates the source data. This list is typically expanded as our characterization progresses, in order to accommodate new characteristics or variables derived from functions, interpretations or predictions involving the source data.

In general, investigative source data are stored as hole or map data structures; interpretation results are stored as volume data structures; and variable prediction results are stored as 3D grid data structures (cf. Fig. 3.1, Chap. 3). In the hole data structure, each set of characteristic and observation values is associated with a defined interval along the line geometry of the structure. In the map data structure, each set is associated with point, line or polygon features referenced to the map plane. The volume data structure only accommodates characteristic values; each value is associated with a component volume. The 3D grid data structure essentially accommodates variable values, associated with grid-cell volumes and located at the cell centroids. However, it also accommodates characteristic values and volumes for spatial control purposes.

4.9 Data Interfacing Considerations

In the previous sections we established a spatial data framework that accommodates all of our information management requirements. In almost every case of characterization we also need to consider interfacing with external computer technologies. Some or all of the source data may exist beforehand in electronic form, typically in spreadsheet, database or GIS format. Moreover, the results of our characterization frequently provide the input for further studies, such as groundwater flow modeling, reservoir production simulation or mine planning (cf. Fig. 3.10, Chap. 3). To satisfy these interfacing requirements we require efficient data transfer (import/export) capabilities for each of the data structures.

As yet there are no suitable industry standards for transferring 3D geoscience information. Most of the accepted standards, such as IGES, DXF, HPGL and various raster image standards, were developed originally for drafting, mapping or graphical applications. They tend to lose information and connectivity (structure) in a 3D context, even when dealing with relatively simple map data structures. They are hopelessly inadequate in the context of volumes and 3D grid data structures. More recent and promising developments in the formulation of spatial data standards have been sponsored by the petroleum industry. These are discussed briefly in Chapter 16. In the meantime we are forced to rely on the next best alternative to standards. This involves ensuring that all import and export files include sufficient intelligence to enable the contents to be deciphered.

This approach to data transfer is generally known as the *self-defining file* approach. At the lowest level it employs the ASCII standard for transfer of character-based information, which is widely accepted and intelligible to all modern computers. All *data records* within a file are preceded by a *format record* that defines their structure. This ensures that the data records can be readily manipulated, or re-formatted, to suit the requirements of the target technology. Most computer operating systems provide data manipulation tools. For example the UNIX operating system provides AWK, a sophisticated, high-level data manipulation tool which any geoscientist involved in computer applications should be aware of. Without self-defining files and data manipulation tools an inordinate amount of time is likely to be spent moving data from one technology to another.

5 Analysis of Spatial Variability

5.1 Objectives of Spatial Variability Analysis

Once we have assembled all of the available investigation information into appropriate data structures, as discussed in Chapter 4, we are ready for the *information review and analysis* step. The principal objective is to obtain an in-depth appreciation of the available information that will dictate the course of the characterization process. The primary tools available for this purpose are *visualization, statistical analysis and spatial variability analysis.* Visualization techniques and statistical analysis are familiar to most of us and are not discussed in detail here; the case studies presented in Part II contain many examples. On the other hand, spatial variability analysis (or *geostatistics* as it is commonly called) is less familiar, particularly to those of us outside the mining sector. This is the principal focus of this chapter.

Since predicting the spatial variation of a variable plays such a large part in geological characterization, a concern as to whether our selected prediction technique is appropriate would be consistent with our objectives. In practice we often apply whatever technique is at hand without giving much thought to whether the algorithms employed are capable of simulating the real, physical variability or not. The commonly available techniques range from simple hand-contouring at one end of the spectrum, through various 2D and 3D computer gridding and contouring algorithms (typically variations on inverse distance weighting), to geostatistical kriging. In Chapter 16 we discuss several new, alternative prediction techniques that have particular advantages. All of these techniques are more or less appropriate in varying degrees to different sets of conditions, and can result in significantly different predictions from the same source data.

The point to be made is that, whichever technique we use, we should have some means of measuring its *appropriateness* relative to the true conditions. To achieve this we require a measure of the true spatial variability; we can then select a prediction technique which is demonstrably appropriate. Unfortunately, we typically never know the true spatial variability, but we can obtain an estimate of it by analyzing the available samples and observations. This process is called *spatial variability analysis*, and the principal analytical tool available for the purpose is the *geostatistical semi-variogram.* By applying semi-variogram analysis, we first of all determine whether the spatial variability is in fact measurable; we perhaps need another round of sampling before we can detect a meaningful result.

Secondly, we obtain measures of how much of the sample variability is *truly random*, how much (if any) is *spatially dependent*, and the *range(s) of influence* over which spatial dependency is in effect. Thirdly, assuming a measurable spatial relationship, we use the results to derive a *prediction model* which is *customized* to the spatial variability of the available samples. The result is the most appropriate prediction model we can get; moreover, we have the means at our disposal to both justify and validate it. The latter are important considerations in any characterization, but particularly so in an environmental contamination context where public health is the prime consideration. Since the extent of remediation is likely to be based largely on our prediction of contamination, it is crucial that we employ all possible means of justifying and validating our results.

This approach is based on the assumption that the available samples and observations are themselves representative of the true conditions. Of course this may not always be the case, however, spatial variability analysis also presents us with a logical basis for establishing whether or not this is true and for controlling our investigation sampling. This aspect is discussed in more detail in Chapter 10.

The principal objective is to quantify the spatial variability of our data. The analytical process typically consists of a series of iterative applications of both semi-variogram analysis and statistical analysis to different subpopulations of the data with different analytical parameters. The purpose is to establish the presence and significance of a number of possible influences in the spatial variability of the data. Those of most concern include geological influences, directional influences, the distribution of sample values (irrespective of location), and the possibility of underlying (determinate) trends in the data. If any of these influences are present to a significant degree, then we need to be aware of them; to a large extent, they determine the appropriateness of the prediction models which result.

To a large extent they also control our interpretation of geological characteristics. Unless we are aware of the geological characteristics that influence the spatial variation of a variable we do not really know which characteristics to interpret. We could be wasting our time performing a detailed interpretation of lithology, only to find that it is of little significance to the end result. In a contamination study there might, or might not, be significant variation differences between different soil types; either way we save ourselves much time and effort by establishing these conditions before we embark on the interpretation step.

To begin our discussion of spatial variability it is convenient to assume that all of these influences are absent from our data. Subsequently, we will discuss their individual significance in detail. The following discussion is aimed at providing a basic understanding of spatial variability and its analysis, its application in practical terms and its advantages and disadvantages. A familiarity with standard statistical analysis techniques is assumed. For those readers who wish to extend their knowledge in geostatistics, there are many detailed, theoretical texts available.

5.2 Measurement of Spatial Variability

Whereas statistical analysis is concerned with value, geostatistical analysis is concerned with *difference in value* and *distance*. An underlying assumption of the semi-variogram is that any spatially determinate trend in the sample values is either nonexistent or has been removed prior to analysis. We assume that differences in the value of a variable are related in some way (within a limited distance), whereas the values themselves are indeterminate. Our objective is to determine a meaningful relationship between difference and distance, essentially a measure of how difference in value changes with distance. We call this measure the *spatial variability* of a set of sample values, each of which is located in space by 3D coordinates. The principal value of semi-variogram analysis lies in its ability to present the results in such a way that we can readily detect any such relationship. Despite the wall of mystique that has been erected around geostatistics, the semi-variogram, which lies at the heart of all geostatistical analysis and prediction, is really a very simple and logical concept, albeit with an unnecessarily complicated title.

The easiest way to explain the analytical process is to follow it through for a given set of samples. Assume we have m samples, each with a value v for the variable of concern and a location defined by three orthogonal NEL coordinates. Suppose we select a pair of samples (cf. Fig. 5.1a) and subtract their variable values (denoted by v_1 and v_2) to obtain the difference. Depending on the order in which we select the pair, the result will be either positive or negative. Negative values present an unnecessary mathematical complication so we square the difference to make life easier. We could perhaps have taken the absolute value instead but difference squared is a more rigorous approach and, as we shall see later, has considerable statistical significance. Let us assume for now that this is a suitable measure of variability and denote it by 2γ (the factor 2 is a mathematical convenience which we accept at face value). Thus, the variability of our sample pair is given by

$$2\gamma_{12} = (v_1 - v_2)^2.$$

The distance s between the pair of samples is the length of the vector between their locations, obtained from a simple 3D hypotenuse expression involving global NEL coordinates. As well as length, this vector has an orientation in terms of azimuth and inclination values (cf. Fig. 5.1a), however, we ignore the orientation at this stage and assume that differences can be measured the same way in any direction.

We now use the textbook approach to determine a relationship between any two variables; we plot them on a graph with γ along the ordinate axis and s along the abscissa. Because we plot γ instead of 2γ, we call the graph a *semi*-variogram instead of a variogram. If we repeat this exercise for a large number of pair combinations, we might be fortunate enough to discern a spatial relationship.

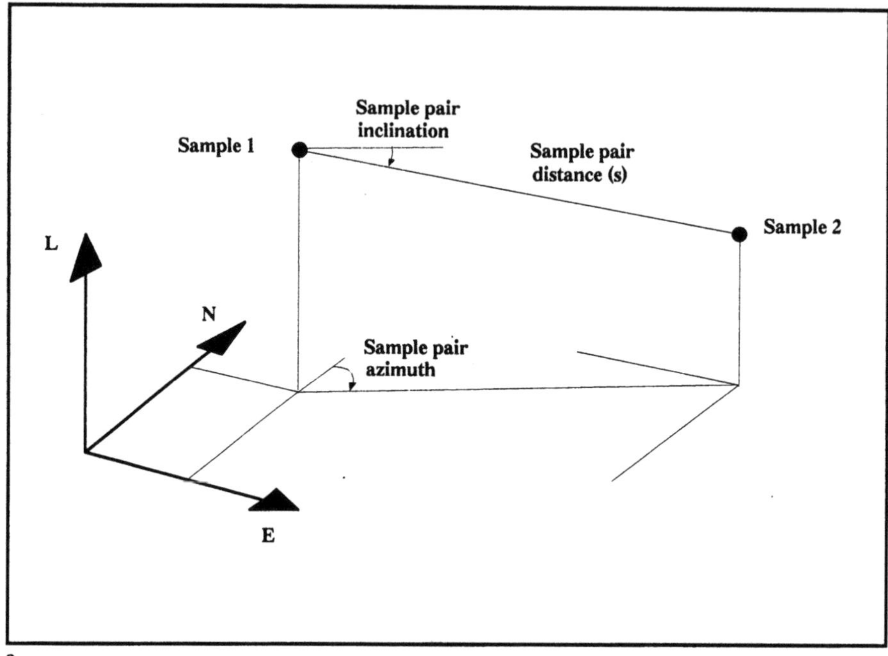

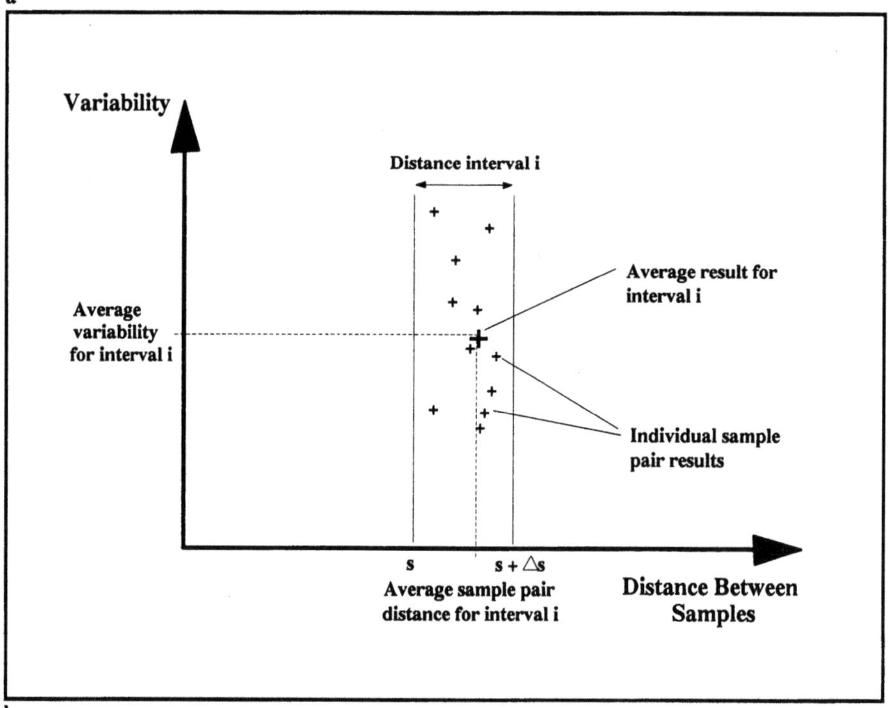

Fig. 5.1   a Sample pair geometry;   b Graphical presentation of the results of spatial variability analysis; the basic format on the semi-variogram

In practice, even a moderately sized sample set has so many possible pair combinations (m! of them) that the result is more likely to be unintelligible. Instead we group the sample pairs into *distance intervals* of width Δs and *average* the γ and s values within each interval (cf. Fig. 5.1b). Providing the selected interval spacing is appropriate (we may have to experiment with different Δs values), a more intelligible result is obtained and we are much more likely to discern any relationship which is present. The average γ value corresponding to an interval, which replaces a number of individual points on our graph, is obtained from the general expression

$$2\gamma_i = \frac{1}{n_i} \sum (v_j - v_k)^2 \quad \text{for} \quad s < s_{jk} \leq s + \Delta s,$$

where $2\gamma_i$ is the average variability for all sample pairs whose distance apart is within distance interval i, between s and s+Δs (cf. Fig. 5.1b). The number of sample pairs in the subset is represented by n_i. The corresponding distance is s_i, the average distance between the sample pairs in the interval. The above expression for $2\gamma_i$ has statistical significance as well. We can show (with several pages of algebraic manipulation) that it is equal to the statistical *variance* of the sample differences for interval i. Thus, the semi-variogram is actually a plot of *semi-variance* versus distance. This has significant impact on the prediction process, discussed in the next section. Of most significance is that we can derive the statistical *standard error* σ_i (also called standard deviation) of the differences for an interval from its semi-variance

$$\sigma_i = \sqrt{2\gamma_i}\,.$$

The advantage of the standard error, assuming a normal distribution of sample differences within the interval, is that multiples of it can be related to different confidence intervals (cf. Chap. 10). For example, with 68% confidence an unknown value is within one standard error of the average; with 96% confidence, it is within two standard errors, etc.. This opens the way for analysis of uncertainty and risk. Because semi-variance has such significance in the analytical process, it is appropriate to use this term with reference to γ for the remainder of our discussion.

In practice the results of a semi-variogram analysis should look similar to Fig. 5.2a, the semi-variogram for benzene concentrations from a set of 146 soil samples from a contaminated site. In this example, the concentrations are expressed as logarithmic values and we have used a distance interval of 5 m. What we are looking for in the results is a relationship γ (s) which relates the semi-variance of sample differences to distance between samples. Such a relationship is already discernible in our semi-variogram example for the benzene concentrations. In general, the γ values increase as distance increases up to a limit of approximately 55 `m, above this they appear to level off and oscillate about a constant value.

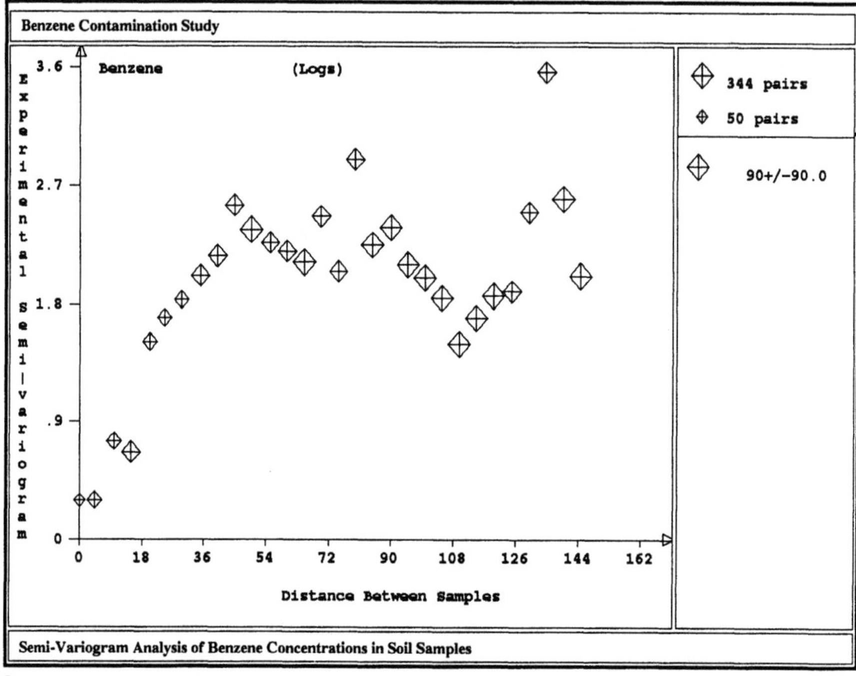

a

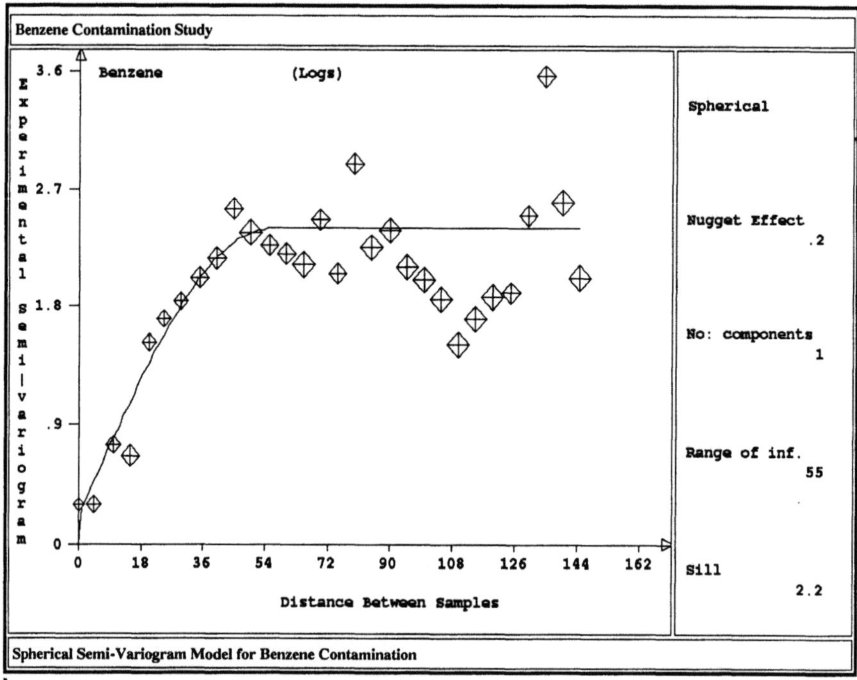

b

Fig. 5.2   **a** Semi-variogram analysis of benzene concentrations from soil samples at a contaminated site;   **b** spherical semi-variogram model fitted to the analytical results

Intuitively, we expect any such relationship to show that semi-variance increases as distance s increases, in other words, a relationship which is positive in a general sense (the reverse is unlikely in practice!). If we think about the semi-variogram for a minute, then there are several other results, besides a positive relationship, which we might anticipate. A positive spatial relationship is not going to be infinite, ever-increasing differences are another practical impossibility. Any relationship which is present is going to be true only over a finite range. At greater distances, as the samples become independent of each other, we expect the semi-variance of the samples to oscillate about some constant value; logically this should be of the same order as the semi-variance of the entire sample population. Theoretically at least, γ should also approach zero as s approaches zero; two samples which are very close together should have almost the same value. In practice this almost never occurs; there is invariably some degree of *background noise* in the sample values, due either to natural randomness, or sampling error, or both.

This approach to the analysis of spatial variability was derived by Matheron 1970, who originally titled it *the study of regionalized variables*, subsequently abbreviated to *geostatistics* by the mining sector. Matheron also derived a theoretical relationship for γ (s) which meets the expectations discussed above. This model (or idealized) semi-variogram relationship is typically referred to as the *spherical model* and is expressed mathematically in general terms as

$$\gamma\,(s) = C_0 \quad \text{for } s = 0$$

$$\gamma\,(s) = C_0 + C\left(\frac{3s}{2a} - \frac{s^3}{2a^3}\right) \quad \text{for } 0 < s \le a$$

$$\gamma\,(s) = C_0 + C \quad \text{for } s > a.$$

In this expression, C_O is the inherent randomness, or background noise of the samples, referred to as the *nugget value*. The distance at which samples become independent of one another is represented by a, referred to as the *range of influence*. The constant value which γ levels off to at distances greater than a is represented by $C_O + C$, where C is referred to as the *sill* of the semi-variogram model, and is generally of the same order as the semi-variance of the sample population. The spherical model was initially developed as a theoretical relationship, based on intuitive expectations of real spatial variability, however, it has been shown to work well in practice. It has since become the same sort of standard to geostatistics as the normal distribution is to statistics. We will discuss a range of prediction models in more detail later in this chapter. An advantage of the spherical model is that it clearly illustrates the concept of a logical relationship which is capable of representing real spatial variability. It is by far the most frequently used model in practice and has been found appropriate to a wide range of spatially dependent variables. The spherical model is defined in general terms in Fig. 5.9; a specific model that has been appropriately customized to the spatial variability of the benzene contamination samples is shown in Fig. 5.2b.

To fit the theoretical model to the observed spatial variability we adjust the values of the model parameters C_0, C, a until we have achieved a good visual approximation of the semi-variogram results. Figure 5.2b is an example of a spherical model which provides a reasonable approximation of the spatial variability of the benzene contamination samples. The initial choice of model parameters is usually based on a purely visual appreciation, later on we discuss various tools for verifying the appropriateness of the model. A factor which influences our choice of parameters is the number of sample pairs in each distance interval, indicated in Fig. 5.2a by the relative size of the plotted symbols. In this case, the largest symbol represents 344 pairs, while the smallest represents only 50 pairs. Obviously, we attach more weight to the larger symbols in our visual appreciation of the results.

5.3 Understanding Geostatistical Semi-Variograms

What do the semi-variogram results and the spherical model presented in Fig. 5.2 tell us about our benzene concentration values? First of all, there is a clearly discernible relationship between sample semi-variance and distance up to a range of influence of 55 m. This result has the most immediate significance, particularly if we are still involved in the site investigation phase. One of the conclusions we can derive from it is that the average spacing of future sampling should be between three and five times less than this range. If we sample closer than this, we are wasting our investigation budget; if we sample further apart, we do not have enough points on our semi-variogram to determine a meaningful spatial relationship.

The results also tell us that there is a background randomness present in the sample population represented by a semi-variance of $C_0=0.2$ at zero distance, which translates to a statistical standard error of approximately 0.6 (from $\sigma = \sqrt{2\gamma}$). No matter how closely we sample our site a background randomness of this order is always likely to be present in the results. We will subsequently discuss the standard error term in considerable detail as it is our primary measure of uncertainty. What it means in the immediate context is that the *real value* of the benzene concentration at any point could, with 68% confidence, be within 0.6 either way of the *measured value* at a sample location immediately adjacent. The underlying assumption here is that the logarithmic benzene concentrations approximate a statistically normal distribution, which they do in this case.

At distances greater than 55 m the sample pairs exhibit a semi-variance that oscillates about a value of $C_0+C=2.4$, which is approximately 2.5 times the semi-variance of the entire sample population. In practice, C_0+C is invariably greater than the population semi-variance because the former excludes sample pairs which are closer together, while the latter includes them.

Now consider the spatial relationship represented by our model semi-variogram in Fig. 5.2b in more detail. This model is an appropriate (or acceptable) approximation of the *measured* relationship between the semi-variance of benzene concentrations and distance. We therefore have a logical basis for *predicting* the semi-variance between values at the sample locations and any other (nonsample) point. Assume for the moment that we are concerned with a point p which is 18 m distant from sample location 1 and 36 m from sample location 2 (cf. Fig. 5.3). We can scale off an estimate of the semi-variance value for a distance of 18 m from the model semi-variogram as shown in the lower part of Fig. 5.3; the result is $\gamma_{p1} = 1.3$ which translates to a standard error of 1.61. This means that with 68% confidence the *predicted concentration* v_p at point p is within 1.61 of the *measured concentration* at sample location 1. Similarly, we can obtain $\gamma_{p2} = 2.1$, equivalent to a standard error of 2.05, which means that the predicted concentration v_p is within 2.05 of the measured value at sample location 2. From these results we can see that sample 1 provides a better estimate of v_p, with less error, than sample 2, which we would intuitively expect since sample 1 is closer. However, we have many more than two samples, a total of 146 in this case, and each of them can be used individually to predict the concentration at point p in this way.

These individual predictions for point p are not particularly useful to us; we need some form of *weighted averaging algorithm* for combining them to obtain a single optimized prediction v_p^*. For this purpose a linear, weighted combination of the sample values that gives more weight to the closer samples is a logical choice, thus

$$v_p^* = w_1 v_1 + w_2 v_2 + \ldots\ldots + w_n v_n,$$

where the weights w_i sum to one. Now we need a rationale for assigning the weights; what better way to assign them than on the basis of minimizing the resulting variance, and hence the error, at point p? *The resulting v_p^* is then the optimum predicted value with minimum associated error and uncertainty and maximum confidence.* In fact, we can achieve this quite easily by substituting the above in the expression for statistical variance for a linear estimator, differentiating with respect to each of the weights w_i in turn, and solving for minimum variance

$$\frac{\partial}{\partial w_i} (\text{variance}) = 0 \quad \text{for } i = 1, 2, 3, \ldots\ldots n.$$

The result is a set of equations in terms of the as yet unknown weights w_i, coefficients obtained from the semi-variogram model, and constants which ensure mathematical rigorousness. The coefficients are the estimated semi-variances between the sample locations and point p and between the samples themselves. These are obtained by scaling γ values off the model for predetermined distances, as we did for point p above. By adding the final condition that the weights w_i must sum to one, we can solve these equations to obtain the optimum set of sample weights, and back-substitute these in the above expression to obtain the optimum value of v_p^* and its associated standard error.

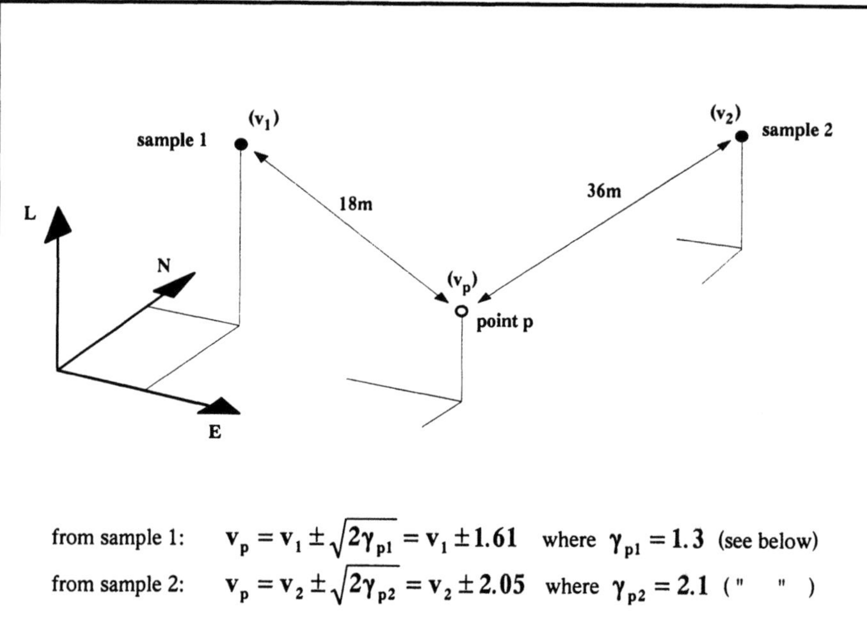

from sample 1: $\mathbf{v}_p = \mathbf{v}_1 \pm \sqrt{2\gamma_{p1}} = \mathbf{v}_1 \pm 1.61$ where $\gamma_{p1} = 1.3$ (see below)

from sample 2: $\mathbf{v}_p = \mathbf{v}_2 \pm \sqrt{2\gamma_{p2}} = \mathbf{v}_2 \pm 2.05$ where $\gamma_{p2} = 2.1$ ("   ")

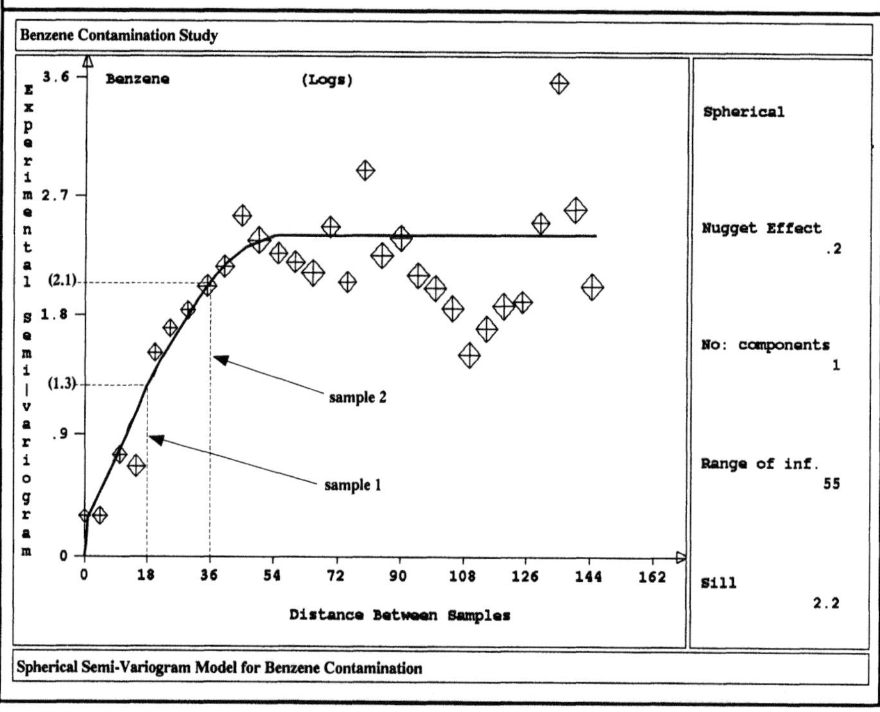

Fig. 5.3 Use of the semi-variogram model of benzene contamination (cf. Fig. 5.2b) with point-to-sample distances and sample values to predict benzene concentrations at points located between the samples

We are not going to show the mathematical steps or the resulting equations here; they require several pages and have already been adequately detailed in several geostatistical texts. For those interested, the general form of the equations is given in Chapter 16. In practice we seldom use all of the available samples to predict the value at a particular point. Typically, we ignore those samples which are at distances greater than the range of influence of the semi-variogram model, since we have determined that they are unrelated anyway.

This prediction technique is called *kriging* after its founder, Danie Krige, who extended Matheron's original work to the prediction of mineral ore grades. It has since been adopted by most of the geoscience sectors and found to be appropriate to the study of many spatially dependent variables. The underlying assumptions of kriging are that (1) the spatial variability of the available samples can be adequately represented by a semi-variogram model, (2) that the sample values (or some suitable transformation of them) approximate a statistically normal distribution, and (3) that there is no spatially determinate trend present in the sample values. Provided these conditions are met, then kriging presents us with a logical, justifiable prediction technique which ensures a result with minimum uncertainty and maximum confidence. This is a far cry from the conventional approaches to prediction which employ a variety of relatively simplistic distance-weighting algorithms. These may be totally inappropriate to the real spatial variability of our variable, and yet provide no measure of the goodness of the result.

This discussion has focused on predicting the unknown *value* of a variable at a particular *point*. In Chapter 8 we also discuss an extension of kriging that allows us to predict the unknown *average value* for a specific *volume*.

In practice, and depending on the complexity of the site conditions, the analysis of spatial variability can take up much of our time. The kind of spatial relationship which is readily apparent in our samples of benzene contamination may, in other cases, require a significant amount of analytical detective work to uncover. We may have to separate our samples into subsets with different *characteristics* because of *geological influences* and analyze them independently, or separate the sample pairs into subsets with particular *spatial orientations* because of *anisotropy* in their spatial variability. Frequently, our samples will have a distribution of values which cannot be approximated by a statistically normal distribution; in these cases, we must first search for and apply a suitable *transformation* to the sample values. Occasionally, the sample values will have an underlying *spatial trend* which is determinate; we must separate the trend component in the values before analyzing their variability. We discuss all of these techniques in the following sections.

Finally, what if there is no measurable spatial variability in our sample values? On these occasions our choice of prediction technique must be based on experience and intuition alone. Kriging is likely to give as good a result as any other technique, and our lack of knowledge of sample variability will be reflected as a high degree of uncertainty in the prediction results.

5.4 Analysis of Geological Influences

A deficiency of all prediction techniques is that they tend to *smooth* discontinuities in the spatial variation of a variable. Geostatistical kriging is better than conventional techniques in this regard, and new stochastic techniques discussed in Chapter 16 are better still. Nonetheless, they all have difficulty with representing discontinuities in a spatial variation unless we integrate them with a capability for recognition of *geological influences.* The reason for this is that the spatial variation is invariably influenced by secondary properties and characteristics of the host materials, which may be physical, chemical or mineralogical in origin. These properties have unique, independent spatial variations of their own, often subject to sudden spatial discontinuity caused by structural deformation; as a result, so does the variable we are interested in. If we measure all of these properties at the same time as we sample the variable of interest and carefully analyze their interdependencies, then we might be able to perform an exhaustive characterization. Limited investigative resources invariably preclude this. Instead, we typically summarize the secondary influences of these properties into one or two *characteristics* that we perceive as generally having a bearing on the end result; soil or rock lithology, or mineralogy are the most commonly applied. These tend to be characteristics that can be easily observed, in most cases simply by visual inspection of borehole cores or exposed surfaces.

This type of influence is illustrated by an example from the characterization of a mineral ore deposit. Figure 5.4 shows the semi-variogram results for copper grades (as a percentage by weight) from two different ore types present in the deposit. Prior to analysis the sample values are sorted and separated by their mineralogical characteristic values. The results in Fig. 5.4a are obtained by analysis of 1138 samples from ore type MBO; those in Fig. 5.4b from 678 samples from ore type PYO. As we can see, they have significantly different spatial variability relationships; they also have different statistical means and standard errors. Ore type MBO has low grades and a pronounced spatial variability with a range of 185 m, whereas type PYO has higher grades, a less definite spatial variability with an apparent range of 150 m and a greater background randomness (or nugget effect). We can only speculate on the mineralogical origins of these differences. Their practical significance lies in the fact that our prediction of copper grades is much more precise if we take these differences into account.

If we were to use a single semi-variogram model, without geological control, to predict copper grades throughout the deposit, then the results would include a gradation of copper grade between the two ore types. In reality, there is a discontinuity in grade variation at their common boundary. To obtain a realistic representation of this condition we must base the prediction of grade within each ore type volume on a semi-variogram relationship (or prediction model) appropriate to the ore type, and on samples from within the ore type. (This statement holds no matter which prediction technique we use; it is not limited to geostatistical kriging.) Thus, the grade prediction for a point on the MBO side of

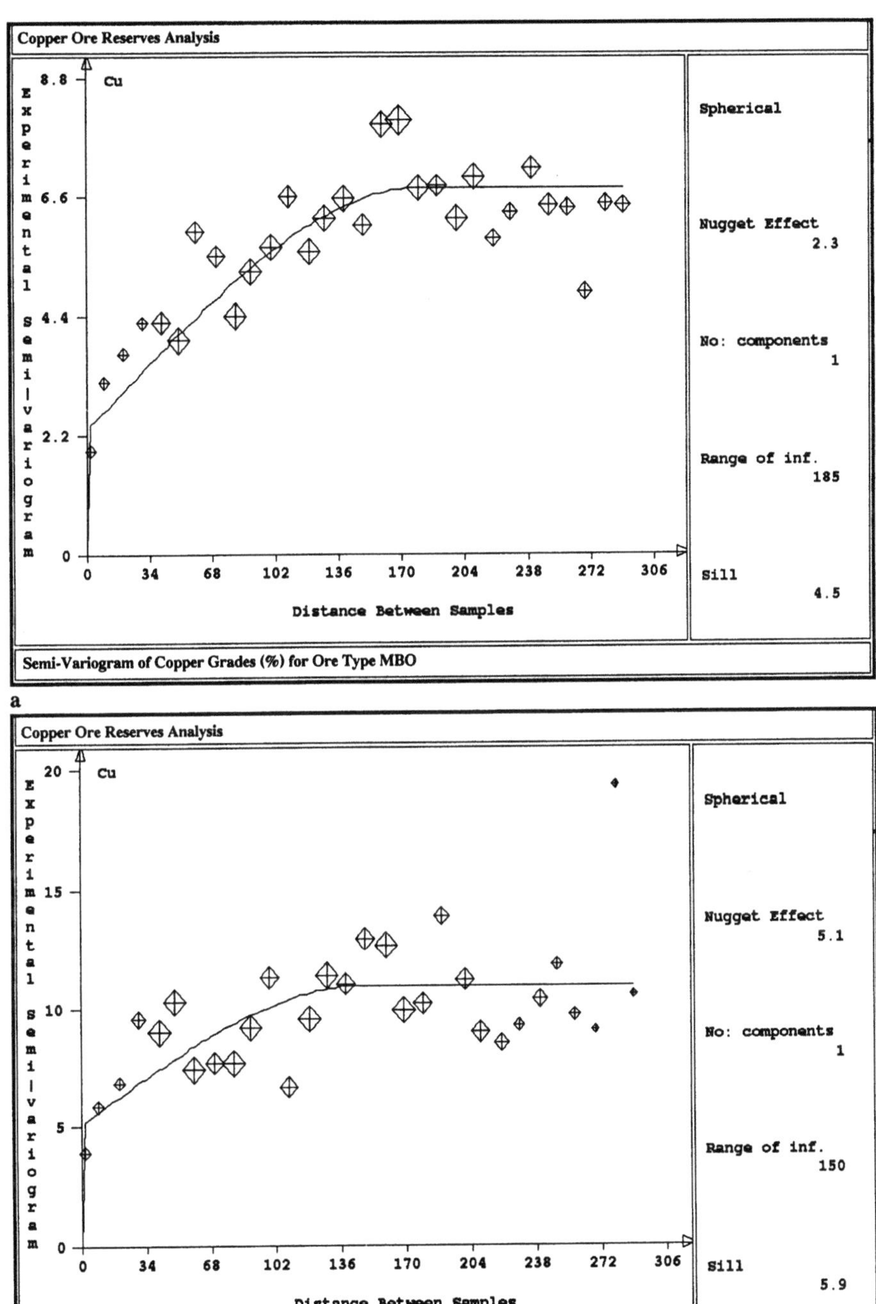

a

b

Fig. 5.4 **a, b** Semi-variogram analyses of copper ore grades from two adjacent ore types within the same mineral deposit, showing significant differences in variability

the boundary is based solely on MBO samples and the MBO semi-variogram relationship. Another point immediately adjacent, but on the PYO side, is predicted from PYO samples with the PYO prediction model. The predictions for both ore types must acknowledge that the ore deposit is surrounded by barren host rock in which the copper grades are effectively zero. This may seem obvious, but many characterizations completely ignore the influence of geological characteristics. The approach presupposes an ability to determine the characteristic value attached to any point for which we need to predict a variable, a topic which we touched on in Chapter 4 and discuss in detail in Chapter 6.

Similar discontinuities in the spatial variation of a variable are likely to occur under a wide variety of conditions. The dispersal and retention of subsurface contamination may vary significantly from one soil type to another. The variation of porosity may be subject to dramatic differences between different lithological units. Similarly, the quality of rock material for excavation purposes is frequently controlled to a large degree by its lithological characteristics.

This type of influence is invariably present to some degree in all geological characterizations, although in some cases it may be small enough to ignore for all practical purposes. A principal objective of the spatial variability analysis step is therefore to determine which, if any, characteristics are likely to influence the prediction of variables. This information dictates the course of the geological interpretation step discussed in Chapters 6 and 7.

5.5 Analysis of Directional Influences

So far, we have assumed that spatial variability is similar, or isotropic, in all directions within a particular volume; this is often not the case. It is quite possible for a variable to have a significantly different variability in orthogonal directions. In many cases, this is due to anisotropy in the host material itself; a sand or silt material deposited in many sublayers may have significantly greater porosity in horizontal directions relative to the vertical porosity. These conditions influence the dispersal of a contaminant within the material and hence its spatial variation. In an oil reservoir the continuity in porosity values along an old river channel may be greater than that of the cross-channel porosity. A wind-blown contaminant may show significantly less variation (and more continuity) in the direction of the prevailing wind. Further, mineral ore deposits frequently exhibit greater continuity in the plane of the deposit compared to that across the thickness of the deposit. For a good example of anisotropic conditions, we return to the benzene contamination example discussed earlier.

Figure 5.5 compares the spatial variability of the benzene contamination samples in the southwest/northeast direction (azimuth 45°) with that in the southeast/northwest direction (azimuth 135°). These results are obtained by sorting the sample pairs by their spatial orientation (cf. Fig. 5.1a) into two principal direction intervals during semi-variogram analysis; in other words, by applying

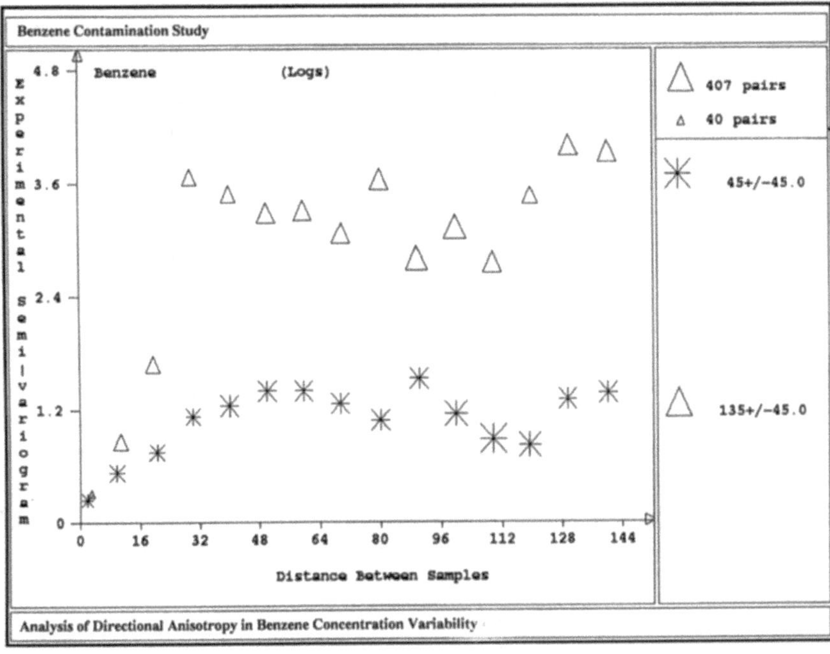

a

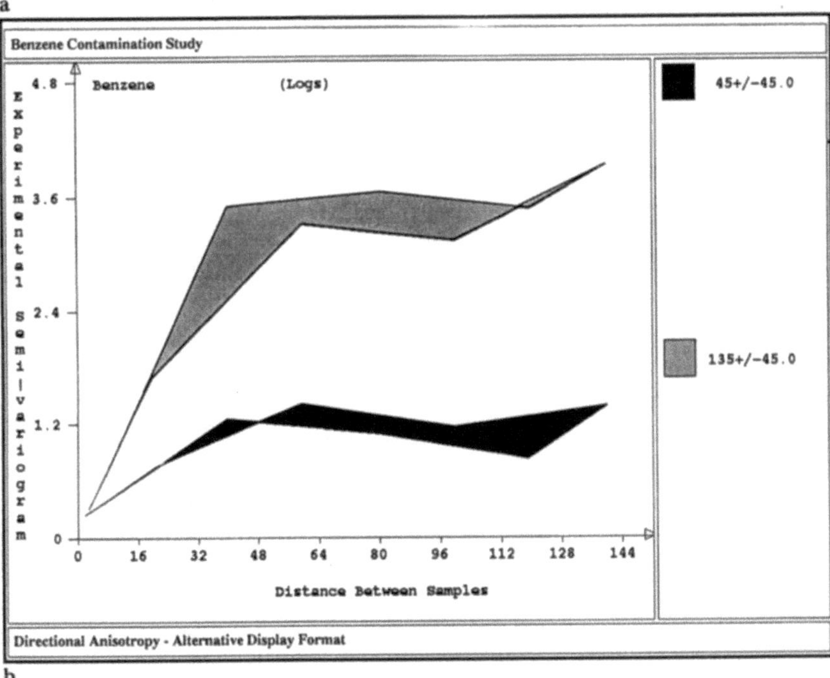

b

Fig. 5.5 **a** Analysis of directional anisotropy in the spatial variability of benzene concentrations in contaminated soil samples; **b** alternative display format for the semivariogram results of Fig. 5.5a

directional control to the analysis. In this case the semi-variogram results are quite dramatically different for the two directions, despite the fact that, in the earlier analysis, we obtained what appeared to be satisfactory results without any directional control at all (cf. Fig. 5.2). It is now apparent that there is much greater continuity and less variability in the azimuth 45º direction, with a range of influence of approximately 55 m, as in the previous analysis, whereas the range in the azimuth 135º direction is approximately 33 m and the overall variability is roughly three times as large. This is an example of second order (two-dimensional) anisotropy in the horizontal plane; things get more complicated when we have third-order anisotropy in the vertical direction as well.

In summary, the analysis of directional influences involves a determination of the *principal directions of anisotropy* and the relevant semi-variogram parameters in each direction. The three orthogonal ranges of influence effectively define an *ellipsoid of influence* around any point in space. For prediction purposes the range is assumed to vary elliptically between the principal directions, and depends on the directional relationship of a sample to the prediction point, whereas for isotropic conditions we assume a sphere of influence with a constant range in all directions.

5.6 Use of Data Transforms

As stated earlier, an underlying assumption of semi-variogram analysis is that the distribution of values in a sample population approximates a statistically normal distribution. In practice, we frequently have to apply some form of *data transformation* to the sample values in order to establish this condition. The benzene contamination example discussed earlier employs a *logarithmic* data transformation. The justification for this is provided by statistical analysis of the distribution of values in the sample set; a frequency histogram shows the set to have a log-normal distribution. This means that the sample population contains a large number of low values and relatively few very high values, a fairly common condition in soil contamination studies.

This condition is better illustrated by another example of soil contamination, in this case, a study of lead concentrations. Figure 5.6a shows the results of a frequency histogram analysis of 553 samples without any transformation. In Fig. 5.6b, a logarithmic transformation is applied; the distribution of values now approximates a statistically normal distribution. Figure 5.7a,b shows the before and after results of semi-variogram analyses applied to the same set of samples. Theoretically we could use the resulting semi-variogram relationship (shown in Fig. 5.7b) to predict the spatial variation of logarithmic values of lead concentrations and then back-transform the results using anti-logarithms to obtain real values. We can also do this in practice, provided we adjust the final results to eliminate a scale distortion of values which is inherent to logarithmic calculations. This correction to the prediction results is discussed in Chapter 8.

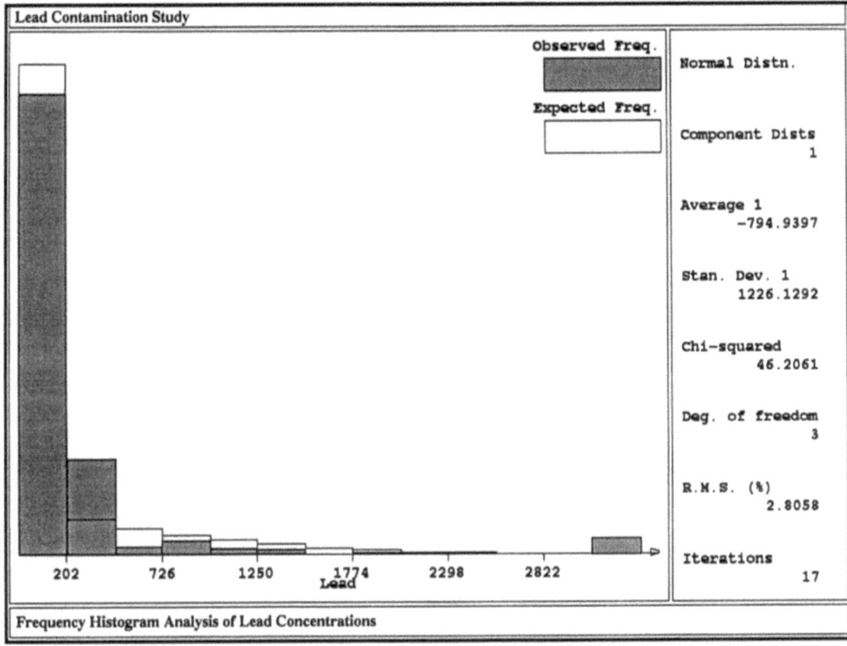

a

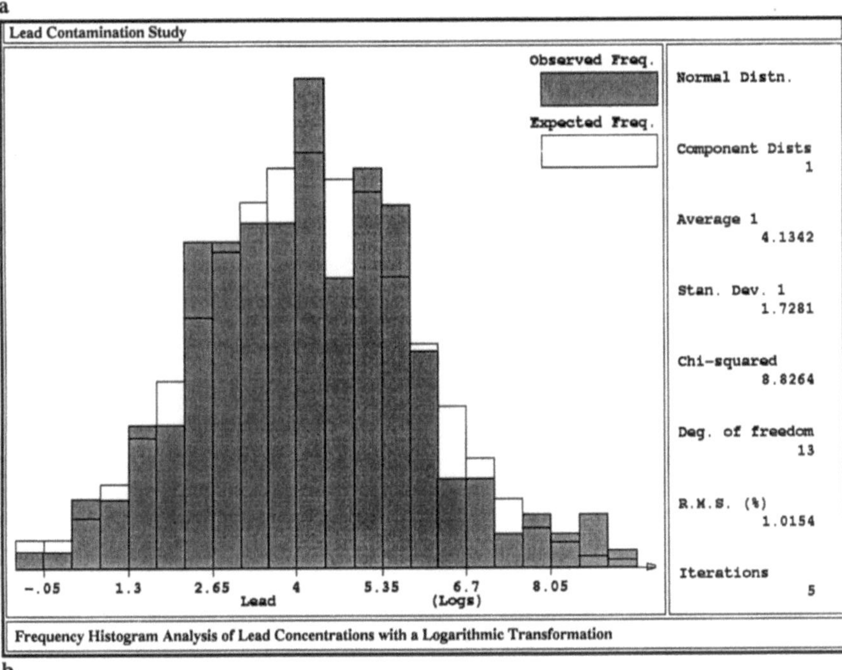

b

Fig. 5.6   **a** Frequency histogram analysis of lead concentrations in soil contamination samples; **b** analysis of the same sample set with a logarithmic data transformation applied; these results approximate a statistically normal distribution

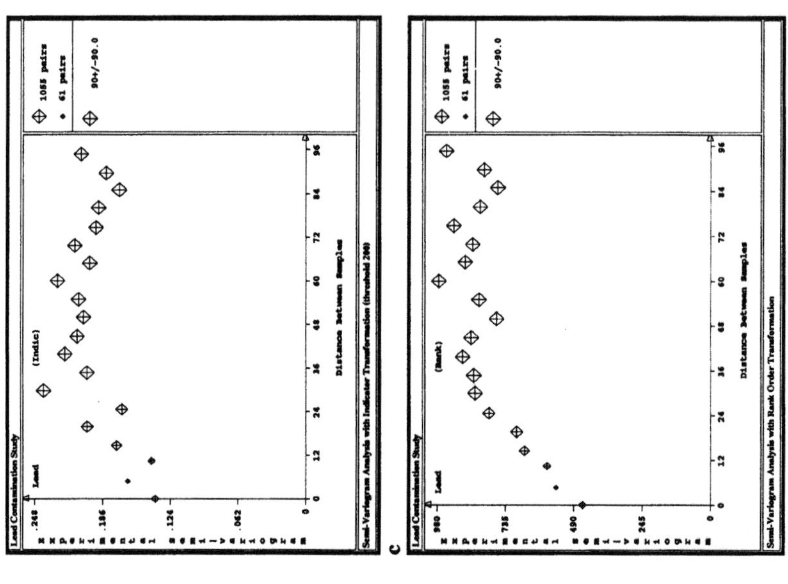

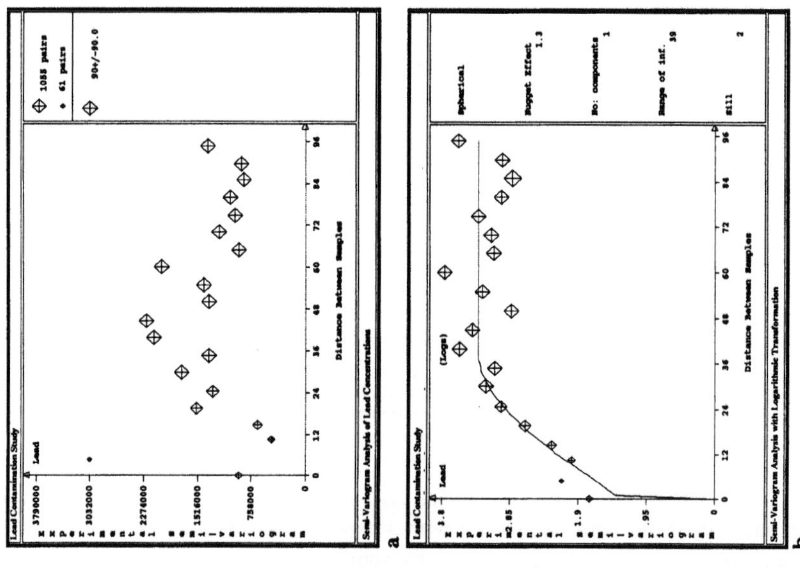

Fig. 5.7 **a** Semi-variogram analysis of untransformed lead concentrations; **b** with a logarithmic data transformation and a spherical semi-variogram model; **c** with an indicator data transformation; **d** with a rank order data transformation

The logarithmic transformation allows us to extend semi-variogram analysis and kriging prediction to a large group of spatially related variables that would otherwise be incompatible with the underlying assumptions. It is the most frequently used of the standard transformations; the others are the *indicator transformation* and the *rank order transformation*. All three of the standard transformations provide us with different ways of *scaling* our sample values to eliminate the distorting effects of differences in value that may span several orders of magnitude. The logarithmic transformation achieves this by converting the values to a logarithmic scale. The indicator transformation converts all values to ones and zeros, depending on whether they are above or below a specified threshold value. The rank order transformation first ranks all the values in ascending order, then converts them to their integer sequence values.

Figure 5.7c,d illustrates the semi-variogram results obtained from indicator and rank order transformations of the lead contamination samples discussed above. All three of these simple transformations are surprisingly effective in helping us to detect and measure the otherwise hidden spatial relationship contained in the sample set. All three indicate a range of influence of approximately 39 m. In this instance, the rank order transformation actually exhibits a stronger relationship than the logarithmic transformation. However, the latter is preferred for prediction purposes since the rank order transformation is limited to semi-variogram analysis. It is not used in the prediction step because back-transformation is impossible. On the other hand, indicator transformation, like logarithmic transformation, is an extremely useful prediction tool.

Consider an individual sample: if its value is greater than or equal to the threshold, then its indicator value is one. The *probability*, on a scale of zero to one, that it exceeds the threshold is also one. Conversely, an indicator value of zero means that there is zero probability that the sample value exceeds the threshold. In other words, the indicator transformations of a sample population are a measure of the probability of occurrence of values greater than the threshold. If we apply kriging prediction to the indicator values, then the resulting predicted value at any point is a measure of the probability of exceeding the threshold at that point. This is an extremely useful result, particularly in environmental contamination studies, since it provides a direct relationship between risk and cost. For example, with 90% certainty, all contamination above a specified threshold concentration is contained within the volume defined by the 0.9 probability isosurface (or 3D contour). We shall discuss indicator kriging in detail in Chapter 8.

At the level of spatial variability analysis there are no clear guidelines as to which data transformation to apply to different conditions. As with geological and directional influences, in practice this form of analysis typically boils down to a process of eliminating alternatives until we find something that works! This may require elimination of a significant number of different combinations of data transforms with a range of possible geological and directional influences.

5.7 The Effects of Spatial Trends

Another underlying assumption of semi-variogram analysis is that there is no spatially determinate trend present in the sample values, or if there is, then we remove it prior to analysis. Semi-variogram analysis assumes that the *differences* between sample values are related (within a certain range) but the *values* themselves are indeterminate. If a significant trend is present, then it is likely to distort the semi-variogram results to the extent that they may be unintelligible. Alternatively, the determinate trend may overwhelm the natural variability to the extent that the latter is undetectable. Situations of this type are likely to arise wherever *natural variability* is superimposed on a relatively strong *determinate trend*. For example, the spatial distribution of a groundwater contaminant is likely to be influenced to a large extent by contaminant dispersion and hydraulic flow in porous media, both of which are determinate processes. An additional component of natural variability is likely to be superimposed on these conditions; this is the component with which spatial variability analysis is concerned.

To illustrate this effect we use a simple 2D data set consisting of measured piezometric heads in a group of groundwater wells. In this case, our analytical variable is the piezometric head. Figure 5.8 illustrates the results of applying semi-variogram analysis to the data set before and after removing a quadratic polynomial trend. The upward-turning, parabolic results in Fig. 5.8a are indicative of a dominant trend in the data. In an analysis of this type we use trend surface (regression) analysis to determine an appropriate linear, quadratic or cubic polynomial trend, subtract this component from the measured values, and apply semi-variogram analysis to the residuals. Figure 5.8b illustrates the semi-variogram analysis of these residuals, a typical spherical relationship is now apparent. This relationship was overwhelmed by the dominant trend prior to the latter's removal. Having determined an acceptable semi-variogram model, we use this to predict the natural variability component of piezometric head at a point and then add back to it the determinate trend component. The approach works equally well in three dimensions.

5.8 Geostatistical Prediction Models

We have already discussed the *spherical* semi-variogram relationship. This fits many natural occurrences of spatial variability and is by far the most frequently used kriging prediction model. Geostatisticians have devised a host of other semi-variogram models to apply to more unusual conditions, and will also on occasion combine several models in piggyback fashion to achieve the desired effect. We concern ourselves only with the less esoteric, more frequently used models here; these suffice for more than 90% of characterization requirements. If we require

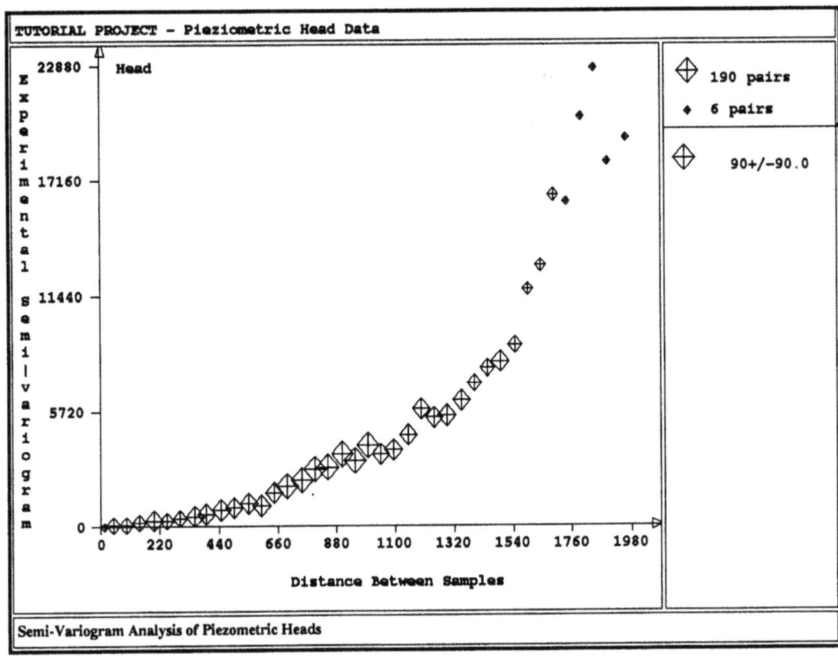

a

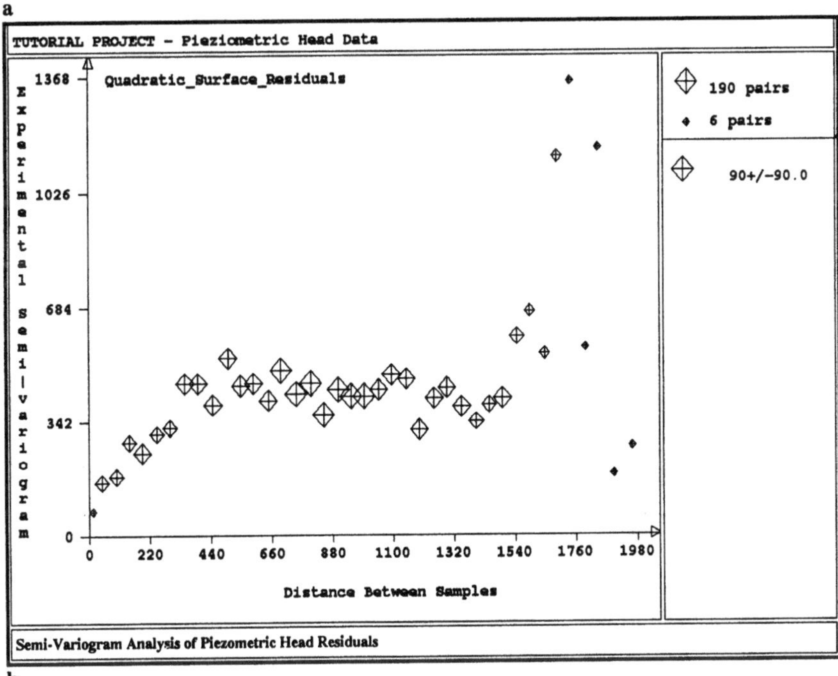

b

Fig. 5.8 **a** Semi-variogram analysis of piezometric heads in groundwater wells, exhibiting a dominant spatial trend; **b** semi-variogram analysis of residual piezometric heads after removal of a polynomial (quadratic) trend; the results exhibit a typical spherical spatial variability

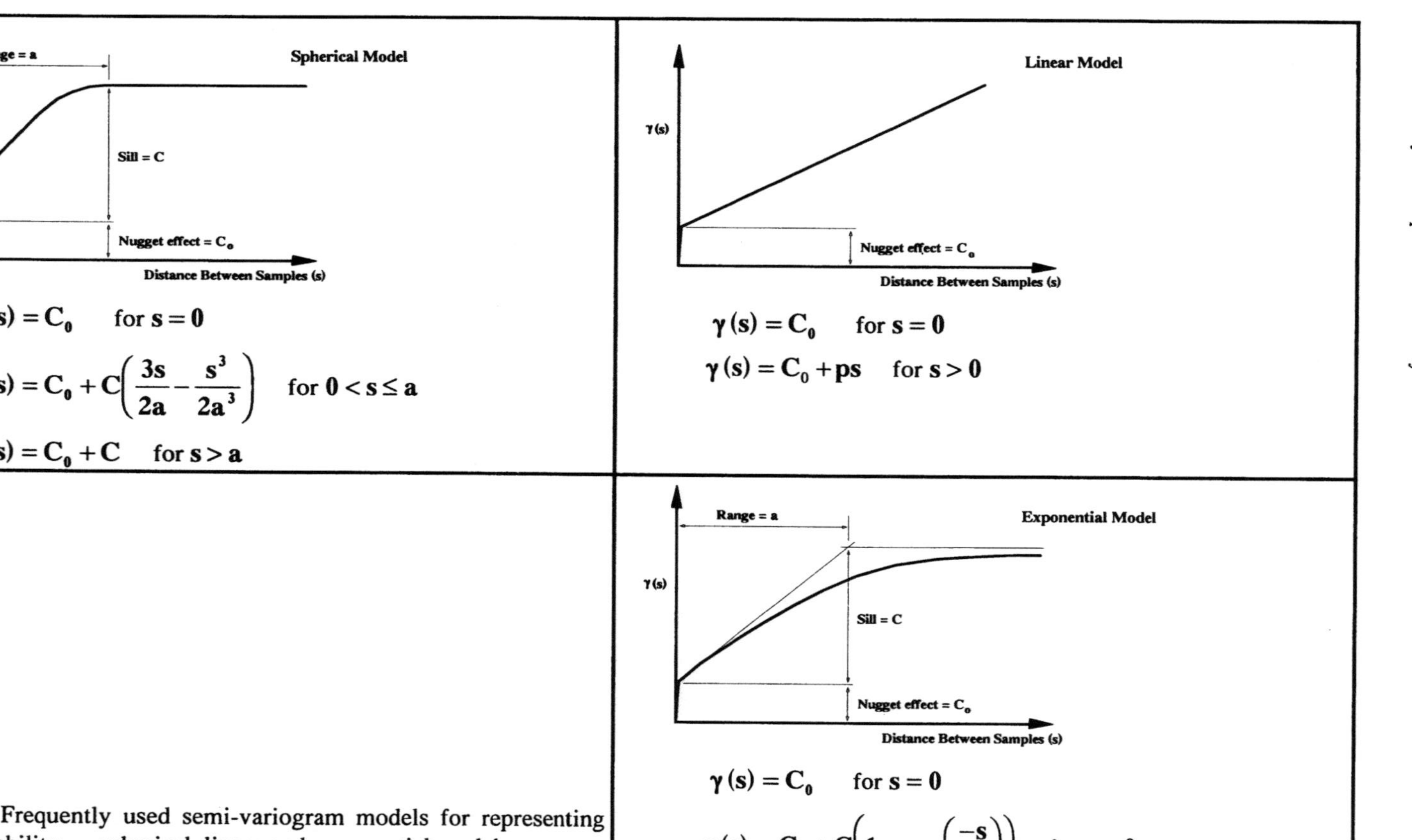

$$\gamma(s) = C_0 \quad \text{for } s = 0$$
$$\gamma(s) = C_0 + C\left(\frac{3s}{2a} - \frac{s^3}{2a^3}\right) \quad \text{for } 0 < s \le a$$
$$\gamma(s) = C_0 + C \quad \text{for } s > a$$

$$\gamma(s) = C_0 \quad \text{for } s = 0$$
$$\gamma(s) = C_0 + ps \quad \text{for } s > 0$$

$$\gamma(s) = C_0 \quad \text{for } s = 0$$
$$\gamma(s) = C_0 + C\left(1 - \exp\left(\frac{-s}{a}\right)\right) \quad \text{for } s > 0$$

Fig. 5.9 Frequently used semi-variogram models for representing spatial variability .... spherical, linear and exponential models

something more sophisticated, then we should be consulting a geostatistical specialist anyway!

There are two other models besides the spherical model that we may have occasion to use. These are displayed in their general form in Fig. 5.9. The *linear* model assumes a linear relationship between semi-variance and distance. It has no range and is theoretically infinite. In practice, it is always appropriate only over a limited, finite distance. The *exponential* model is quite similar in form to the spherical. It is designed to handle conditions of exponential decay with distance of the influence between samples. The model becomes asymptotic to the sill value for large distances.

The definition of semi-variogram models may seem unnecessarily complex at first glance. We should not lose sight of the fact that they are nothing more than convenient relationships for approximating the way in which sample semi-variance changes with distance. To obtain a prediction model that is appropriate for a particular set of semi-variogram results we merely choose the model with the most appropriate form (or general shape) and customize its parameters until it best approximates the results.

5.9 Cross-Validation of Prediction Models

A selected prediction model is an approximation of the spatial variability of the *observed or measured* conditions. In turn, the observed conditions may or may not be representative of the *real* conditions. There is no way we can verify this since we do not know the real conditions. The only way we can obtain an indication of the degree to which a set of samples represents the real conditions is to perform a round of infill sampling. If the semi-variogram results of the combined sample population are significantly different, then the first set obviously is *not representative* on its own. As a general rule, the average sample spacing should be three to five times less than the range of influence. If we have achieved this condition, then there is generally not much to be gained by any further sampling.

There remains the question of how well our prediction model approximates the observed conditions. We can obtain a measure of this by a process called *cross-validation*. Suppose we remove a sample from the population and use the prediction model, together with the surrounding samples, to predict a new value and a standard error at the sample location. The predicted value should be close to the omitted sample value. Ideally, the ratio of their difference to the predicted standard error should be less than or equal to one. It is unlikely to be zero since the sample value itself is not included in the prediction process. This ratio is called an *error statistic*. By repeating this exercise at each of the sample locations in turn and statistically analyzing the results, we obtain a measure of the *appropriateness* of the prediction model with respect to the sample population. If our model is a good approximation, then the error statistics should have a statistically normal distribution, with an *average* close to zero and a *standard error* close to one. Of

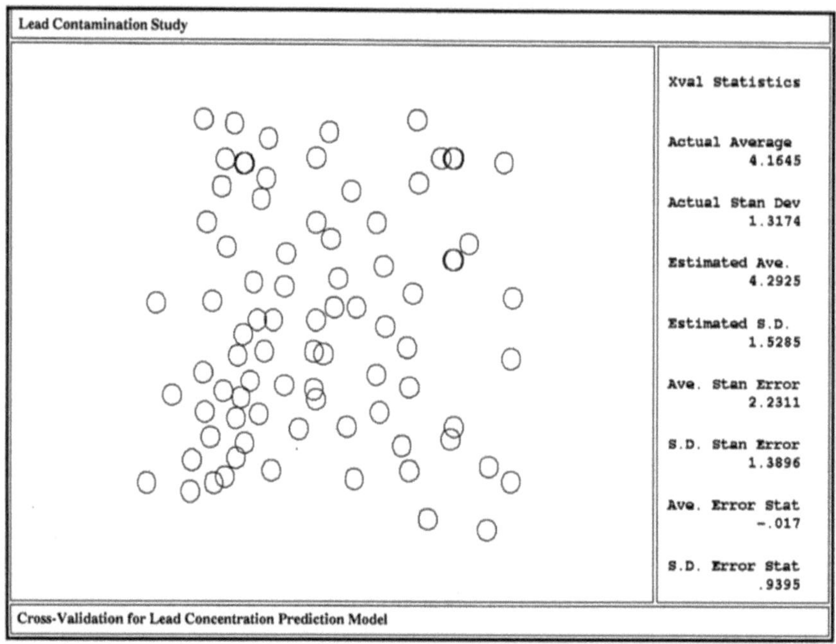

a

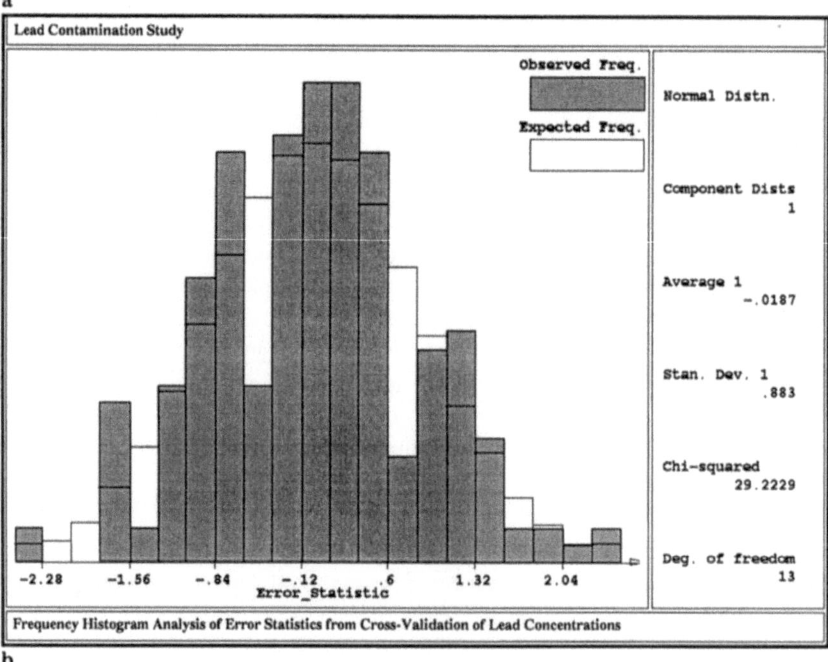

b

Fig. 5.10 a Cross-validation results for the semi-variogram model for lead concentrations shown in Fig. 5.7b; **b** frequency histogram analysis of the cross-validation error statistics

course, the average and standard error of the actual predicted values should also be close to those of the sample population.

To illustrate this process we return to the lead contamination example. Figure 5.10a summarizes the cross-validation results; it also shows the relative locations of the samples. The predicted average and standard error of 4.29 and 1.53 compare with the equivalent sample values of 4.16 and 1.32, respectively. The average error statistic is -0.017 (close to zero) and the standard error of the error statistic is 0.94 (close to one). Figure 5.10b shows the results of a frequency histogram analysis of the error statistics, which approximate a normal distribution. Despite the obvious discrepancies, in practice these are considered reasonable results. We should bear in mind that any subsequent prediction of contamination will be based on *all* of the samples and will result in a closer approximation of the real conditions.

6 Geological Interpretation and Modeling

6.1 Interactive Geological Interpretation

Geological interpretation, which is by nature an iterative, interactive process, typically requires definition of large quantities of spatial information. Ideally, what we need for this frequently difficult task is a kind of *3d CAD system for geologists*, one that allows us to define and visualize highly irregular geological volumes quickly and efficiently. This type of approach to interpretation is the subject of the following discussion. Of course, there are significant differences between it and normal CAD systems, such as accommodating the kind of data structures we discussed in Chapter 4, working with highly irregular volumes instead of cubes, cylinders and spheres, and most importantly, allowing us to attach geological attributes to those volumes. It should also accommodate, in a convenient way, the familiar methods of interpretation that geologists have developed with experience.

When we interpret geological structure and stratigraphy on paper we are, in essence, characterizing subsurface space in two dimensions, in the plane of a section, by lithology, or mineralization, or structure, or geological sequence. In a manual exercise we simply draw boundaries that represent our interpretation, or *educated guess*, of the intersection of various geological entities with the section plane. We then color the areas between the boundaries to distinguish the geological entities by their characteristic value. All of this is done with visual and spatial reference to the observations, or characteristic data, which are relevant to the section. Of course, we seldom achieve a satisfactory result the first time, and much erasing and editing are typically required.

This procedure is relatively easily and efficiently duplicated on a computer. We can orient our computer screen, defined earlier as a *viewplane*, in terms of three-dimensional coordinates and rotations, so that it corresponds to the orientation of our interpretation section. Further, we can display on the viewplane a scaled projection of the geological data that is spatially relevant, i.e. within a specified distance or range of the section. This is achieved by transformations of data structure geometry to the local coordinate system of the viewplane, as discussed in Chapter 4. We call this *background information* since it is displayed for visual reference purposes only. Background information can take many forms, depending on what is relevant to our interpretation. Projections of color-coded lithology intervals are frequently used, perhaps with viewplane intersections with previously modeled topographical and fault surfaces, or surfaces derived from reduced

Fig. 6.1 **a** Interpretation of several intersections of a geological unit on a viewplane section, with topography, fault surfaces and boreholes displayed in the background; **b** a three-dimensional fence model of geology consisting of a number of completed sectional interpretations at different viewplane orientations

geophysical surveys, or intersections with other interpretation sections.

We can then interpret the boundaries in similar fashion to a paper interpretation. Instead of a pencil we use a graphics device (such as a mouse) to position a cursor on the viewplane and activate it to define points on the interpretation boundary. Each boundary consists of a series of such points joined by line segments. Since we control the number of points in each boundary, we also control the degree of geological detail incorporated in the boundary. This interactively interpreted boundary is called *foreground information.*

We store the completed boundary as a component in a volume data structure, as described in Chapter 4, initially with zero thickness and the three boundaries of the data structure equal and coincident. The viewplane becomes the component plane; they share the same origin and orientation. We assign a code to the component that designates its characteristic value and controls the color in which the geometry and its intersection with the viewplane are displayed. Figure 6.1a shows the interpretation of a single, faulted geological unit consisting of several boundaries on a viewplane section with borehole information, fault traces and the topographical surface displayed in the background. An interpretation of fault surfaces and a triangulated representation of topography have been determined beforehand and stored as volume data structures.

We can continue in this fashion until all the geological intersections of interest have been interpreted on this section, and then repeat the interpretation exercise on a number of other sections at various orientations (cf. Fig. 6.1b). The result, viewed in three-dimensional perspective, is a fence model of geology. Compatibility between intersecting sectional interpretations is ensured by displaying their viewplane intersections in the background for reference purposes during the interpretation process.

6.2 Interpretation of Geological Volumes

We are now ready to embark on a full three-dimensional interpretation of geology, using these simple two-dimensional sectional interpretations as our starting point. To illustrate the process let us consider the interpretation of a single geological unit (cf. Fig. 6.2a), and suppose that we have interpreted the same unit on two adjacent viewplane sections, parallel to the first but displaced from it. On these sections our interpretation boundaries have somewhat different configurations (cf. Fig. 6.2b).

The first step is to extend the sectional interpretations normally, so that they meet at common planes between the interpretation sections (cf. Fig. 6.2c). This is achieved by assigning fore- and back-thickness to the volume data component. The result is equivalent to the simplifying assumption, implicit in many conventional interpretations, of assuming uniform geology from mid-way between a pair of

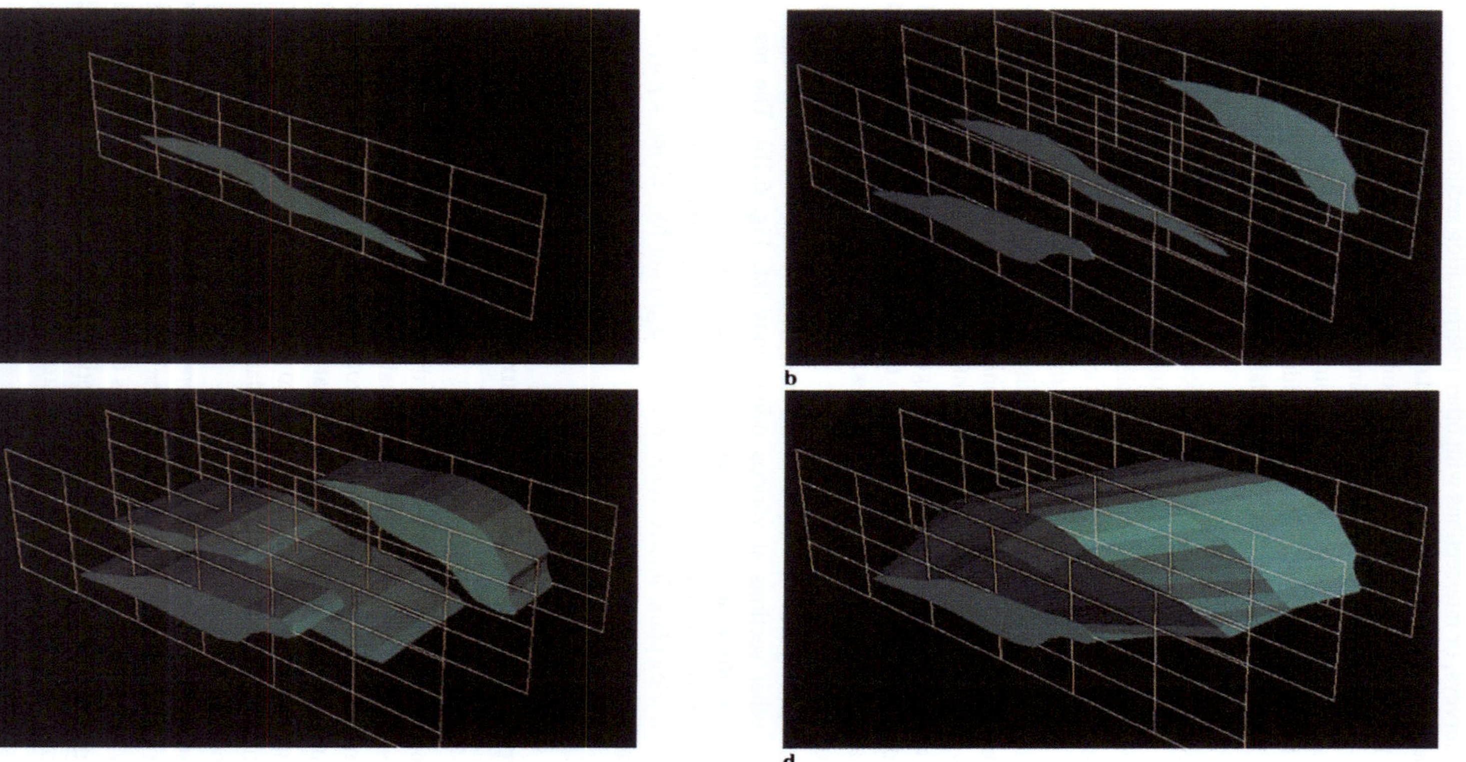

Fig. 6.2 a Interpretation of the intersection of a geological unit on a viewplane section; **b** interpretation of the same unit on two adjacent sections; **c** extension of the boundaries to common planes between the sections; **d** interpretation by boundary matching on the common planes

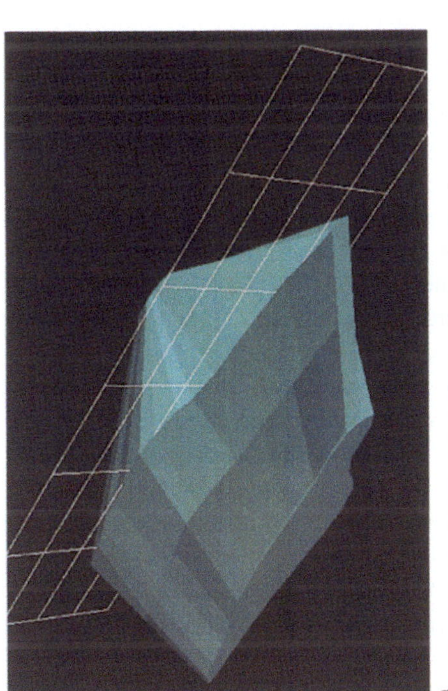

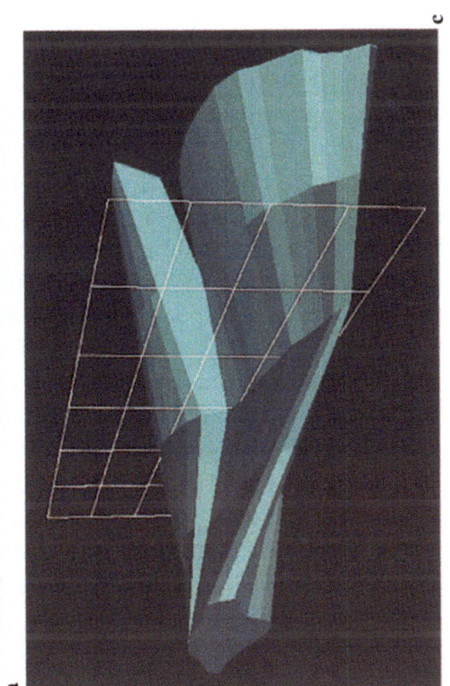

Fig. 6.3 (cf. Fig. 6.2) **a** alternative interpretation of a geological pinch-out between sections; **b** interpretation of a bifurcation condition; **c** interpretation of a fault condition

sections to mid-way between the next pair. However, with the aid of the computer we can enhance our interpretation considerably beyond this point.

A boundary matching algorithm can be applied so that the pairs of boundaries are modified and matched on the common planes to provide a continuous three-dimensional interpretation of the geological unit (cf. Fig. 6.2d). This is equivalent to a linear interpolation of the boundary profiles between each pair of sections. Our initial two-dimensional interpretation boundaries have been *stretched and modified* in the third dimension so that they now define an irregular volume, equivalent to the volume of the geological unit between the sections.

Of course, our example is a simple one and reality might be considerably more complex. In many cases, linear interpolation of boundaries is inadequate, or the geological unit might pinch-out, or bifurcate, or be faulted between the sections. To cater to this type of complexity an alternative technique that allows interactive modification of the extended boundaries is provided. This allows us to *shrink* the extended boundary to an irregular line, or a point, in the case of a pinch-out (cf. Fig. 6.3a); or to match one boundary to two boundaries in the case of a bifurcation (cf. Fig. 6.3b); or to interpret the displacement of a fault discontinuity (cf. Fig. 6.3c).

We can continue this process throughout the region represented by our geological model until the unit is fully defined in three dimensions. The final representation consists of a number of contiguous subvolumes, each represented by a component in the volume data structure (cf. Fig. 6.5a). Each component defines a portion of the volume of the geological unit and its geological characteristic. There is no limit to the number of components used to define a unit, thus the degree of geological detail incorporated in the model depends on the diligence (and patience!) of the interpreter.

In a similar fashion we can perform three-dimensional interpretations of all the geological units to provide a complete geological model of the region (cf. Fig. 6.5b). Various facilities are provided to ensure that the boundaries of adjacent units match each other; any spatial *gaps* that we might inadvertently leave in the model will later be considered *undefined space*. An equally important consideration is that the initial sectional interpretations are not confined on parallel planes. Individual interpretations can be performed on the viewplane orientation that is most appropriate to the geological structure, and volume modeling components can be defined at any orientation. Thus, in the case of a complex geological structure, gaps in our three-dimensional model of geology that might remain after an initial 3D interpretation from parallel sections can subsequently be *filled in* with appropriately shaped components defined from appropriately oriented viewplanes.

The result is a geometric characterization of subsurface space by geological characteristic values, a three-dimensional extension of the equivalent conventional interpretation on two-dimensional sections.

We might argue that it would be more efficient, in computational terms, to represent each geological unit with one continuous geometrical definition, rather than a number of components. This might indeed be the case for relatively simple geological conditions. However, the flexibility and efficiency offered by the multi-component volume modeling approach, in terms of defining, representing and editing complex interpretations, far outweigh the computational advantages of the alternative. As we have seen, it also provides an efficient avenue for the geoscientist to impose interpretational control on the result.

In fact, the multi-component approach to volume modeling works equally efficiently for conditions involving simple stratigraphy, wherever we have sufficient information to define the contacts between units as triangulated surfaces, as discussed in Chapter 7.

6.3 2D and 3D Geological Sections

The ability to cut and visualize arbitrary viewplane sections through our geological model is of paramount importance to characterization as a whole, and particularly to the interpretation process. It is provided by the unique geometric characterization of irregular volumes implicit to the volume data structure described in Chapter 4. The geometry of a typical volume model component is illustrated by Fig. 6.4a,b. It consists of three boundaries, comprising the initial interpretation boundary and the two extended and modified boundaries, and links that join the boundaries. These boundaries and links form the *edges* of the surface that encloses the volume of the component.

All of this geometrical information is stored in local XYZ coordinates relative to the viewplane from which the component was defined, i.e. the sectional interpretation viewplane. This viewplane is in turn defined by global NEL (real world) coordinates and rotations, and hence the component geometry is fully defined in three-dimensional space by an appropriate coordinate transformation of the form described in Chapter 4.

Using the inverse form of the coordinate transformation algorithm, the component geometry can now be redefined in terms of the local coordinates of any other viewplane, for example, a new viewplane that intersects the component at some arbitrary orientation (cf. Fig. 6.4c). After the transformation, the geometry of the component is defined in terms of the local X and Y coordinates in the plane of the new viewplane and the Z coordinate, normal to the viewplane, that defines the offsets of the points contained in the geometry. Each boundary segment and link in the component geometry can now be inspected to determine whether or not they intersect the viewplane. This occurs wherever the end points of a link or segment in the component geometry satisfy the condition

$$\frac{Z_1}{|Z_1|} \neq \frac{Z_2}{|Z_2|}$$

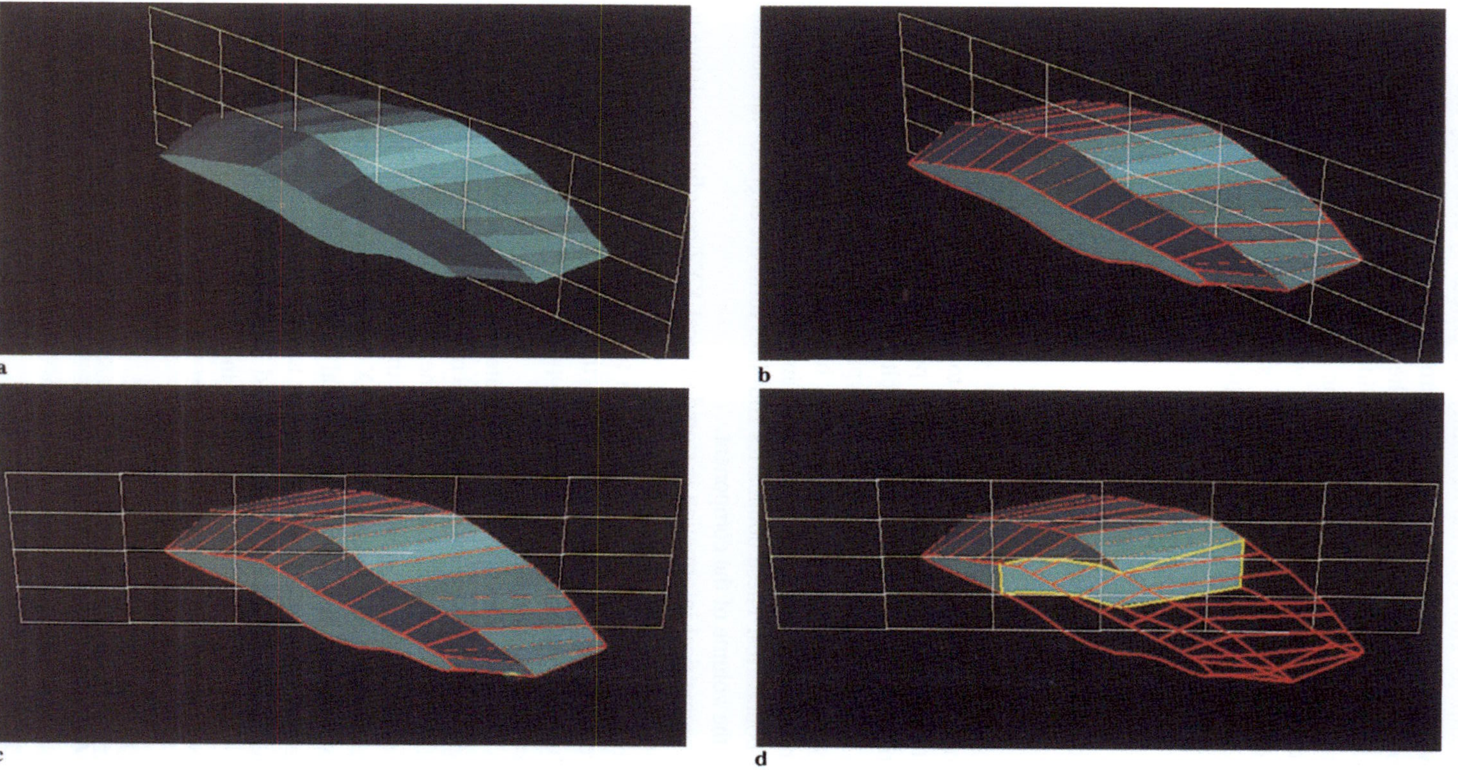

Fig. 6.4 a A single volume modeling component which represents a portion of a geological unit; **b** the geometry of the links (edges) which define the local geometry of the component; **c** intersection of the component by a new viewplane at an arbitrary orientation; **d** the area of intersection with the new viewplane

If they do, then the X, Y coordinates of the point of intersection with the viewplane are readily determined from expressions of the form

$$X_i = X_1 + \frac{(X_2 - X_1)}{(Z_2 - Z_1)} \cdot Z_1$$

$$Y_i = Y_1 + \frac{(Y_2 - Y_1)}{(Z_2 - Z_1)} \cdot Z_1 .$$

These points of intersection, together with line segments that connect them, collectively define the boundary (or boundaries) of intersection of the component with the new viewplane (cf. Fig. 6.4d).

If each of the components comprising a geological volume is transformed and analyzed in this fashion, then the collective intersection boundaries represent the intersection of the volume with the new viewplane, displayed in the appropriate color. The intersection with the entire geological model is obtained in a similar fashion. The process may sound laborious, but in fact takes only a few seconds on a modern graphics workstation, even for models containing several thousands of components. We have thus achieved a methodology for rapidly cutting sections at any orientation through our three-dimensional model of geology (cf. Fig. 6.5c), and the same method can be employed for three-dimensional cut-aways in perspective views (cf. Fig. 6.5d).

6.4 Analysis of Geological Volumes

At this point we introduce the subject of volumetrics analysis. Reflect for a moment on the intersection of a volume model component by a viewplane of arbitrary orientation, as illustrated in Fig. 6.4, and suppose that we duplicate this process on a number of parallel, equally spaced planes throughout the volume of the component (cf. Fig. 6.6a). The two-dimensional area of intersection of the component with the plane k is readily determined by applying a simple algorithm for numerical planimetry to the points in the intersection boundaries

$$A_k = \frac{1}{2} \sum_{j=1}^{j=M} \sum_{i=1}^{i=N_j} (X_{i+1} - X_i) \cdot (Y_i + Y_{i+1})$$

$$\text{where} \quad N_j = \quad \text{number of points in the boundary } j$$
$$M = \quad \text{number of intersection boundaries.}$$

If we multiply each of these intersection areas by the spacing of the parallel planes, and sum the resulting sectional volumes for all the planes that intersect

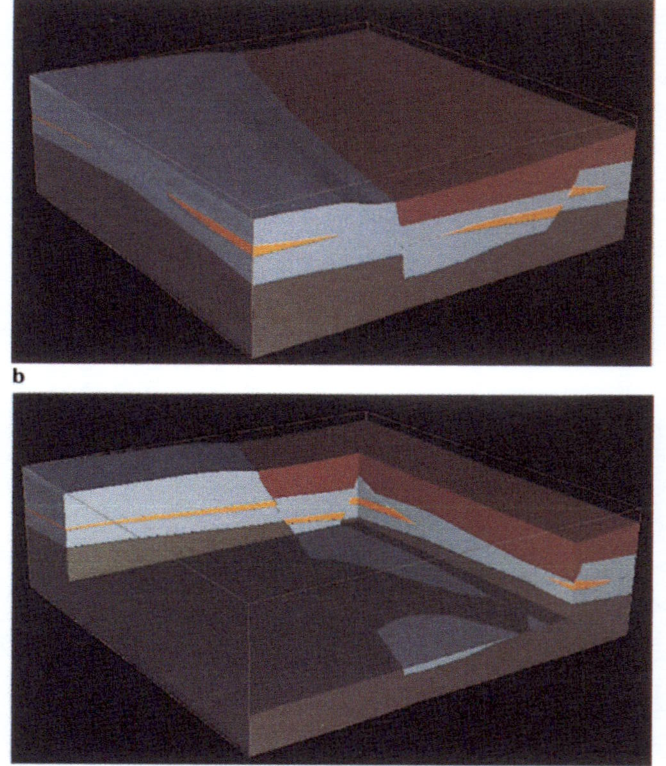

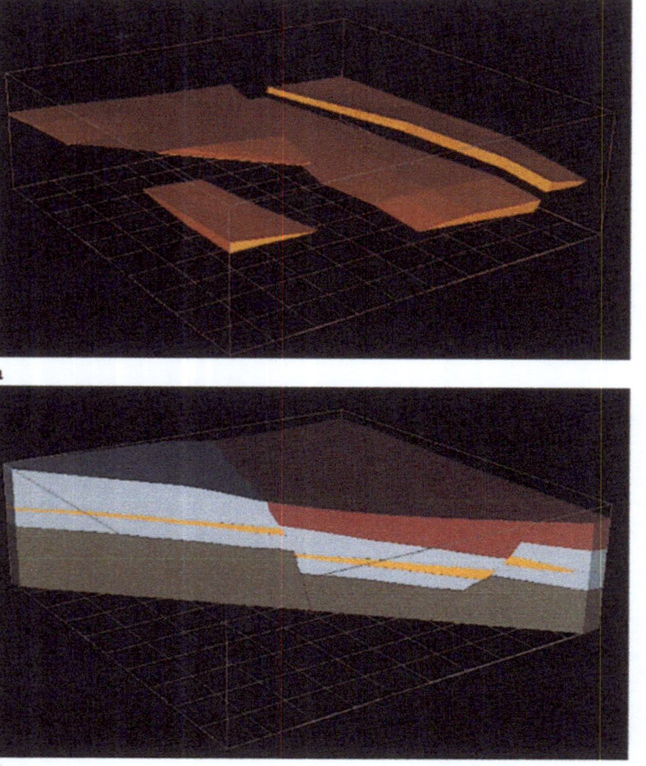

Fig. 6.5   **a** Three-dimensional interpretation of a faulted, discontinuous unit; **b** a complete volume model representation of geology; **c** a geological section at an arbitrary orientation through the model; **d** a three-dimensional cut-away section through the same model

the component, then the result is a measure of the total volume of the component (cf. Fig. 6.6b)

$$V = \sum_{k=1}^{k=I} \Delta s \bullet A_k$$

$$\text{where} \quad \Delta s = \text{ spacing of intersection planes}$$
$$I = \text{ number of intersection planes}$$

If we reduce the spacing, thereby increasing the number of intersection planes, then the result approaches the true volume of the component. We thereby achieve a methodology for determining the volume of a component (with controllable precision), and hence, by extension, the volume of an entire geological unit. Whatever the orientation of the intersection planes with respect to the unit (or the individual components), the result always approaches the true volume as the intersection spacing decreases. The process is analogous to the conventional approach for determining volumes by hand-planimetering profiles on individual sections, but much more efficient, flexible and precise.

6.5 Volumes of Intersection

The above is a potentially useful result, however, the ability to determine the volume of a geological unit hardly satisfies our requirements for volumetrics analysis. To derive a suitable extension of the algorithm let us consider two overlapping components, as illustrated in Fig. 6.6c. For purposes that will become apparent, we will initially assume a simple rectangular configuration for the second component.

A plane that intersects both components has a boundary of intersection with each of them. The two boundaries can be readily analyzed on their common plane to determine whether they overlap and, if they do, the boundary of the area that is included by both. This is the *area of intersection* of the two components on the common intersection plane. If we repeat this exercise on a set of parallel, equally spaced planes, multiply the areas of intersection by the spacing and sum the results, then we have a measure of the *volume of intersection* of the two components (cf. Fig. 6.6d). As before, the volumetric precision is controllable and the results are repeatable for any orientation of the intersection planes. The second component is obviously not limited to a rectangular configuration and any irregular component geometry is accommodated by this technique.

This is a more useful result, since it allows us to determine, for example, the volume of intersection between a set of components representing an excavation volume and a set representing a geological unit. What is more important, it provides us with an avenue for applying geological control to the process of

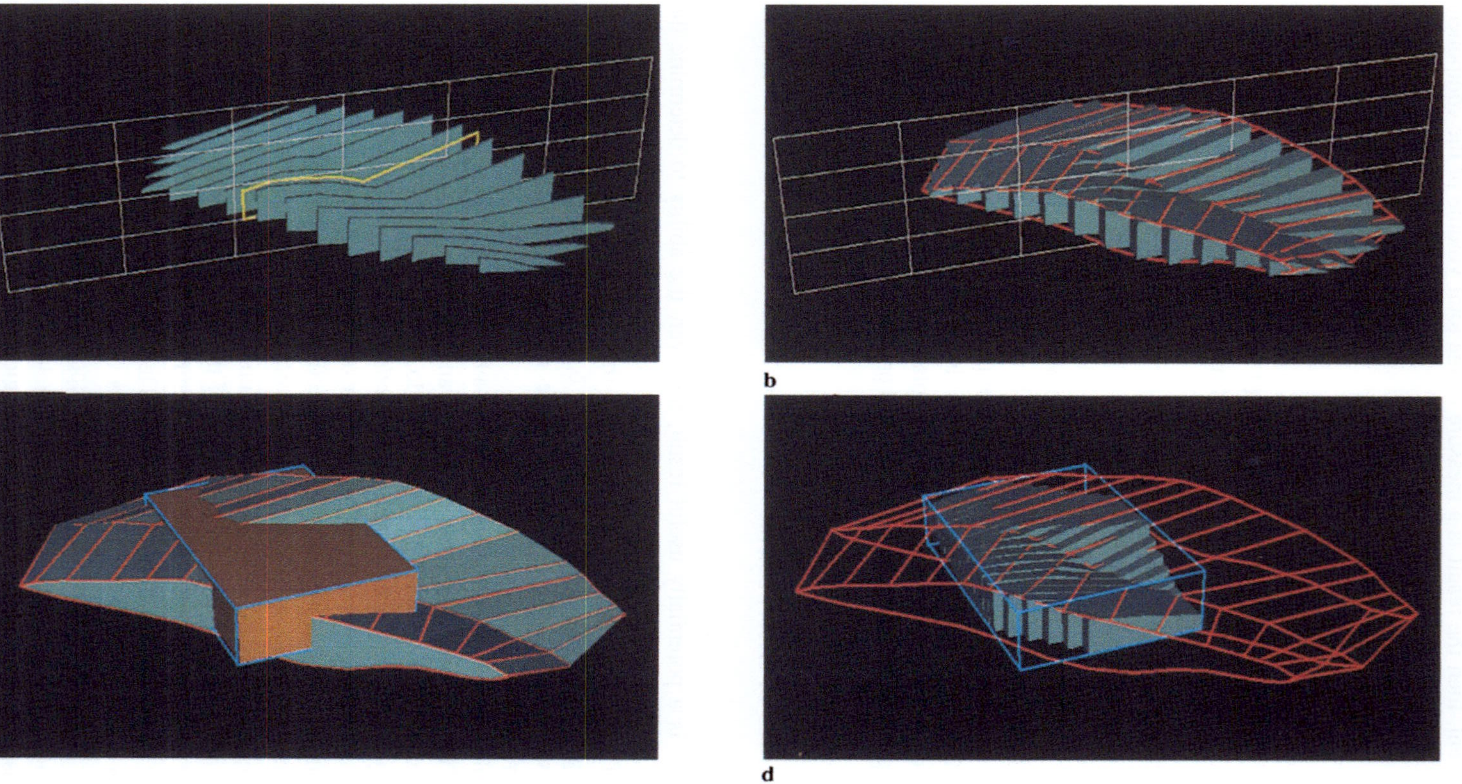

Fig. 6.6 **a** Intersection of a component by a set of parallel, equally spaced planes; **b** determination of the component volume by integration of planar sections; **c** an intersection of two components; **d** the volume of intersection by integration of common sections

predicting the spatial variation of a variable. It allows us to integrate the volume data structure with the 3D grid data structure.

Figure 6.7 shows a portion of a 3D grid model superimposed on a volume model component. Our objective is to determine the volume of intersection between the component and any one of the grid cells. If we simulate each grid cell by a rectangular component of appropriate dimensions then we can determine the intersection volume by applying the process described above. If we repeat this process for every grid cell and every component in our volume model of geology, and accumulate the results in the grid model data structure as described in Chapter 4, then we know precisely which grid cells intersect which geological units, and by how much. This knowledge gives us the ability to apply geological control to the variable prediction process, as discussed in Chapter 8.

6.6 Spatial Integrity in the Modeling Process

It is important to remember that the modeling process we have described is primarily concerned with the representation of *geological volumes*, as opposed to surfaces. This is compatible with the objectives of geological characterization; physical geology *consists* largely of discrete volumes. As well as providing an efficient medium for interactive interpretation of geology, the process also allows us to define these volumes *directly*. This is quite different from the surface-based modeling techniques of the past. These require geology to be determined initially in terms of *geological surfaces* that are only subsequently used to derive volumes. Anyone who has attempted to apply surface-based approaches to complex conditions with significant discontinuities is well aware of the procedural difficulties involved.

The difficulties created by discontinuities are largely eliminated by the volume modeling approach of using as many irregular components as necessary to represent any complex shape. Discontinuities at worst require interpretation of a few more appropriately shaped and appropriately oriented components. These advantages have a cost, however, that can be measured in terms of the extra effort required to ensure the spatial and geological integrity of our interpreted volumes. When dealing with discrete, independent volumes it is relatively easy to create unintentional voids between them, particularly when we are interpreting adjacent volumes in the third dimension. The integrity can be verified quickly and efficiently by cutting viewplane sections at appropriate orientations, and rectified by interactively editing the component boundaries. However, it is obviously preferable to ensure integrity from the start.

We achieve this by making use of interactive interpretation tools. These allow us to incorporate any portion of an existing boundary into the current component definition, or to *snap* points in the current definition to adjacent boundary points. The thickness of a component can also be modified interactively while we are extending it in the third dimension and matching it to adjacent boundaries. If we

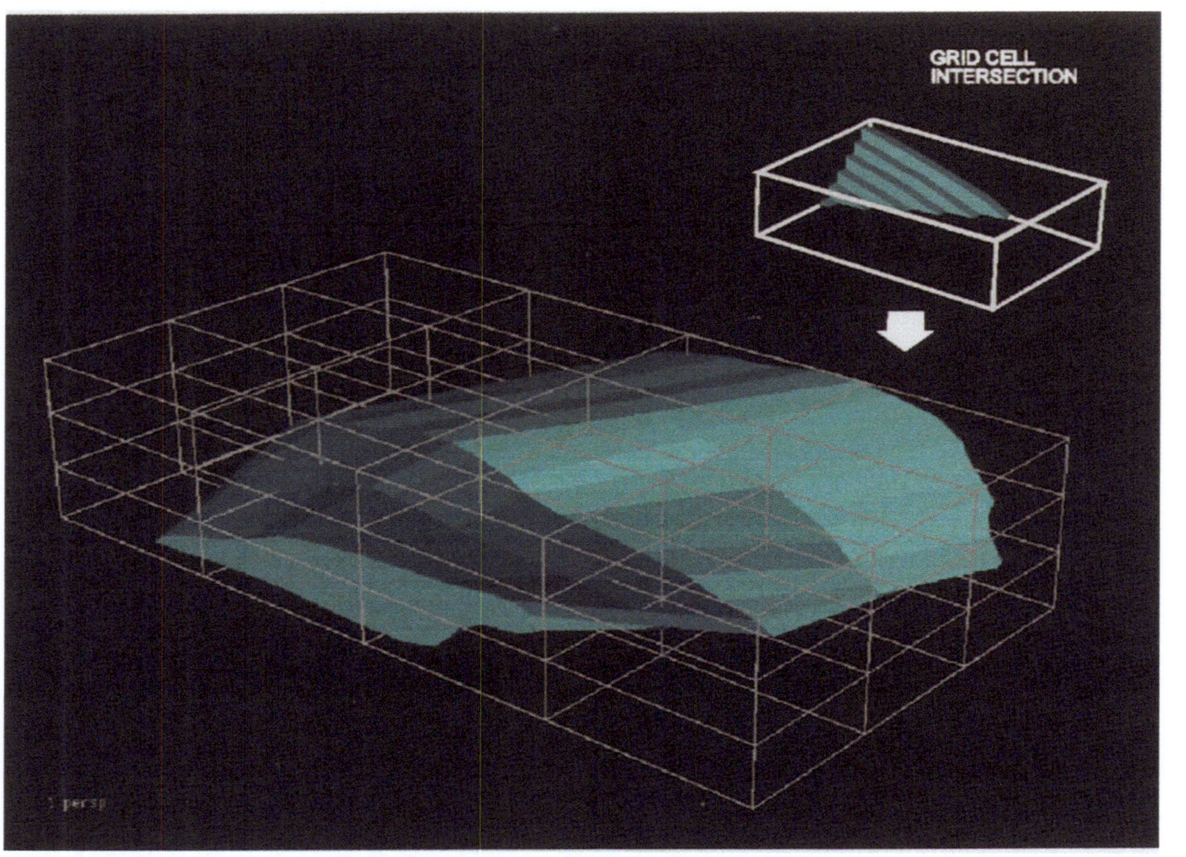

Fig. 6.7   Spatial integration of a component with intersecting grid cells from a 3D grid data structure; the volume of intersection is stored for each grid cell (inset)

need additional definition in a component boundary, then new points can be entered at any stage. For automatic interpolation between boundaries, we can establish *interpolation links* that maintain the integrity of prominent features. Tools such as these make interpretation of complex conditions as simple and efficient as possible, and ensure the integrity of the result. They also provide another advantage over surface-based approaches in that they make editing of volumes, to incorporate new information for example, a relatively simple operation. With the availability of interactive tools such as those described we are able to both interpret and represent complex geological conditions. This capability is illustrated in Fig. 6.8, an example of folded, faulted and tilted geological structures.

6.7 3D Modeling Considerations

We have spoken at length about data structures, viewplanes and interactive editing tools that make the process of 3D interpretation as simple and efficient as possible, however, we should not underestimate the work involved. Interpretation of complex geological conditions can be a difficult, tedious, repetitive and time-consuming process, no matter what medium we choose to do it in or which methodology we use. 3D interpretation is a quantum leap removed from the 2D equivalent in terms of complexity and level of input. If our characterization involves considerable geological complexity, then we should be aware from the start that the process of 3D interpretation is going to require a significant effort on our part.

This brings us to the next point: the tailoring of our interpretation to the available information and to the objectives of our characterization. There is absolutely no point in interpreting unsubstantiated detail into our model of geology. If the interpreted level of detail cannot be justified by either the source data or our knowledge of the region, then we are wasting our time. This wasted effort will most likely be compounded by having to reinterpret geology later. In a similar vein, there is no justification for incorporating detail that will have little measurable effect on the outcome of our characterization. All of this may seem obvious, but it is surprising how frequently we begin a characterization project by interpreting far too much detail, only to run out of time half way through and have to finish it in a hurry. A characterization project benefits significantly from planning and scheduling, just like any other project.

Finally, a few words on the advantages of interpretation in a 3D context are appropriate. The ability to move easily from one section to another to view the effects of what one has just defined or interpreted in the third dimension enhances the process considerably, as does the ability to ensure that geometries correlate in all directions. Because inconsistencies in an interpretation are so glaringly obvious when viewed on another section or in 3D perspective, the result is generally an interpretation that is significantly more realistic, and geologically more plausible, than the conventional 2D equivalent. A set of 2D interpretations, that may appear

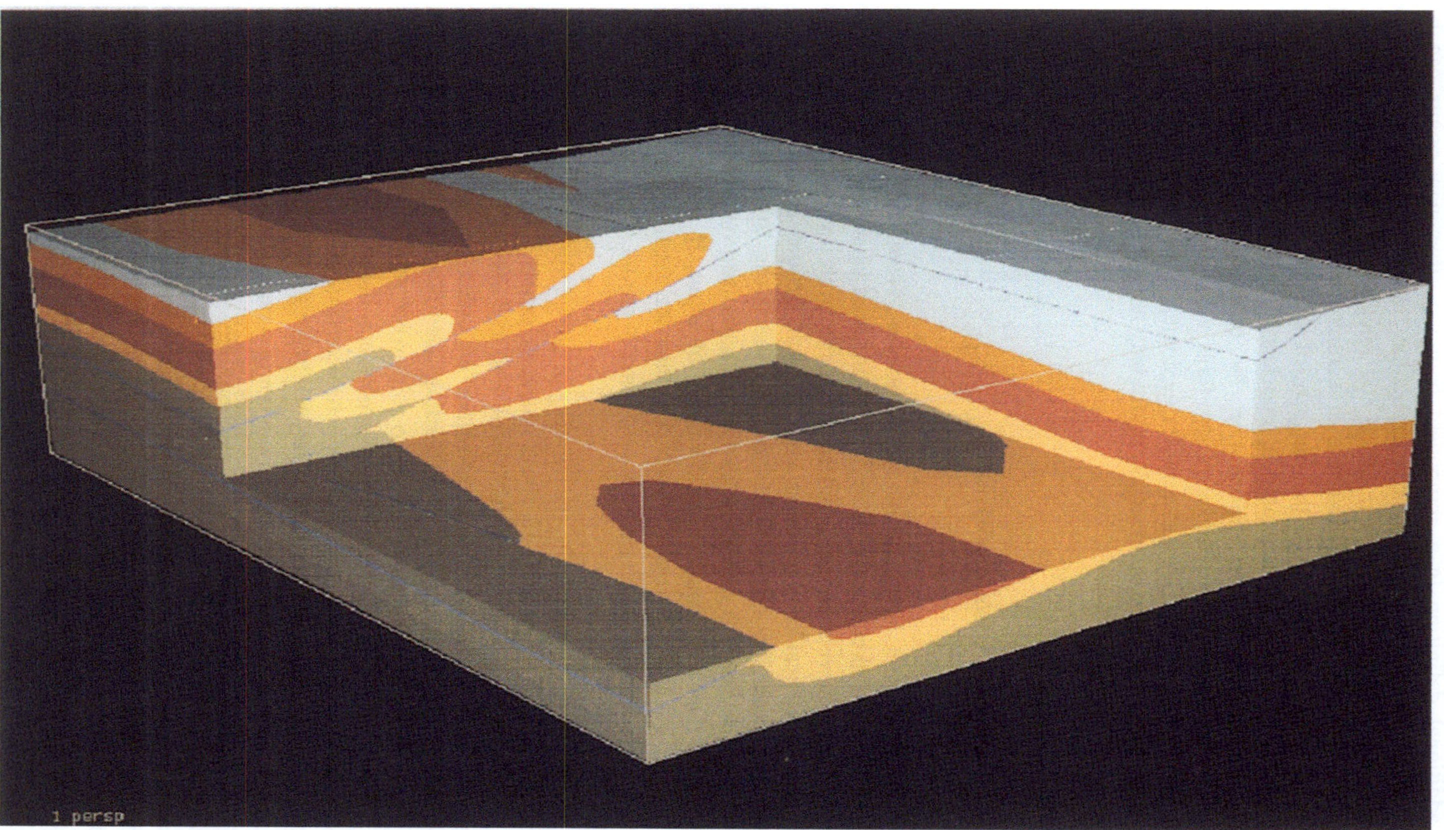

Fig. 6.8   3D volume model representation of folded, faulted and tilted geological structures, with a 3D cut-away section applied

perfectly reasonable on individual sections and plans, invariably undergoes significant change as a result of being subjected to 3D interpretation. The extra effort involved is always justified by the more rigorous interpretation that results.

7 Geological Modeling from Surfaces

7.1 Advantages and Disadvantages of Modeling from Surfaces

For many applications of layered or stratified geology, the interactive interpretation techniques discussed in Chapter 6 are more than we need. For these relatively simple situations there are much less input-intensive ways of achieving a 3D characterization of geology, provided we have sufficient information to define the principal geological surfaces. We can then instruct the computer to fill the spaces between surfaces with suitably shaped volume model components.

The underlying concepts of this approach are twofold. Surfaces are initially represented as triangulated networks, with the known data points as the triangle nodes. Triangulated volume model components (cf. Fig. 4.3, Chap. 4) that match the dimensions of the surface triangles are then generated between the surfaces.

The principal advantages of this approach are that it is much faster - the computer does most of the work - and the integrity of the model is ensured, provided we take the necessary precautions. There is less risk of leaving unintentional voids in our volume model representation of geology. In those cases where detailed surface information is available, the surface-based approach actually provides an enhanced representation compared to the sectional approach discussed in the previous chapter. The disadvantages are mostly concerned with its limited application to relatively simple layered conditions, although techniques that help to alleviate this limitation and make the approach more generally applicable are also discussed below.

7.2 Representing Surfaces as Triangulated Networks

Many of the conventional approaches employ rectilinear 2D grid data structures to represent surfaces, typically with a horizontal orientation. This severely limits their functionality with respect to the representation of inclined surfaces. It also

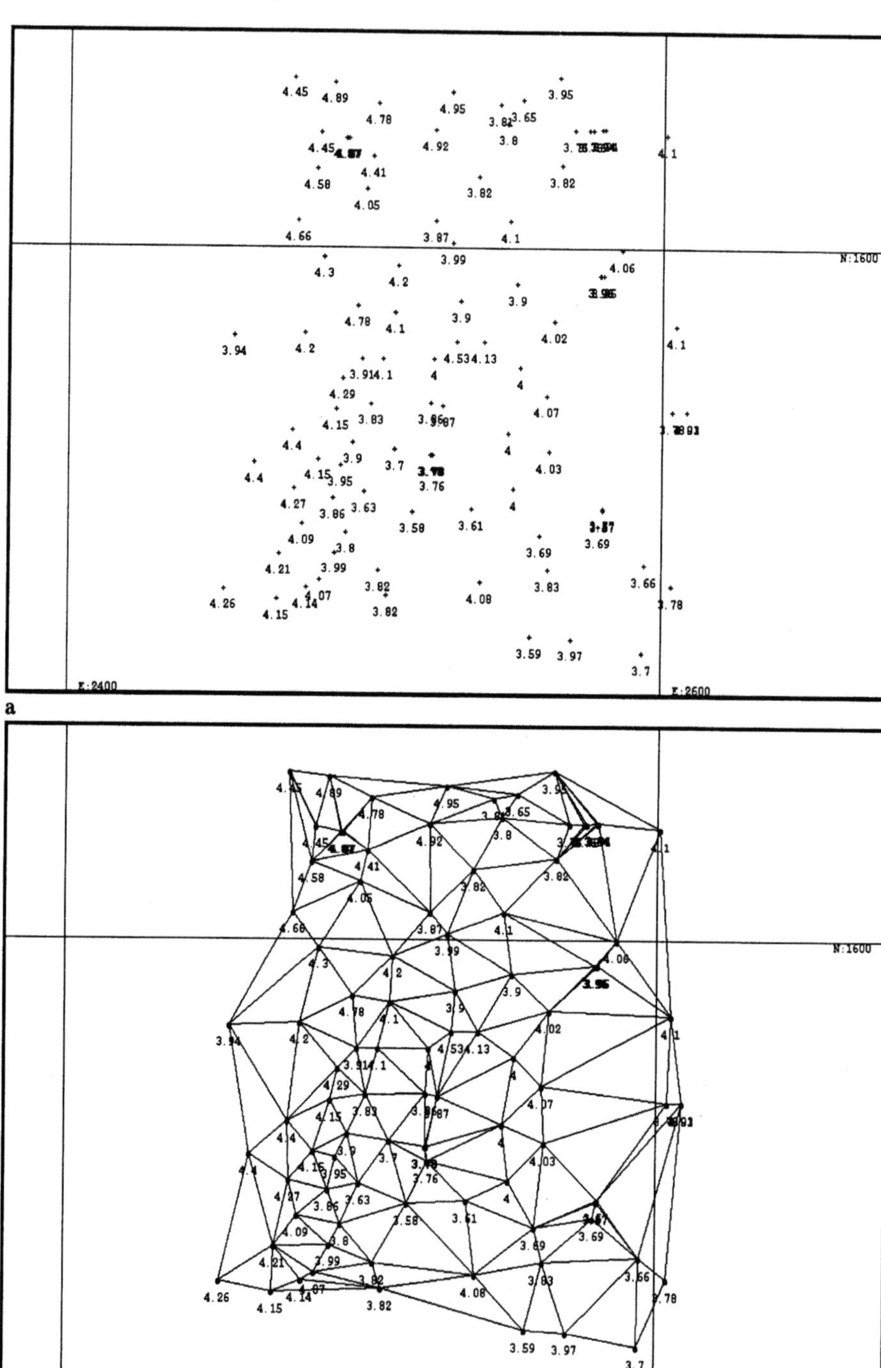

Fig. 7.1   **a** A set of points with elevations displayed as local Z values; **b** a triangulated network (TIN) that includes and connects all of the points

implies a *smoothed* representation, since the elevation values at the grid nodes must be interpolated from the observation points, which are typically randomly spaced.

A much more general approach is to represent the topology of a surface by an optimized *triangulated network* (TIN) that connects the XYZ coordinate locations of the known observation points. A primary advantage of the TIN approach is that the observed 3D locations of the points are honored, no smoothing or interpolation is necessary. The resulting surface always includes the observed points, a feature that is generally compatible with the representation of *hard* geological surfaces. Another advantage is that the observed density of information is maintained; small triangles occur where the point density is high and correspondingly large triangles where the density is low. In contrast, the gridded surface approach assumes a uniform density of interpolated (or predicted) information throughout, which is acceptable provided the prediction results include a measure of the prediction uncertainty. We discuss geostatistical prediction of surfaces at the end of Chapter 10. The TIN approach is sufficiently general that it can deal with any degree of information density, irrespective of whether it is uniformly or randomly spaced. A third advantage concerns the representation of inclined surfaces; the TIN approach can be applied from any orientation that is approximately parallel to the surface considered. Horizontal, inclined, vertical and overturned surfaces are handled with equal ease.

The final advantage of the TIN approach concerns its ability to represent discontinuity in the extent of a surface. There is no necessity for a surface to be continuous throughout the modeling region, as there is with many of the gridded approaches. Moreover, the ability to represent an irregular edge or surface extremity is enhanced.

There are two criteria for successful implementation of the TIN approach to representation of surfaces. The first concerns an efficient algorithm for generation of triangles, efficient in terms of speed and ability to produce a triangle set that is representative of the topology of the physical surface. The most commonly applied algorithm is *De Launay Tessellation*. It results in the optimum triangle set with triangles that are as well-ordered as possible in terms of their *equilateralness*. The second criterion concerns the ability to manipulate the information associated with one or more TINs in order to define volumes rather than surfaces.

The triangles are generated on the basis of the XY locations of the points. The local Z coordinates of the points provide the topological relief. This means that generation and manipulation of TINs must be performed relative to a local reference plane that is approximately parallel to the surfaces considered. This is readily achieved by defining an appropriately oriented viewplane and then transforming the source information for the surfaces to the local XYZ coordinate system, as described in Chapter 4. The viewplane becomes the local TIN reference plane. In order to generate volumes from a TIN we require additional information; this is most conveniently expressed in terms of a *thickness* (T) at each point (cf. Fig. 7.5). TIN manipulation is generally concerned with derivation of the Z and T values for volume definition from one or more surfaces.

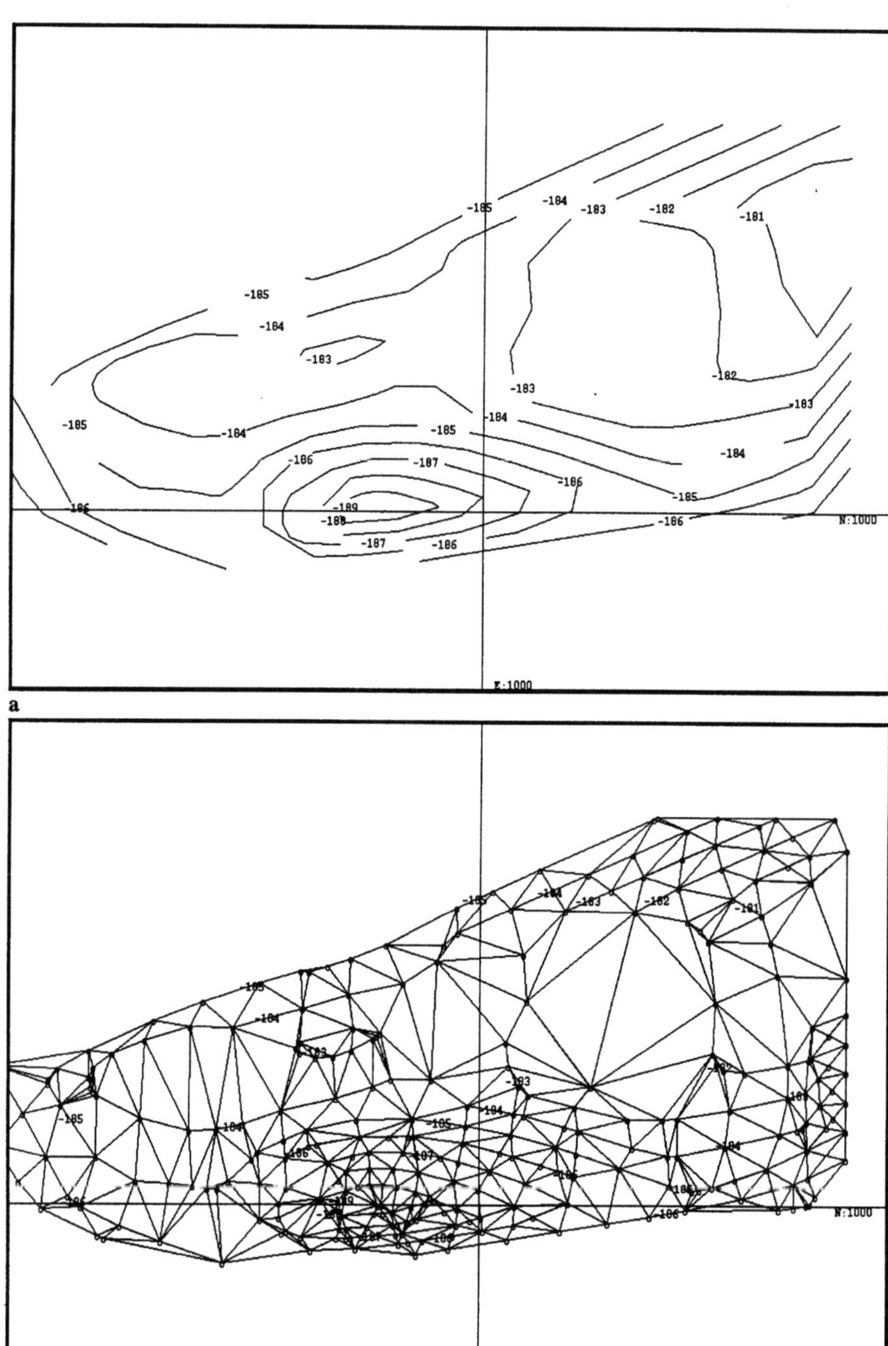

Fig. 7.2 **a** Elevation contours of the top of a reservoir formation; **b** a triangulated network (TIN) that honours the contour features

In reality, a TIN is yet another data structure with its own unique definition of local geometry, just like the principal data structures discussed in Chapter 4. Because they are quick to generate, and seldom the end objective of a modeling exercise, it is convenient to treat them as transient, or temporary, data structures.

Much of the source information for surfaces comes in the form of randomly spaced point information. Its origin might be an analysis of borehole intersections with a geological unit, or a topographical survey. Figure 7.1 provides an example of the triangulation of this type of information. In this case, the viewplane has a horizontal orientation at zero elevation. The positive Z values displayed in Fig. 7.1a indicate that the surface is above the viewplane. If the available point information has already been interpolated to a regular grid, then it is handled in a similar fashion, with the exception that the triangulation is much quicker and easier since the triangles are regular.

However, in many cases, the source information is provided in a line format rather than a point format. Surface contours, survey traverses and geophysical survey lines are examples. With this type of information we have an additional consideration: these line features frequently contain topological (or connectivity) information over and above the XYZ locations of the observed points. It is generally undesirable that the topology of the TIN that connects the points contained in these features should violate this connectivity. We must therefore have the ability to force the triangulation algorithm into honoring the line features where this is necessary. Figure 7.2 provides an example; inspection of Fig. 7.2b confirms that none of the resulting TIN triangles cross any of the contour features from which they were generated.

To make the TIN approach more generally applicable we require a range of surface handling tools in addition to TIN manipulation. These include the ability to interactively modify the triangle set of a TIN, to clip a TIN internally or externally to a polygonal boundary, to map one TIN onto another to ensure a common point set, to determine the intersection of two TINs, and to truncate one TIN above or below another. These tools and their applications are discussed in detail in the following sections.

7.3 Structural, Stratigraphic and Topographical Surfaces

The interactive 3D interpretation process discussed in Chapter 6 is frequently enhanced if we first define the controlling surfaces and discontinuities. These generally include topography, and may include structural surfaces, such as faults, and stratigraphic horizons or contact surfaces as well. The ability to define a viewplane at any appropriate orientation and have the intersections of these controlling surfaces displayed in their true locations is of significant advantage to us during interpretation. Figure 6.1 provides an example of this type of viewplane display. Representation of these surfaces as volume model data structures is usually the first step in any 3D interpretation of complex conditions.

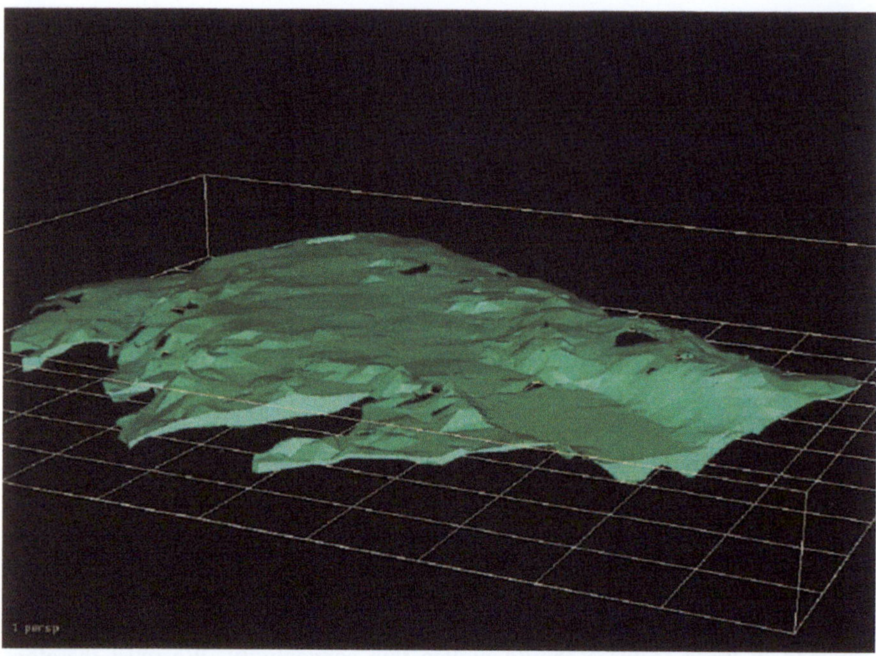

a

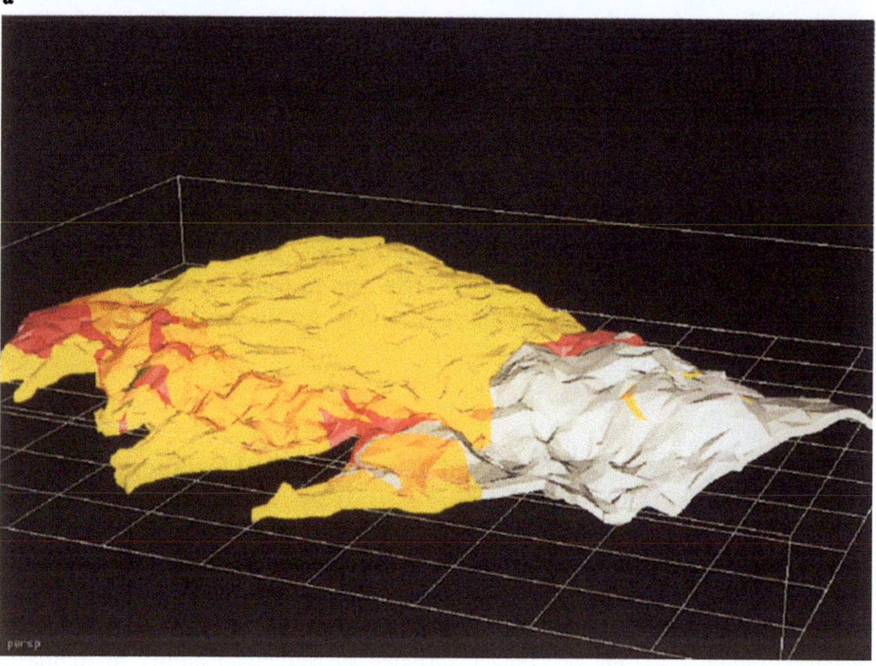

b

Fig. 7.3   **a** 3D view of a surface-based model of overburden thickness between bedrock and topography; **b** a model of the (geophysically determined) bedrock profile that incorporates regional geology characteristics

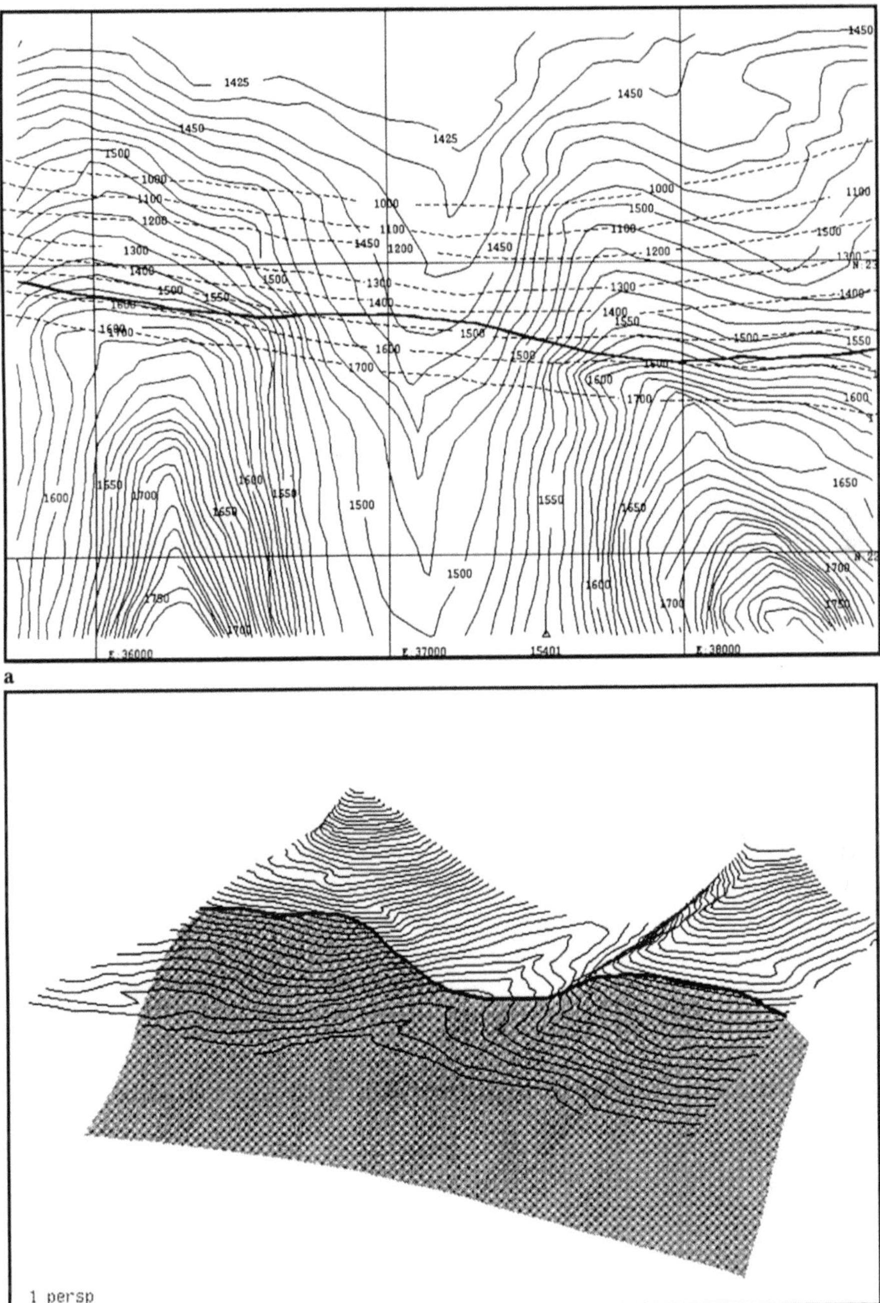

Fig. 7.4   **a** Contours of topography (*solid line*) overlaid on contours of an inclined fault (*dashed line*), with the resulting line of interserction (*heavy line*); **b** perspective of the fault surface truncated at topography, overlaid with topography contours

This requirement is readily satisfied by generating a TIN representation of the surface, assigning a nominal thickness (T) value to all the points contained in the TIN, and then generating triangulated volume model components of uniform thickness, as defined in Fig. 7.6a. The thickness value is usually selected so as to provide an appropriate display thickness, with respect to our working scales, on intersecting viewplanes. Figure 7.3 provides examples of applying the surface-based approach to representations of surface topography, overburden thickness and bedrock geology. In these examples, the original densely gridded information has been *thinned* to remove points that contribute little to the vertical relief of the surfaces. That the thinned information is of variable density is of no concern with the TIN approach, which matches the surface to whatever information is available.

The nature of the source information frequently requires an ability to clip or truncate surfaces, for example, we might wish to clip the representation of a fault surface so that it does not project above topography. To achieve this we generate TINs for both surfaces from the same viewplane, then inspect their Z values and the triangle connectivities to determine the boundaries of intersection between them. These boundaries are used to clip the fault surface TIN before generation of volume model components. Figure 7.4 provides an example of a clipped intersection between two surfaces and the resulting volume model representation.

7.4 Manipulation of Surfaces

To apply the TIN approach to representation of geological volumes we require thickness (T) values at all the points, in addition to the Z values, which are generally assumed to refer to the upper surface of a geological volume (cf. Fig. 7.5a). In some instances, we may have direct access to both types of information, for example, as contours of the upper surface elevation together with isopach contours of thickness. By generating a primary TIN of the upper surface and a secondary TIN from the isopach contours, we can merge the two to generate a single primary TIN with both Z and T values, and progress to a volume model representation. The best possible representation is always obtained from a viewplane that is approximately parallel to the surface, since Z and T values are measured normal to the plane. This does not mean that the process cannot be applied from a viewplane that is oblique to the surface; the integrity of the result depends on the degree of relief in the surface.

The thickness values are not always immediately available, however, in many instances, we have to generate this information by manipulating two or more surfaces. Figure 7.5b is a schematic of this manipulation process. Ideally, the points in both TINs share the same XY locations, as in the case of point i in the figure. The thickness at each point is obtained simply by subtracting one Z value from the other. However, XY coincidence of both sets of points is frequently not the case, as in the case of point k in Fig. 7.5b. This means that the points in the primary TIN must be *mapped* onto the secondary TIN to obtain interpolated Z

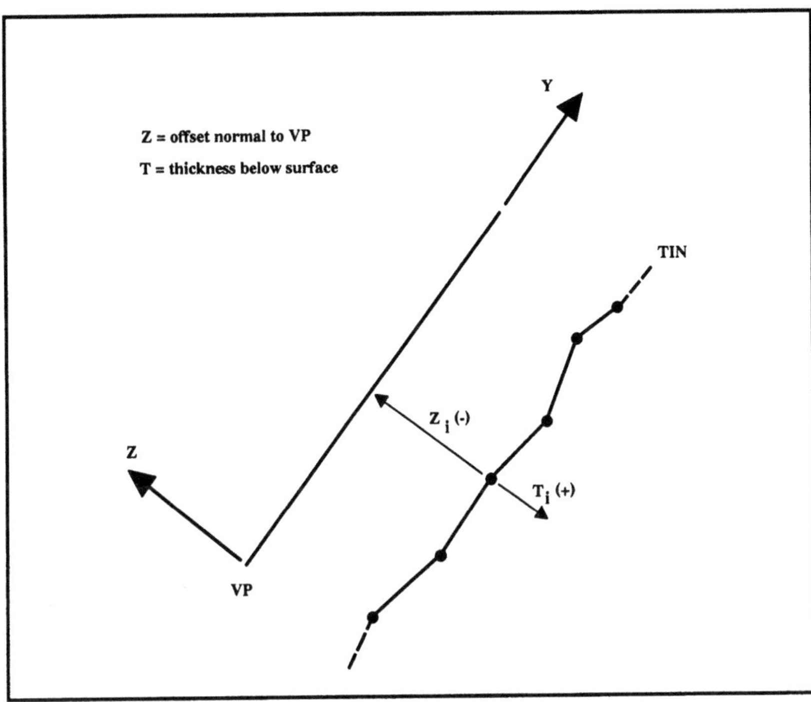

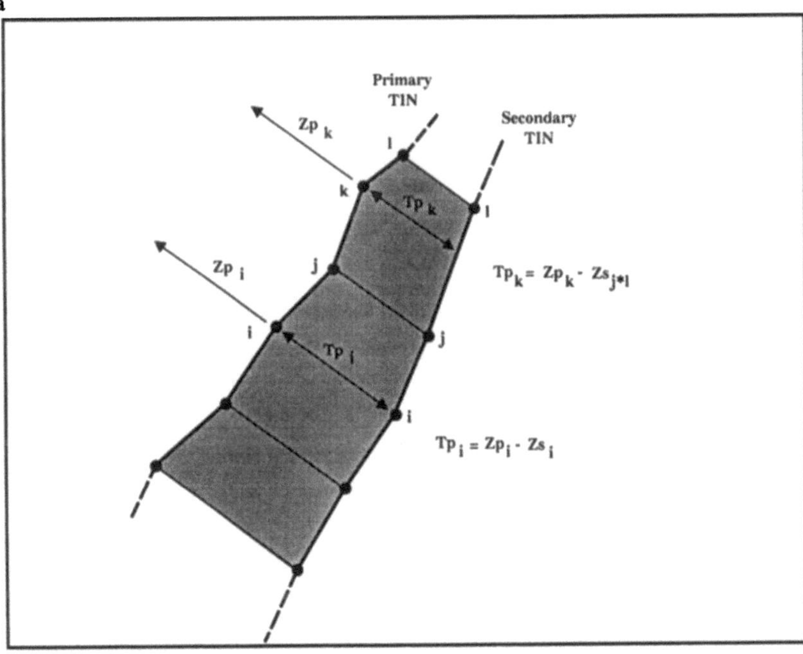

Fig. 7.5   a Cross section through an inclined TIN showing its relationship to the viewplane; b manipulation of two TINs to obtain thickness values for volume modeling purposes

values at appropriate locations before the thickness calculation can be performed. This is conveniently achieved with an algorithm for linear interpolation within a triangle and is a standard function of TIN manipulation.

The *primary TIN* is an underlying concept of the entire approach to surface-based modeling. It provides the spatial framework for both the manipulation of Z and T values and the derivation of component geometry for the final volume model representation. It is important that we recognize the significance of the latter function. To ensure the integrity of the volume model representation of a sequence of stratigraphic units, the same primary TIN, or subsets of it, should be used throughout the modeling process. If TINs generated from different point sets are used to model two adjacent units, then their resulting volume model representations will not be 100% contiguous, voids and overlaps will exist at the contact between the units. This is because the points in the two TINs do not coincide at the common surface between them. This loss of integrity is eliminated by using the same primary TIN for every unit. To generate a primary TIN that contains sufficient detail to obtain a good representation of all units we may have to merge several point sets into one.

7.5 Derivation of Volumes from Surfaces

The geometry of a triangulated volume model component is defined in Fig. 7.6a. The component geometry is obtained from a triangle in the primary TIN; when viewed from the reference (view)plane, the triangulated boundaries of the upper and lower surfaces are coincident. The Z values of the upper surface are obtained from the primary TIN Z values. The Z values of the lower surface are obtained by subtracting the primary TIN T values from the Z values. The edges joining the vertices of the two triangulated boundaries are normal to the reference plane. The result is a simple geometrical configuration that can be rapidly generated by the computer from the information contained in the primary TIN.

Figure 7.6b shows a volume model representation of three stratigraphic horizons that define the contact surfaces between a sequence of soil layers. In this case, an analysis of borehole intersections provides the source information for generating the surfaces. Figure 7.7a illustrates a portion of the TIN geometry of the surfaces and generation of a single volume model component for one of the soil layers between the horizons. Figure 7.7b shows the final volume model representation. The soil geology is very simple in this case, however, the objective of creating a geological model for spatial control of contaminant prediction is achieved quickly and with minimal effort. In the following section, we discuss application of the TIN approach to more complex examples involving geological discontinuity.

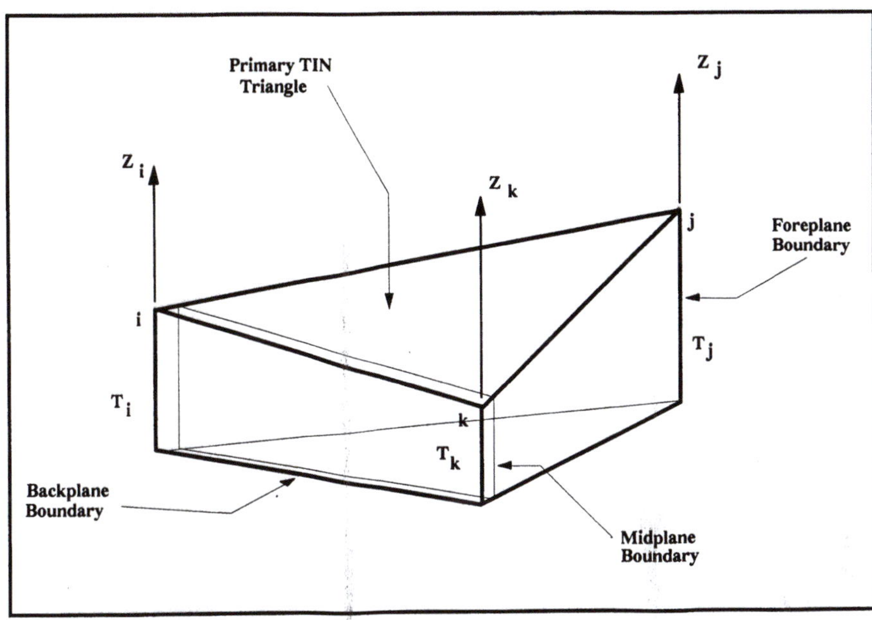

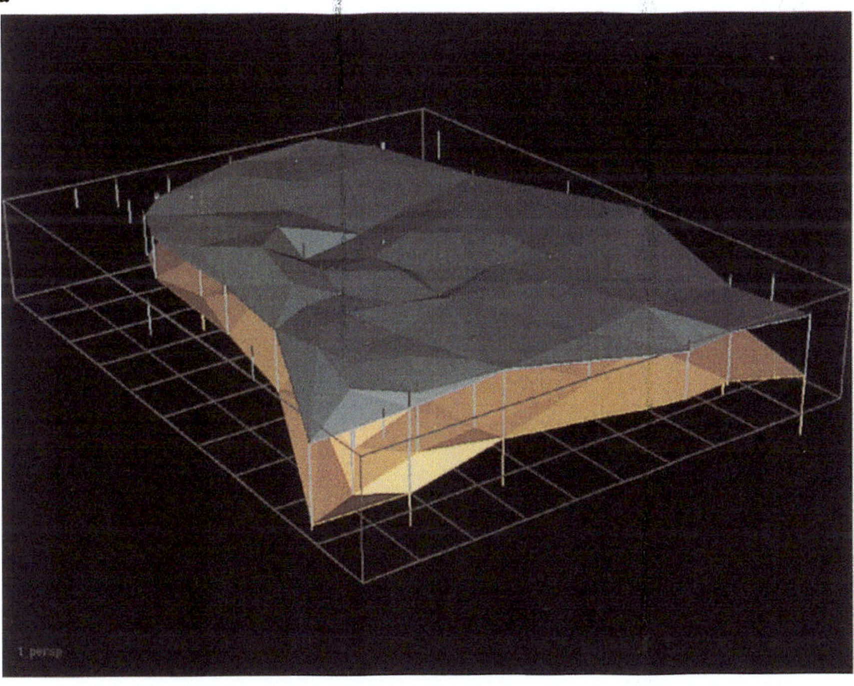

Fig. 7.6   **a** Geometrical definition of a triangular volume model component, derived from TIN manipulation; **b** surface representation of stratigraphic horizons with triangular components generated from borehole intersections

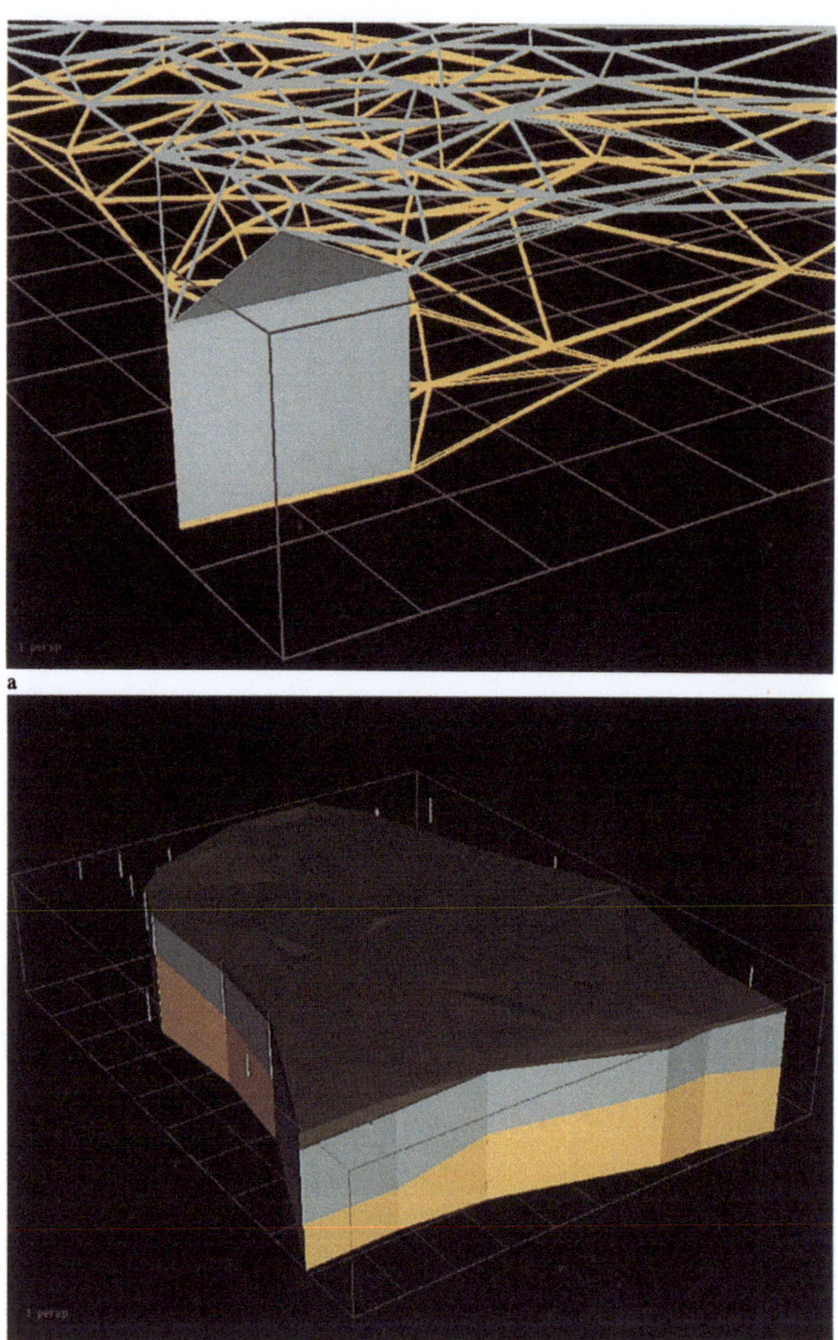

Fig. 7.7   **a** A triangulated, computer-generated volume model component between two of the surfaces shown in Fig. 7.6b;   **b** complete volume model representation of the stratigraphic units between the surfaces

7.6 Modeling Stratigraphic and Structural Discontinuities

The previous example involved a straightforward representation of irregular, but continuous, soil stratigraphy. If the geological unit we are concerned with pinches out, or is cut by erosion channels, or is discontinuous for any other reason, then the modeling process becomes more complicated. Stratigraphic discontinuities of this type are typically dealt with by a combination of surface intersection and clipping to obtain an appropriate primary surface representation, followed by TIN manipulations similar to those discussed earlier.

For example, consider the case of a sequence of stratigraphic units that outcrop at topography due to erosion and are also truncated by a fault surface in the foreground (cf. Fig. 7.10). The solid lines in Fig. 7.8a are contours of topography and the dashed lines are contours of the fault surface and stratigraphic horizons. The heavy lines represent the intersections of the various surfaces, obtained by analyzing the intersection of their TINs. The intersection analysis simply involves linear interpolation of the zero Z difference line segment within each triangle for each pair of surfaces. These intersection boundaries are stored as traverse features, with variable X, Y and Z values at each point, in a map data structure. By merging these intersection features with the source information we can use them to *clip and paste* surface information to derive *composite surfaces* that are suitable for modeling purposes.

Figure 7.8b illustrates the composite surface for the uppermost stratigraphic unit in the example sequence, comprising clipped topographical contours and intersection features between topography and the bottom of the unit. In this case, the clipping and merging operations are conducted directly on the contours, represented as map data structures, whereas in the previous example (cf. Fig. 7.4) clipping is applied to the TIN representation. The resulting composite surface is used to generate a primary TIN for modeling the volume of the unit. The thickness values, which taper to zero at the intersection features, are derived by TIN manipulation with the stratigraphic horizon that represents the bottom of the unit.

Modeling the intermediate unit is more complex since it is defined by four different surfaces: topography, two stratigraphic horizons and the fault surface. Figure 7.9a illustrates the composite surface for this unit, comprising a number of intersection features and appropriately clipped portions of the individual bounding surfaces. Figure 7.9b shows the equivalent composite surface for the lower stratigraphic unit. Once these composite surfaces have been derived, the final volume model representation (cf. Fig. 7.10) is generated rapidly and efficiently using standard TIN techniques.

The use of composite surfaces in conjunction with the TIN approach provides a powerful tool for handling more complex conditions of layered (stratified) geology. As demonstrated by the above example, it allows us to represent discontinuities due to erosion, pinch-outs and inclined faults, all of which present severe modeling problems for the conventional gridded approaches. The TIN approach can also be used in conjunction with the interactive interpretation

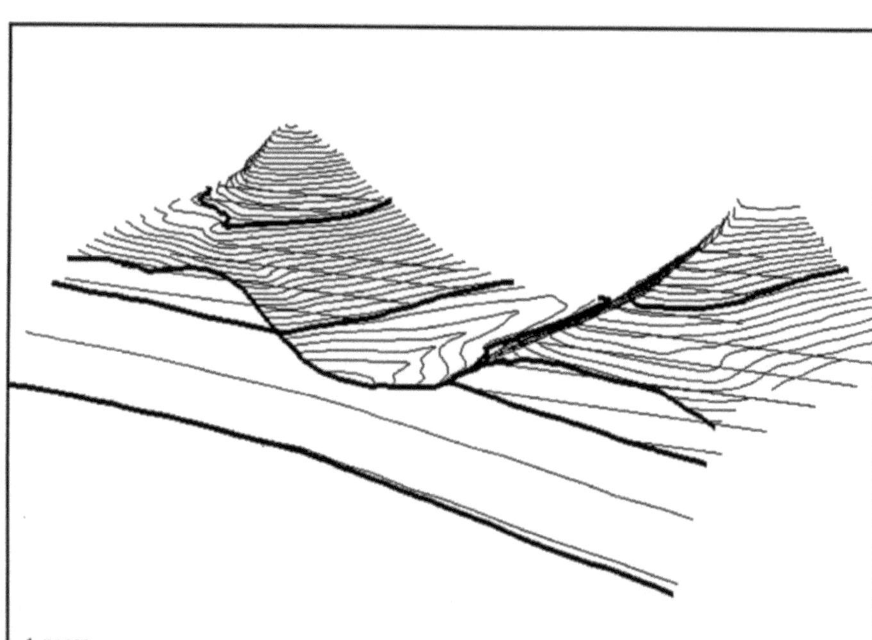

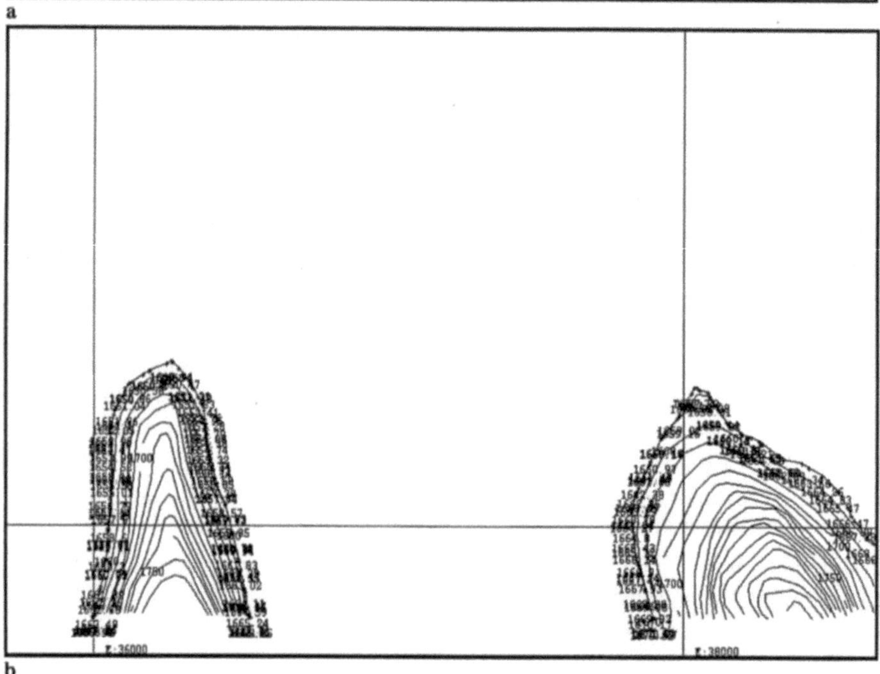

Fig. 7.8   a Contours of topography, fault surface and stratigraphic horizons, with surface intersections shown as *heavy lines*, (cf. Fig. 7.10); **b** composite surface for modeling the uppermost stratigraphic unit, consisting of topography and intersection traverses

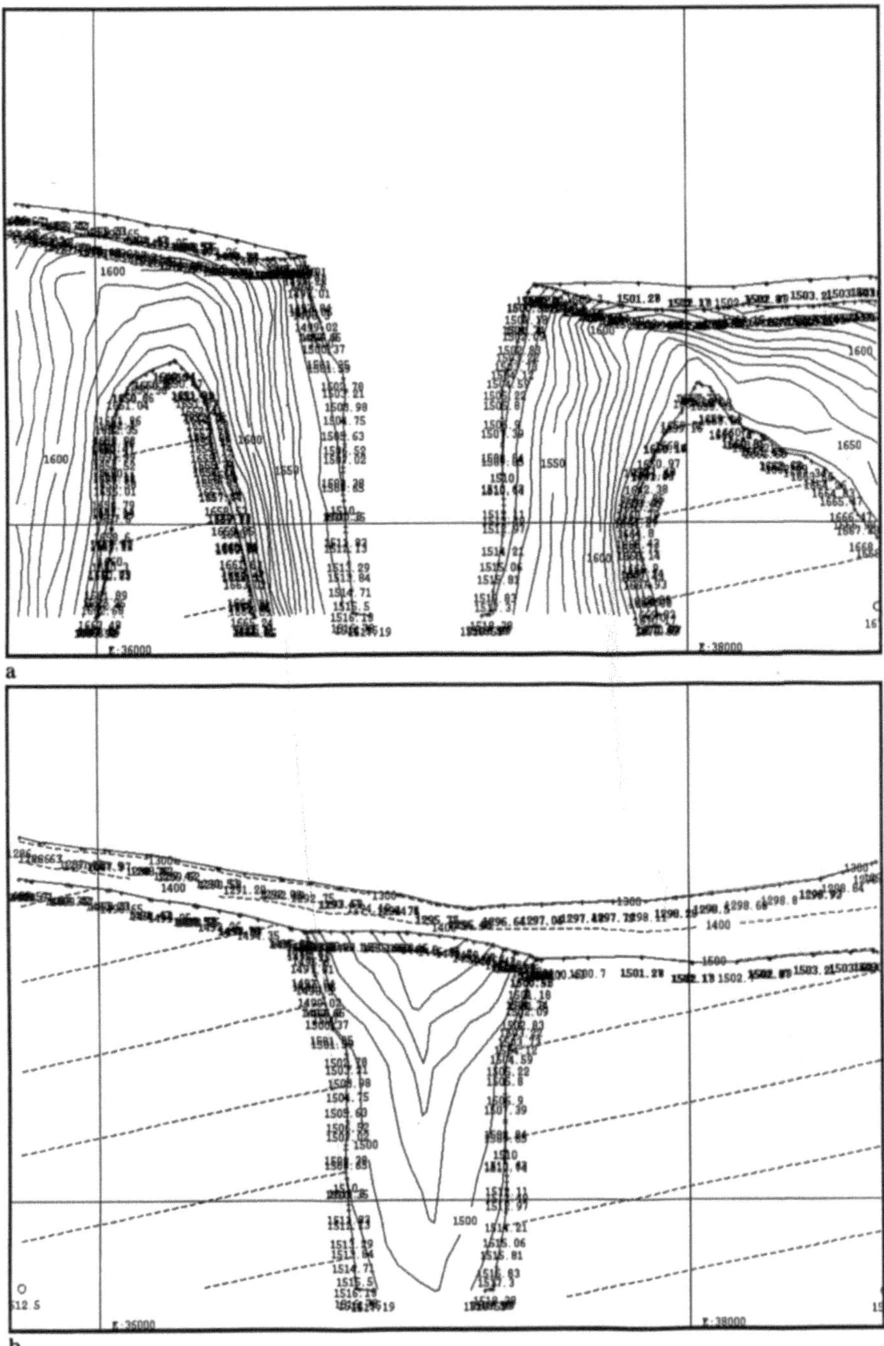

Fig. 7.9 a Composite surface of topography, fault and stratigraphic contours and intersection traverses for modeling the intermediate stratigraphic unit in Fig. 7.10; **b** composite surface for modeling the lower stratigraphic unit

Fig. 7.10 Volume model representation of faulted, eroded stratigraphic units that outcrop at topography, truncated by a near-vertical fault surface in the foreground; generated from the composite surface information shown in Fig. 7.8 and Fig. 7.9

techniques discussed in Chapter 6. This allows us to add detailed information, where necessary, to achieve our final geological characterization.

In summary, use of the surface-based TIN approach to modeling of relatively simple, stratified geological conditions is entirely complementary to the interactive interpretation approach to modeling discussed in Chapter 6. Combination of the two approaches extends the modeling capabilities to efficient representation of a wide range of geological conditions.

8 Geostatistical Prediction Techniques

8.1 Spatial Prediction Process

Our objective here is to predict the spatial variation of one or more variables of interest. The data structure, or vehicle, for prediction is the 3D grid data structure discussed in Chapter 4. The methodology comprises whatever prediction technique, prediction model(s) and parameters we select as a result of our analysis of spatial variability, as discussed in Chapter 5. Spatial control of prediction is provided by our interpretation of relevant characteristics, discussed in Chapters 6 and 7.

An underlying assumption of the prediction process is that a continuous measure of spatial variation is acceptably represented by values at discrete points at regular intervals in three dimensions. The discrete points are the grid-cell centroids of the data structure; their spacings in orthogonal directions are generally selected with due consideration of the corresponding spatial variabilities. Another implicit assumption is that the value at the centroid of a cell is representative in some way of the values throughout the volume of the cell, hence, the analogy to raster-like information. If this is true, then we can interpolate between the predicted values at grid-cell centroids to achieve even greater precision. As we shall see, there are several ways in which the predicted values can be representative of their associated cell volumes, each is appropriate to different objectives and conditions.

One of the principal advantages of the 3D grid data structure is that it, and its variations, are employed by many external technologies. This means that we can readily export our predicted spatial variations for further analysis. For example, we can export predicted mineral grades for mine planning and scheduling purposes, predicted porosities and oil saturations for an oil reservoir simulation, or predicted contaminant concentrations and aquifer transmissivities for a groundwater flow and dispersion study.

In this chapter, we focus initially on the implications for spatial prediction of integrating the vector-based volume data structure with the 3D grid data structure. Although this approach is considerably more precise than the conventional approaches discussed earlier, there are some residual approximations involved that we should be aware of. The remainder of the chapter is devoted to discussion of the more common forms of geostatistical kriging and their variations, with a brief digression into alternative techniques. The case for kriging as the spatial prediction technique of choice has already been made in Chapters 1 and 3.

The chapter closes with a discussion of some of the more common pitfalls of spatial prediction and a summarized performance review of a selection of prediction techniques.

8.2 Implications of Spatial Discretization

In Chapter 6 we discussed at length the volumetric analysis techniques that allow us to intersect a 3D grid data structure with a volume model representation of a controlling characteristic, e.g. lithology or mineralogy. The result of this intersection is stored in the data structure as the volume of each characteristic value that occurs within each grid cell, for convenience we refer to these volumes as *characteristic intersections*. This knowledge allows us to limit the spatial prediction of a variable to only those cells that intersect a particular characteristic. Many cells are totally enclosed by a geological volume and have only one characteristic intersection. Depending on the spatial complexity of our interpretation, and on the selected dimensions of the grid data structure, some cells invariably have two or more such intersections. Figure 8.1a is an example of such a cell; it has two intersections with characteristic values A and B, respectively. In the prediction process we predict values of a variable at the cell centroid for both characteristic intersections, in each case using the appropriate prediction model and sample subset. This introduces an approximation, since the cell centroid is not, generally speaking, a point that is geometrically representative of the volume of a partial characteristic intersection. This discrepancy is illustrated in Fig. 8.1b; the centroid location of the predicted value for characteristic intersection A actually lies outside its intersection with the cell.

This discrepancy is not as negative as it seems at first glance. First, the characteristic intersection A is generally used only for controlling variable prediction within the cell, and not for any subsequent spatial analysis. Second, the predicted value at the cell centroid is based on the measured spatial variability of the variable for characteristic value A. The only approximation involved is caused by the implicit assumption in the prediction process that characteristic intersections A and B are both centered on the cell centroid. Third, if we require a more representative predicted value, then all we have to do is perform a volume-weighted average calculation for the cell, discussed under spatial analysis techniques in Chapter 9.

This type of cell-scale approximation is inherent to any raster-like approach that involves a discretization of space into uniform, regular subunits. By integrating the volume model representation of characteristics with the gridded representation of variables the precision with which we can predict spatial variations is advanced significantly beyond other computer approaches, which in many cases assume uniform conditions throughout a cell. The precision of the result is obviously influenced by the grid-cell dimensions. We discuss a logical basis for selecting appropriate dimensions below. Provided we adhere to this, and bear the above

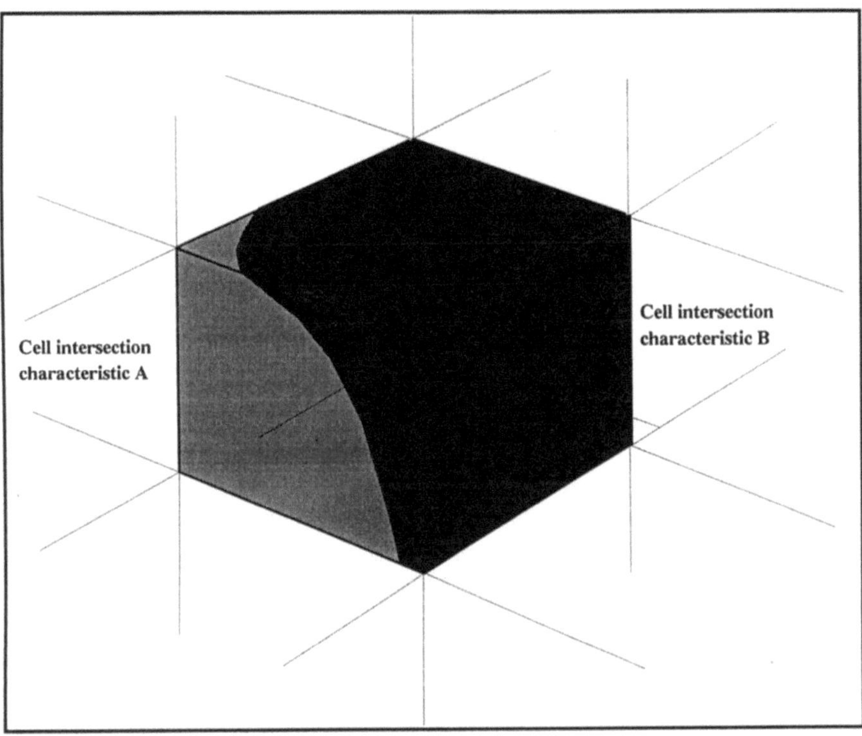

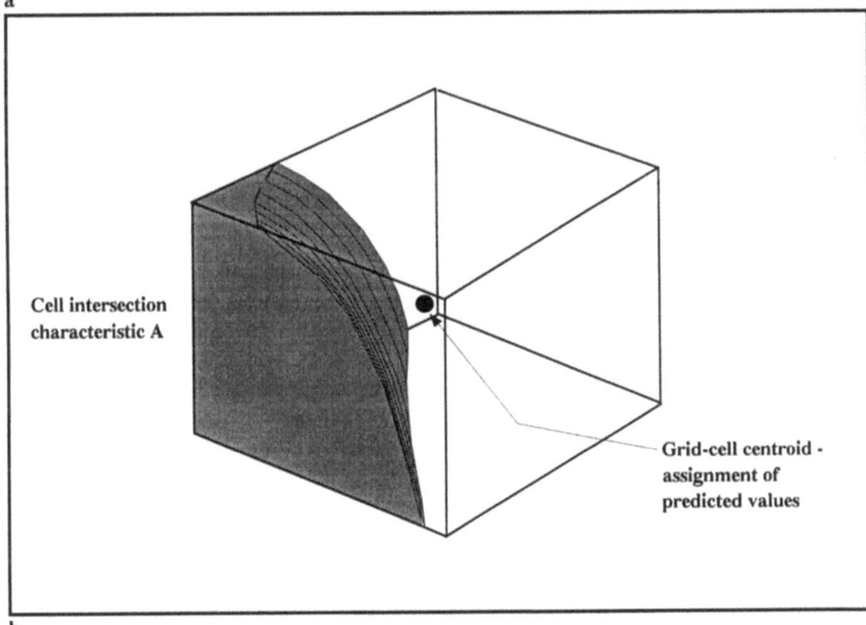

Fig. 8.1   **a** Grid-cell intersection at the boundary of two geological volumes; **b** spatial relationship of the grid-cell centroid to a partial geological volume

considerations in mind, then the integrated approach is capable of representing extremely complex conditions with acceptable precision.

The overall extent of the selected grid data structure should obviously encompass the region of interest. To avoid working with an unnecessarily large grid we typically also align it with the region of interest in terms of azimuth and inclination rotations. Figure 8.2a is an example of a 3D grid for an ore deposit characterization that has been aligned with the strike (azimuth) of the deposit. In cases of anisotropic spatial variability, we might also consider aligning the grid axes with the principal directions of anisotropy. All sample data relevant to prediction, whether in hole data or map data structures, must then be transformed to the local coordinate system of the grid data structure, as discussed in Chapter 4.

The selection of appropriate grid-cell dimensions is typically more subjective and is based to a large extent on our analysis of spatial variability, as discussed in Chapter 5. Ideally, the cell dimensions in each orthogonal direction should be between five and ten times less than the corresponding range of influence. This means that the cells are rectangular (rather than cubic) if the spatial variability is anisotropic. If the measured spatial variability has a small range of influence, then the corresponding cell dimension should be small and vice versa.

There are several other factors that also have a bearing on the selection of cell dimensions. There is generally little to be gained by selecting a cell dimension much less than a fifth of the average sample spacing in a particular direction. The cell dimensions may also be dictated to some extent by excavation considerations. In a mining context, for example, we may select dimensions that are related to the precision of the proposed method of excavation. On the other hand, in contrast to the conventional approaches, the cell dimensions are typically independent of any complexity present in our interpretation of geological characteristics. Figure 8.2b is an example of a grid that has been tailored to the anisotropic spatial variability of a soil contaminant. The grid-cell dimensions are 5 m x 5 m in plan but only 1 meter vertically. The measured horizontal range of influence is 35 m and the average sample spacing is 12 m. The vertical range is 2.5 m, which would normally dictate a cell dimension of 0.5 m, however, in this case 1 m precision in the vertical direction is considered adequate for remediation planning purposes.

8.3 Point Kriging to a 3D Grid

In Chapter 5 we discussed the concepts involved in predicting the value of a variable at a point by geostatistical kriging. We now relate these concepts to the practical requirements of geological characterization. First, we assume that any complicating influences such as anisotropy and underlying trends can be ignored. We consider the prediction of a variable whose spatial variability for a particular value of the controlling characteristic is represented by the spherical semi-variogram prediction model shown in Fig. 8.3b.

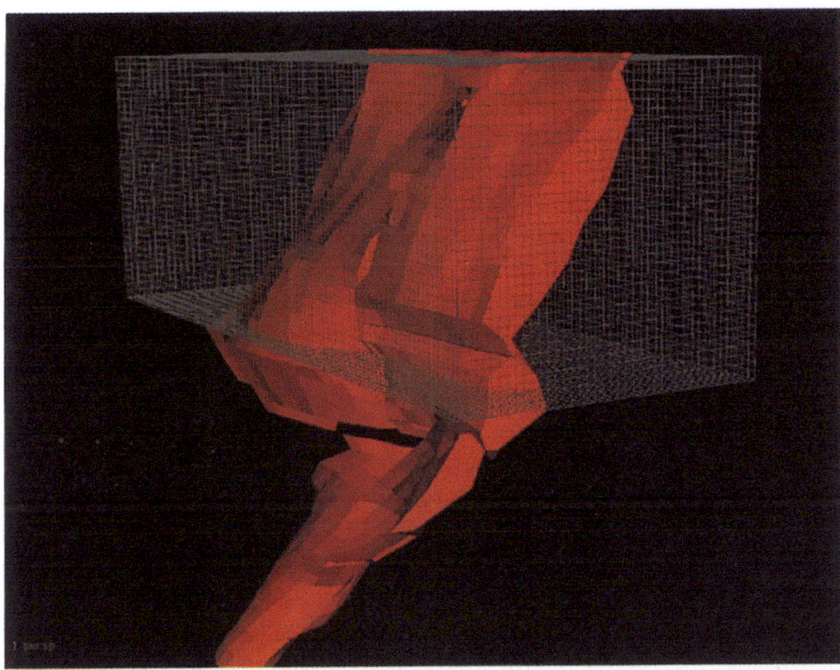

Fig. 8.2 **a** 3D grid model geometry for representing grade variation in a steeply inclined mineral deposit; **b** 3D grid model geometry for representing relatively shallow soil contamination variation

We visit each grid cell in turn to determine whether it has a characteristic intersection of the right value. If not, the variable is left undefined at the cell centroid location and we move on to the next cell. If it has an intersection, then we search for the samples that are relevant to the centroid of the grid cell. To qualify a sample must have the right characteristic value, a defined value for the variable of interest, and fall within a *search volume* that is centered on the cell centroid. In this case the search volume is a sphere with a radius equal to the range of influence of the prediction model (cf. Fig. 8.3a). We discuss several variations in the definition of the search volume in subsequent sections. For each sample that falls within this volume we determine its distance from the cell centroid and then scale the semi-variance value from the prediction model as shown in Fig. 8.3b. The semi-variance values are used to compile the set of equations that define the minimum variance for the cell centroid location in terms of the unknown sample weighting coefficients. Solution of the equations by numerical inversion gives the optimum set of weighting coefficients; and substitution of these and the sample values into a standard linear estimator (discussed in Chap. 5) results in a predicted value for the cell centroid. The principal advantage of this approach to prediction is that the semi-variance (and hence the uncertainty) at the cell centroid is minimized, based on the surrounding sample values and the measured spatial variability. Any other predicted value, irrespective of how it is obtained, would have a greater associated uncertainty (see Chap. 10). The semi-variance of the predicted value at the cell centroid is also obtained from the prediction process (for subsequent convenience we convert this as discussed in Chapter 5 and store it as the statistical standard error), our primary measure of prediction uncertainty.

We repeat the above process for every grid cell and for each value of the controlling characteristic. For those cells that have more than one characteristic intersection, we end up with a corresponding number of predicted values and uncertainties for the variable of interest. Figure 8.4a,b illustrates the results of separate contamination predictions for two adjacent soil layers. Cells that intersect the boundary between the layers have two predicted values, one for each layer.

At this stage it is appropriate to digress for a minute to discuss the implication of *undefined values* and how we manage them. Characteristics are typically defined in terms of *alpha strings* and an undefined value is readily dealt with by assigning it a null (zero length) string. However, in the context of *numeric values* computers are designed to deal with real numbers and zeros, the concept of something being undefined is inherently foreign and incompatible. On the other hand, undefined values are things that we have to deal with all too frequently in the real world. They signify that we do not know something, probably because we do not have enough samples in the area for the variable of interest. The nearest equivalent to a null alpha string in the context of numeric values is a zero, however, this is frequently a real, measured value and thus cannot be used to signify that a variable is undefined. The only solution is to arbitrarily select a value that is unlikely to occur in practice, for example, $-1E33$, and to use this for internal computer representation of all undefined numeric values, so that we can recognize them in

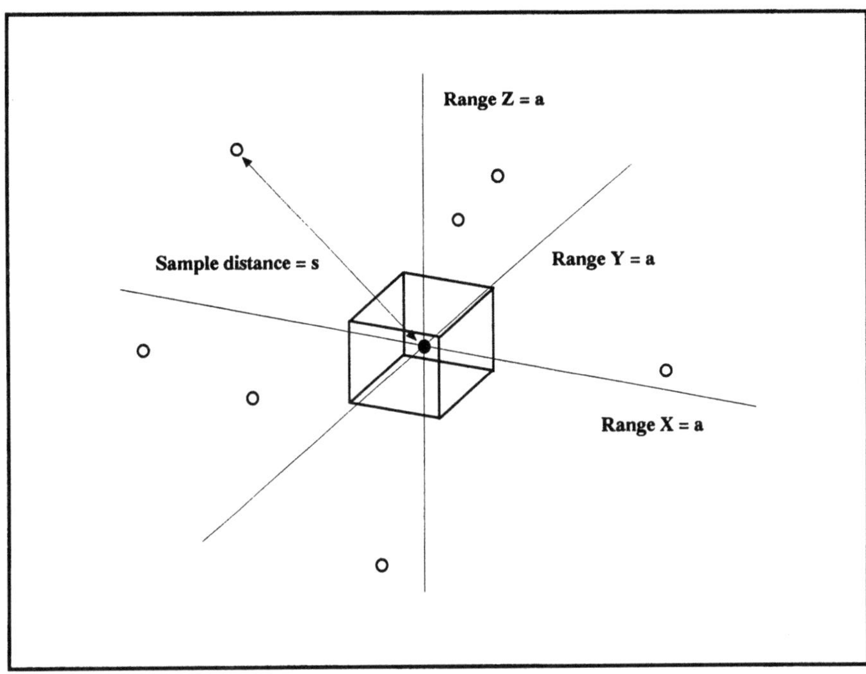

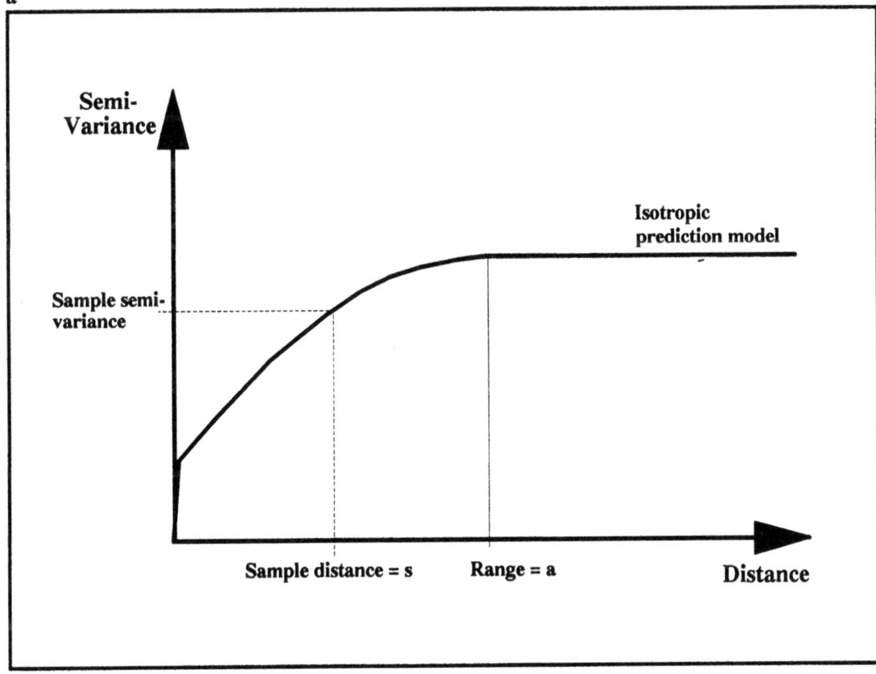

Fig. 8.3 **a** Sample and grid-cell geometry for isotropic prediction of a variable; **b** scaling of sample / prediction cell semi-variance from the isotropic prediction model semi-variogram

any subsequent analysis. This is a consideration in any characterization. It is just as important to know that a variable is undefined in a certain region as it is to know the uncertainties associated with predicted values of the variable. We revisit the concept of undefined values in our discussion of spatial analysis techniques in Chapter 9.

8.4 Sample Selection Controls

We have already mentioned several aspects of sample selection, including characteristic value control and search volume control. The former has already been discussed in detail. However, the latter requires further discussion. The sample search volume can have a significant effect on the prediction results. There are also several useful refinements to sample control that we should be aware of.

As we discussed in Chapter 5, samples that are at a greater distance than the range of influence from the point to be predicted are considered to be unrelated in terms of the measured spatial variability. Nonetheless, we can still use them to predict a value at the point if necessary, for example, when there are too few other samples within range and we are determined to predict a complete measure of the spatial variation. The disadvantage of doing this is that the predicted value has a correspondingly high uncertainty. This is a logical result since the available samples are so far away that the semi-variance between them and the prediction point is the maximum measured value. The obvious advantage is that we can still obtain a predicted spatial variation in regions where the sample density is low. To achieve this result we merely extend the radius of the search volume beyond the range of influence, a multiple of two times the range is fairly common. To stretch the search volume beyond this is less common, and is equivalent to indulging in geological fiction!

A useful refinement we may wish to apply to sample control allows us to ensure that values are predicted only at points that have an acceptable distribution of samples around them. This is usually done by specifying the minimum number of search volume quadrants that must be occupied by at least one sample before prediction occurs. If we have no concerns regarding sample distribution then, in the worst case, one occupied quadrant is adequate. If we want to be conservative, then we should specify three or four.

A second refinement is more concerned with the efficiency of the prediction process. If there are many samples within each quadrant of the search volume, then those furthest from the prediction point are going to have minimal influence on the predicted value. The efficiency of prediction (in terms of computer processing) is increased by only considering the closer samples. In many cases, we limit the number of samples considered in each quadrant to the five or so closest samples.

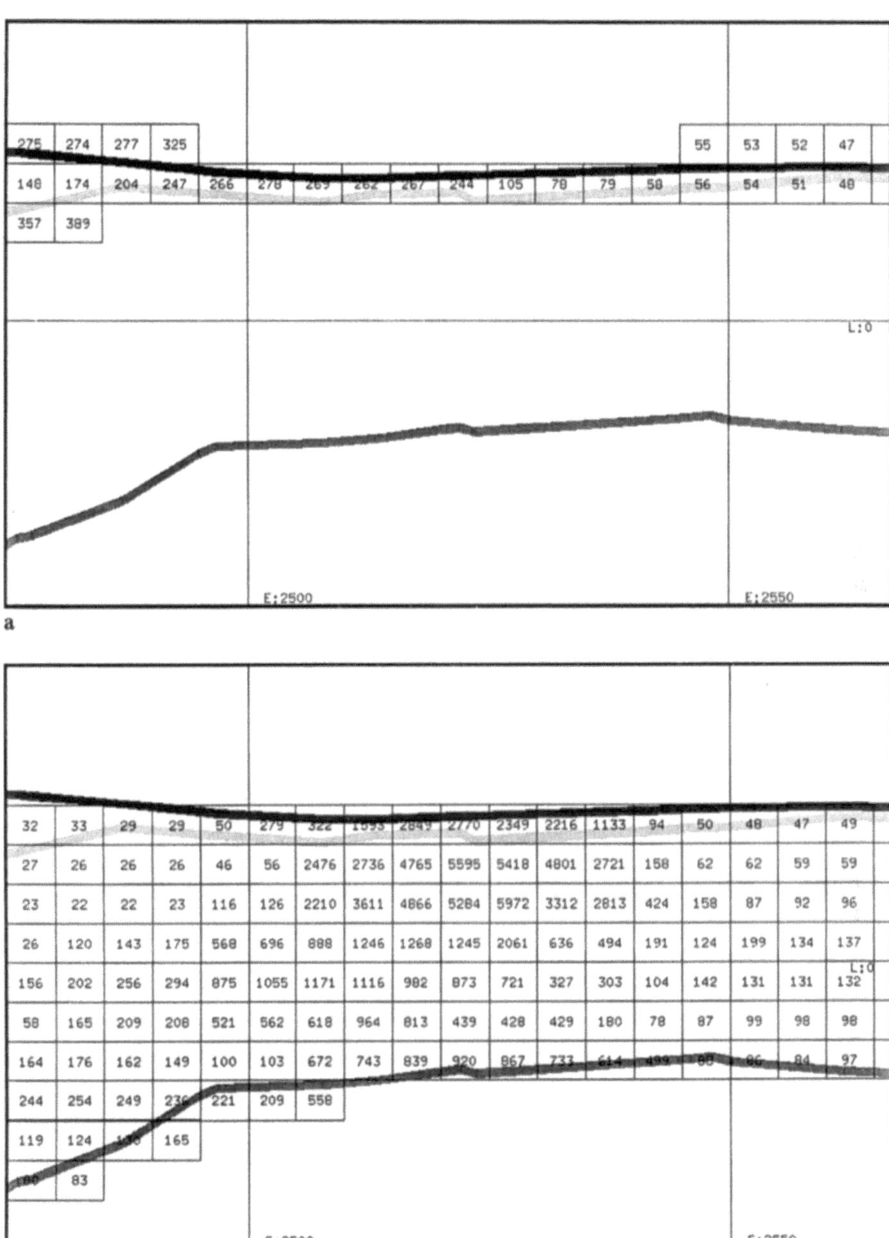

Fig. 8.4   **a** Predicted values of copper contamination for the top soil layer on a vertical section; undefined cells are not displayed; **b** predicted copper contamination for the intermediate soil layer

8.5 Management of Spatial Anisotropy

Anisotropy in the spatial variability of a variable occurs frequently in practice, as demonstrated by some of the examples we considered in Chapter 5. The principal manifestation of anisotropy is a difference in the semi-variogram ranges of influence in orthogonal directions. We accommodate this in the prediction process by assuming that the range of influence of the prediction model has an elliptical variation about the prediction point, or grid-cell centroid, as shown in Fig. 8.5a. The principal radii of this ellipse coincide with the ranges of influence and principal directions of anisotropy.

This has considerable implications for the prediction process. First, it means that our prediction model is now represented by a *semi-variogram envelope* (cf. Fig. 8.5b), rather than a simple line model (cf. Fig. 8.3b). The extremes of the envelope are defined by the different ranges of influence in the principal directions. Second, we now have to determine both a distance and an *orientation relative to the cell centroid* for each sample considered. Knowing the orientation, we can calculate the corresponding range of influence from the expression for the influence volume ellipsoid (cf. Fig. 8.5a). This, in turn, defines the appropriate prediction model, from which we can scale off the semi-variance value corresponding to the distance between the sample and the prediction point, as shown in Fig. 8.5b. This means that the semi-variance values in the sample weighting equations are determined from a number of different semi-variogram models, all of which fall within the model envelope. In this way the anisotropy in the measured spatial variability is taken into account in the determination of the sample weighting coefficients.

The third implication is that the ellipsoidal influence volume must now be oriented in terms of azimuth and inclination so that its ranges of influence (and principal radii) correspond to the principal directions of anisotropy. These changes obviously implicate the sample search volume as well; it also is now ellipsoidal in shape and must be oriented properly in spatial terms.

Further inspection of the anisotropic prediction model shown in Fig. 8.5b reveals that, although the range of influence varies with sample:centroid orientation, the remaining semi-variogram parameters (nugget effect and sill values) remain constant. In practice, the measured spatial variabilities in the principal directions of anisotropy are quite likely to exhibit different nugget and sill values, although they should always be of the same order. The standard approach to this inconsistency is to take the maximum observed values of each of the parameters and to use these as constants for the prediction model. There are techniques for accommodating nugget and sill values that vary with orientation, however, these are complicated in application and should be left to the geostatistical specialist. Unless the observed values vary dramatically with orientation, the differences in the final results are insignificant in most cases.

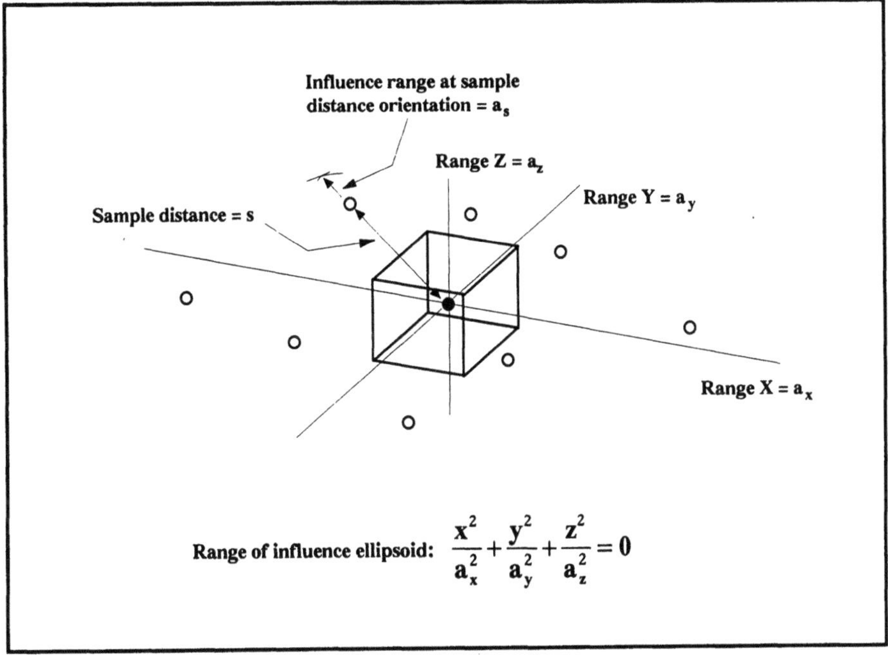

Range of influence ellipsoid: $\dfrac{x^2}{a_x^2}+\dfrac{y^2}{a_y^2}+\dfrac{z^2}{a_z^2}=0$

a

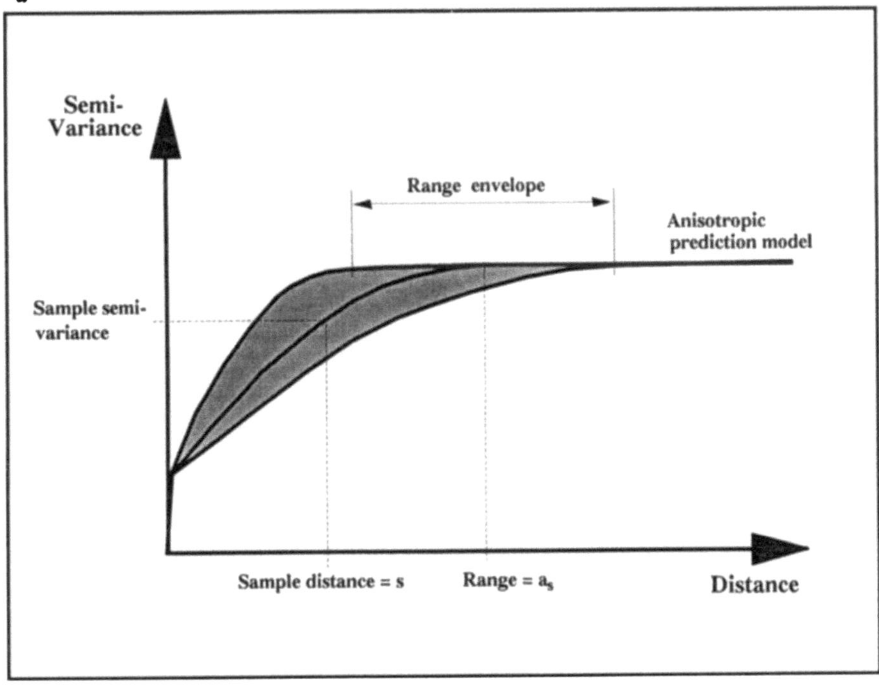

b

Fig. 8.5 a Sample grid-cell geometrical relationship for anisotropic prediction; **b** semi-variance scaling from the anisotropic prediction model

8.6 Volume Kriging to a 3D Grid

All previous discussion has focused on prediction of a value for a *point* at the location of a grid-cell centroid. The result is independent of the volume of the cell. If the dimensions of the cell are expanded or reduced, the predicted value remains the same, assuming the centroid remains in the same location. We now introduce a variation on the point kriging technique which allows us to predict the *average value for the cell volume*, called *volume kriging*. Due to the geometrical definition of the grid data structure, we continue to store the result at the cell centroid, although this may not coincide exactly with the true location of the cell average. The differences in predicted results between the two techniques depend largely on the relationship between the cell dimensions and the prediction model, since the average value is obviously dependent on the cell volume. If the dimensions are relatively large and the spatial variability is high, then the differences are likely to be significant on a local scale, however, the large-scale spatial variations are always similar. The most significant benefit of volume kriging is that the semi-variance (and hence the uncertainty) is always less than that predicted by point kriging. It is an intrinsic property of kriging that, for a given set of samples, the larger the volume we predict an average value for, the less the associated uncertainty. If we think about it, this is a logical outcome; the predicted average value for a volume is always going to be a more stable entity than the value at a specific point, and the average becomes more stable as the volume increases. In Chapter 16 we discuss a new, generalized approach that extends the volume kriging technique to prediction of an average and an uncertainty for any large, irregular volume.

How do we decide which technique is most appropriate to our requirements? The answer depends on our characterization objectives. If we are going to use the predicted values to determine isosurfaces and contours of the variable of interest, or for visualization purposes, then strictly speaking, the point kriging technique is more correct since the predicted values are at their true locations. On the other hand, if we are more concerned with analyzing volumes and minimizing uncertainty then volume kriging is more appropriate. In the mineral sector, mine planning is invariably based on the results of volume kriging (cf. Fig. 8.6b); the objective is to predict the contents of mining excavations with maximum precision and minimum uncertainty. Similarly, hydrogeologists concerned with predicting transmissivities for a groundwater study should use volume kriging, since transmissivity is based on volumes and average conductivities.

We are not going to discuss the use of volume kriging to predict grid-cell values in detail here since this is covered in a more general sense in Chapter 16; a summary is sufficient at this stage. Essentially, it involves a discretization of the cell volume into a set of points at regular intervals, as shown in Fig. 8.6a. By some clever mathematical manipulation, the kriging equations are modified to include the point-to-point semi-variance relationships within the cell as well as the point-to-sample and sample-to-sample relationships. In a similar fashion to point

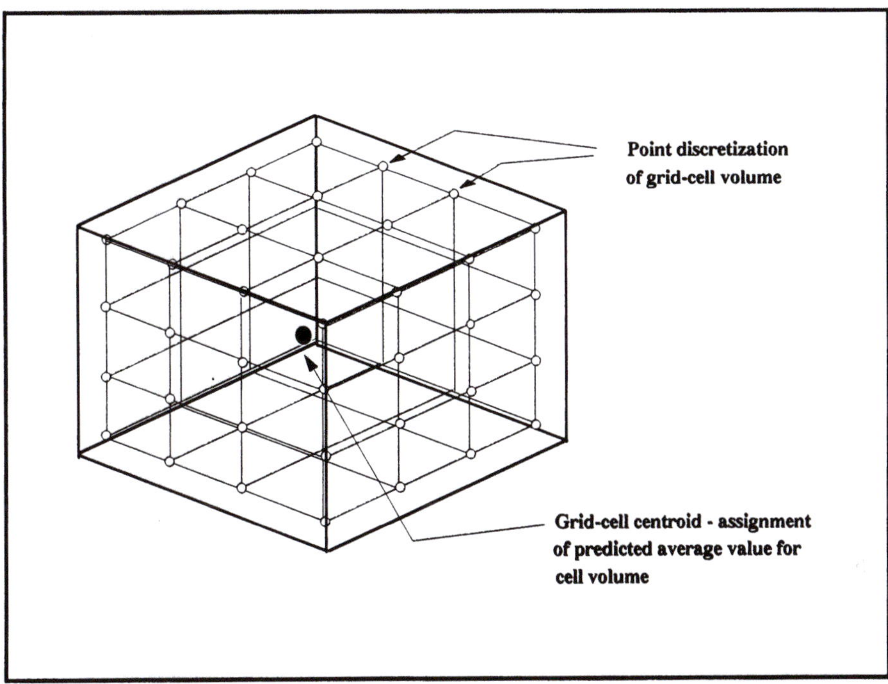

a

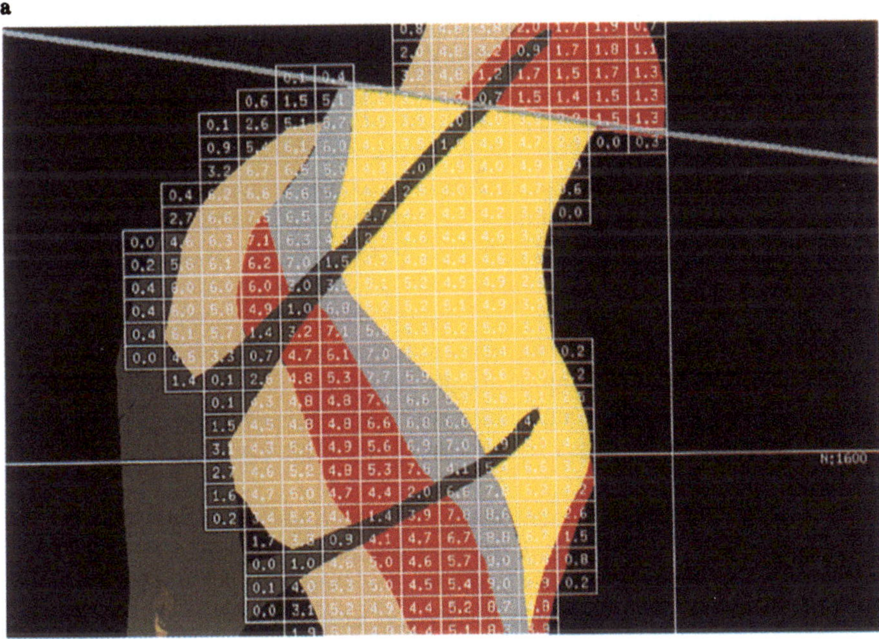

b

Fig. 8.6   **a** Discrete point representation of a grid-cell volume for volume kriging purposes;
b predicted copper grades for a mineral deposit overlaid on a section through a volume model
characterization of mineralogy

kriging, the result is a set of weighting coefficients that directly provide the average value for the cell with minimum uncertainty.

8.7 Kriging with Data Transforms

There are many occasions when spatial variability is not measurable unless we first apply some form of data transformation. This is because the untransformed samples do not exhibit a statistically normal distribution of values, which is a precondition of semi-variogram analysis. As discussed in Chapter 5, there are several transformations that work well in practice; *logarithmic* and *indicator* transforms are the most frequently used. We focus on logarithmic transforms here, as indicator transforms fall into a class of their own and constitute a more *risk-oriented approach* to characterization which we discuss in a later section.

The concept of applying a logarithmic transform to our sample values, measuring the spatial variability, selecting a prediction model, using one of the kriging techniques discussed above to predict the spatial variation, and finally reversing the transformation to obtain the variation in terms of natural values seems straight-forward enough. However, there is a complication in this process that we need to be aware of: inherent to the logarithmic calculations is a tendency to overestimate both the predicted values and their uncertainties by allocating too much weight to higher sample values. We must apply a correction factor to the predicted values to bring them back into line. Provided we recognize the necessity for this, the logarithmic transform is an extremely useful and reliable prediction tool.

An acceptable approximation of this correction factor is obtained by dividing the average of the original sample values by the average of the predicted values. This is equivalent to rescaling the predicted values so that the averages of their value distributions coincide. A more precise correction factor is obtained by taking into account that the predicted spatial variation comprises a spatially uniform density of information, whereas the density of sample information may be highly variable. Strictly speaking, we need to apply an area or volume-weighting algorithm when calculating the sample average, in order to achieve a value that is more comparable to the predicted average.

On occasion we may lack the necessary computer tools to perform this type of back-transformation. In this case, there is an alternative, known as the *empirical method*, that can be applied directly to the original (untransformed) sample values. It involves using the prediction model parameters obtained from analysis of spatial variability with an appropriate data transform to derive an *equivalent prediction model* that works with the original values. This is achieved by fitting a model, as best we can, to the semi-variogram results of the original values that has the same range of influence and the same nugget effect to sill ratio (C_o:C). This approach is justified by the observation that the range of influence is the controlling parameter in the prediction process. In theory, and generally in practice as well, the range of influence is constant whether we apply a data transform or not (cf. Fig. 5.7, Chap. 5).

8.8 Kriging with Data Trends (Universal Kriging)

Determinate data trends in the sample values occur frequently enough in practice to warrant a brief discussion. The prediction technique for handling a trend (commonly called *universal kriging*) is similar to applying a data transformation, as discussed above. In this case, the results of spatial variability analysis include details of a polynomial (linear, quadratic or cubic) expression that defines the spatial variation of the determinate portion of the variable value. The semivariogram prediction model defines the variability of the indeterminate, residual portion only.

The prediction process begins with adjustment (transformation) of the sample values by subtracting the trend portion of the value. Predicting the spatial variation of the residual portion follows the normal kriging steps, and once this is complete we add back the trend portion. Like the logarithmic transformation discussed above, this is a means of extending kriging prediction to variables and sample sets that would otherwise be incompatible with the underlying concepts and assumptions of the technique.

8.9 Indicator Kriging and Probability Estimation

All the prediction techniques we have considered up to now are concerned with predicting the spatial variation of a variable in terms of its *value* at discrete points in space. In contrast, indicator kriging is primarily concerned with predicting the spatial variation of the *probability* of the variable meeting a specified criterion; for this reason the technique is also referred to as probability estimation. The criterion is usually defined as the variable being equal to or greater than a *threshold value*, less frequently as the variable being less than a threshold value. We introduced the concept of the indicator transform in Chapter 5. It involves conversion of the sample values to ones or zeros depending on whether or not they meet the criterion. The transformed (indicator) values are equivalent to the probabilities, on a scale of zero to one, of the variable meeting the criterion at each of the sample locations. A sample location with an indicator value of one has a 100% probability of meeting the criterion; an indicator value of zero signifies 0% probability. Any predicted value at another (nonsample) location that is based on the indicator values therefore represents the probability of the variable meeting the criterion at this location, and will have a value between zero and one.

Depending on the objectives of our characterization, this can be an extremely useful technique, particularly if we are concerned with risk analysis. In a mineral

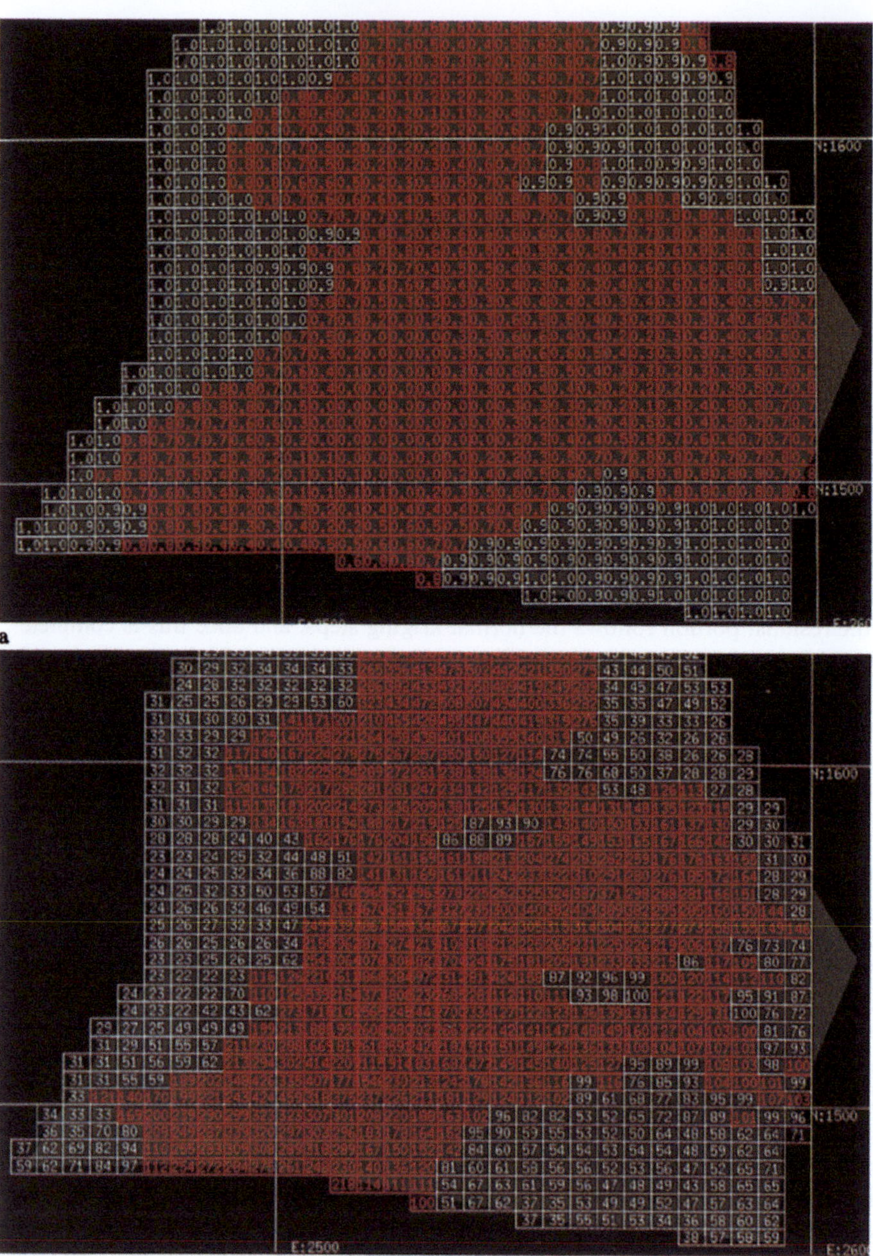

Fig. 8.7   **a** Plan section of probability prediction for copper contamination less than a threshold of 100 mg/g, based on indicator values; colored red below 0.85 probability;   **b** copper contamination predicted by normal kriging based on sample values; colored red above a threshold of 100 mg/g

deposit characterization, for example, our criterion might be that the mineral grade exceed an economic grade threshold (or cutoff). By using spatial analysis techniques (discussed in Chap. 9) on the predicted probability values, we can readily extract the volume of ore enclosed by the 0.9 probability isosurface (or any other appropriate value). This would represent the volume of ore that meets our criterion with 90% confidence (or 10% risk), and the volume would decrease as the stipulated confidence level increased. The more confidence we require in the result, the smaller the volume of ore and, hence, the smaller the potential return from mining the deposit. In an environmental contamination study we would typically reverse the criterion so that the sample indicator values become one wherever the contaminant concentration is less than a stipulated threshold (or action level) and vice versa. In this case, the volume increases as the confidence level increases; the higher the stipulated confidence, level the more material we have to remediate and the greater the cost.

Both these examples illustrate the use of the technique to relate confidence (or risk) to a spatial analysis of volumes, and hence to value or cost as well. A distinct advantage of the technique is that the results are presented in a way that emphasizes the concept of a characterization having an associated (measurable) confidence, and the fact that we always have to deal with a residual risk. Like the other prediction techniques discussed, kriging with indicator transforms also predicts the standard error of the probability value at each point. However, since we are already dealing with probabilities, these values do not have much significance and we generally ignore them.

A useful feature of indicator transforms is the reduced scale of the values we have to deal with. No matter what the scale of the original values, our prediction is solely concerned with indicator values between zero and one. If the original sample values vary by several orders of magnitude, then their predicted spatial variation is likely to be overly sensitive to our choice of prediction model parameters. For example, small changes in the range of influence of the prediction model may result in significantly different spatial variations between the samples. In turn, any subsequent spatial analysis of volumes based on these results is likely to be overly sensitive to our choice of threshold value, particularly if the threshold value is at the low end of the scale.

In these cases, the prediction results are likely to be considerably more stable and less sensitive if we use indicator transforms instead. This has the disadvantage that we have no means of transforming the predicted probabilities back to the value scale from which they originated. However, this is not always a disadvantage. In many cases a knowledge of the volume and extent of the material that meets a specified criterion is all we need. The actual values of the variable within the volume are of no particular concern. This is particularly true of many studies of environmental contamination, where the measured sample values frequently span three or four orders of magnitude and the stipulated threshold (or action level) may be close to the low end of the scale. A knowledge of the volume of contaminated material that exceeds the action level is often adequate for remediation planning purposes. Knowing that the contaminant concentrations within this volume may be

10, 100, or even 1000, times the action level is redundant information if it has no effect on remediation planning and costs.

On introduction to indicator transforms, most of us are probably inherently distrustful of the ability of a technique that converts all sample values to simple ones and zeros to predict a representative spatial variation. As an example of the use of indicator transforms, consider the comparison of prediction results for copper concentrations from a contaminated site characterization shown in Fig. 8.7. Figure 8.7a shows a plan section through the spatial variation of the probability of the concentrations being less than a threshold of 100 mg/g, colored red below a probability threshold of 0.85. These have been generated by normal kriging prediction based on an appropriate indicator transformation of the sample values. For comparison, the results shown in Fig. 8.7b have been obtained by normal kriging based on the untransformed sample values. These reflect the predicted spatial variation of the actual copper concentrations and are colored red wherever they exceed a threshold of 100 mg/g. A comparison of values between the two sets of results is meaningless since they are based on different value scales. What is important is that the spatial variations (the red areas) exhibit a close correspondence for equivalent threshold values; contaminated isosurface volumes determined from the results would show a similar correspondence.

In summary then, indicator transforms are an extremely useful tool for relating confidence and/or risk to value and/or cost. They also present an alternative, more stable prediction technique when we are forced to work with extreme ranges of value and low threshold values. They should not be used wherever we are concerned with the spatial variation of real values.

8.10 Alternative Prediction Techniques

Despite the almost generic applicability of geostatistical prediction techniques, there are occasions when our samples have no measurable spatial variability at all and we cannot justify fitting any sort of prediction model to the results. In Chapter 16 we discuss several new approaches to predicting spatial variations, including a stochastic technique that caters to any degree of random variability, and an extension of volume kriging that caters directly to irregular volumes of any shape. However, these are still emerging techniques and are not yet widely available.

The most frequently employed alternatives to geostatistical kriging are variations of the *inverse distance weighting technique*, a conventional contouring algorithm. Like kriging (cf. chap. 5), these employ a linear weighted combination of sample values that gives more weight to closer samples. Thus the predicted value is still obtained from a linear estimator expression of the form

$$v_p^* = w_1 v_1 + w_2 v_2 + \ldots\ldots + w_n v_n.$$

Unlike kriging, however, the weights are not optimized in any way. Instead, they are determined directly from the expression

$$w_i = \frac{1/s_i^x}{(1/s_1^x + 1/s_2^x + \ldots\ldots + 1/s_n^x)}$$

where s = sample distance
x = power coefficient.

This expression ensures that the weights always sum to one. The only means of controlling this prediction algorithm is to vary the power coefficient x. Values of one and two are used fairly frequently, three or higher less frequently. The expression always gives more weight to closer samples and less to distant samples, however, higher values of x emphasize the weighting of closer samples relative to more distant ones.

What value of x to use in any given situation is generally an intuitive, seat-of-the-pants choice. A value of one tends to smooth the prediction results, a value of three tends to emphasize local high and low values, two is the usual compromise. We also have to make arbitrary choices regarding the extent and shape of the sample search volume. These choices would obviously benefit from a semi-variogram analysis of spatial variability, but if this is measurable, then we should be using kriging as the prediction technique of preference anyway. In many cases the prediction parameters are iteratively adjusted until the resulting spatial variation looks *intuitively reasonable*. Other obvious disadvantages include the lack of any measure of the uncertainty associated with the results, and the inability to accommodate anisotropic spatial variability. Spatial control of the prediction process by interpretation of geological characteristics and sample selection control are applied in the same way as discussed earlier for geostatistical kriging.

Of course, there are other alternative prediction techniques that can be, and are, used for geological characterization purposes. The list includes *trend surface techniques* (used in isolation), *minimum tension* (another variation on 3D contouring) and *sample averaging* (the average of all samples within the search volume). Yet geostatistical techniques are the only ones that allow us to tailor the parameters of the prediction algorithm to the measured spatial variability of the samples and observations.

8.11 Cross-Checking of Prediction Results

Whatever prediction technique is used, we should always cross-check the results to establish their *reasonableness and validity*. The most obvious and logical check is a visual one and relies on the availability of a range of sophisticated graphic display capabilities. These allow us to cut sections and plans through a predicted spatial variation, or to generate 3D perspective views, on which we can overlay the original samples and observations for direct visual cross-checking for reasonableness. Any unanticipated inconsistencies in the predicted variation are frequently detected by visual checking, thus forcing us to either justify them or to revise the prediction parameters.

The most convenient checks are statistical ones. Analysis of the basic statistical properties of the predicted variation, such as average, standard error and normality, quickly confirm whether or not these results are similar to those of the original samples and observations, which they should be. Of course, this just means that the predicted values are in the right ball park; it does not mean that the right values are in the right spatial locations, which is why we should always combine statistical and visual cross-checking for completeness.

8.12 Common Problems in Spatial Prediction

This is not intended to be an exhaustive list of all the things that can go wrong with the prediction of spatial variations. It includes some of the more common problems and hopefully is of assistance to users of these techniques in recognizing where and why things have gone wrong.

The first of these problems can be termed *unconfined prediction*. It is manifested by a much larger than expected distribution of high values. The most likely source of the problem is a lack of samples in the outlying areas of our characterization. All of the prediction techniques discussed above extrapolate as well as interpolate information. Outlying samples of low value are required to moderate and confine the prediction of high values. This condition is a frequent result of oversampling *hot spots* in the characterization of contaminated sites. Instead, we should be concentrating our investigative efforts more on the outlying areas, and if necessary extending the area of characterization, so that the boundary between contaminated and uncontaminated material is better defined. A partial solution is to use tighter sample selection control, as discussed earlier, so that prediction is confined to locations that have samples distributed all around them. Another is to introduce dummy zero-value samples in appropriate locations, assuming we can justify this somehow. The best (but most expensive) solution is, of course, to return to the site and take more samples.

The second class of problems is related to the *sample search volume*. If the search volume is too small, or conversely the samples are too widely distributed in some areas, then the resulting spatial variation consists of *elliptical blobs* of defined values centered around the sample locations. Intermediate locations are *undefined* because they are out of range of the available samples. The solution is to increase the radii of the search volume sufficiently. This means that the intermediate locations are now assigned a predicted value that has the maximum associated uncertainty (which is a logical outcome) since the closest sample distance is greater than the measured range of influence. The opposite condition also occurs in practice; if the search volume is too large, then there is likely to be too much *smoothing and uniformity* in the predicted variation. Ideally, the search volume radii should be equal to the corresponding ranges of influence, however, this is not always a practical solution.

Another possible cause of undefined regions is that the controlling characteristic itself is undefined within the region, or has an incorrect value. In this case, we should revisit our geological interpretation of the characteristic.

The last class of problems to be discussed is concerned with *variable scale effects*. As before, the manifestation is an unexpectedly wide distribution of high values. In these cases, however, the cause is the scale of variability in the sample values. This condition is particularly prevalent in environmental contamination studies; measured contaminant concentrations frequently range over three or four orders of magnitude, whereas the threshold of concern is just as frequently at the low end of the scale. The prediction process tends to be dominated by the exceptionally high values, resulting in exaggerated contaminated volumes. The solution is to use indicator transforms based on the threshold value, as discussed earlier, generally resulting in a more stable result.

8.13 Performance Review of Prediction Techniques

We have discussed a wide range of options concerned with geostatistical prediction: different semi-variogram models, logarithmic transforms, indicator transforms, anisotropy, trends, search volume definition and geological control. There are no clearly defined rules as to which technique, or alternative technique, to use for a particular application. The primary considerations are first, obtaining an acceptable measure of spatial variability and second, choosing whatever best meets the objectives of our characterization.

A number of exhaustive comparisons have been completed on the performance of different prediction techniques with a variety of data sets. In particular, Englund (1990) came to the following tentative conclusions after a comprehensive study of the application of a number of techniques to environmental contamination data. The measures of prediction quality applied to the results include the standard statistical properties as well as the following measures, which are largely influenced by environmental considerations:

- *False positive/negative count*: the number of values incorrectly predicted as being above/below a threshold.
- *False positive/negative deviation*: a summation of the amounts by which these falsely predicted values deviated from the true values.
- *Selection efficiency*: the ratio of the total contaminant represented by the n highest predicted values to that of the n highest true values.

· *Prediction efficiency*: the ratio of the total contaminant represented by the n highest predicted values to that represented by the same n true values.

The results of this study are summarized as follows:

· *Trend surface* is generally a poor spatial prediction technique compared to kriging.
· *Sample mean* (or average) is also a poor predictor.
· *Ordinary kriging* is a relatively consistent, good predictor even for skewed (non-normal) distributions.
· *Logarithmic kriging* is generally a good predictor for highly skewed distributions, but the back-transformation details are critical.
· *Indicator kriging* gave very good results, but was only used once in the tests and this finding is not conclusive.

No single spatial predictor was best or worst consistently for all of the measures of prediction quality. The only meaningful conclusion to be drawn is that all of the variations on geostatistical kriging that were included in the study performed consistently better than the alternatives.

9 Spatial Analysis Techniques

9.1 Spatial Analysis Objectives

In Chapter 3 we briefly discussed the generic requirements of the spatial analysis step of the geological characterization process. In a *quantitative context* these requirements are invariably concerned with volumes of one type or another and their contents in terms of variable values. The analytical volumes may be defined by geological boundaries, or an isosurface of a variable, or an excavation geometry, or various combinations of all of these. We briefly discuss the application of volume modeling techniques to excavation design at the end of this chapter, and accept for the moment that this is a viable process.

The primary information sources for quantitative spatial analysis are interpretations of geological characteristics represented by volume data structures, predicted spatial variations of variables represented by 3D grid data structures, and excavation geometries, again represented by volume data structures. The spatial variations are, in turn, the information source for generating isosurfaces of variables. At the low end of the complexity scale we may wish to determine the volume of a particular geological unit; at a more complex level, the volumes of intersection between a number of geological units and an excavation volume, or the partial volumes of geological units in which the value of a particular variable exceeds a specified threshold (an isosurface intersection). At the most complex level we may have a combination of all of these, for example, the partial volumes of all geological units that are both contained within an excavation volume *and* have a variable value that exceeds a threshold. The last represents a three-stage interrogation of geological space. All of these variations of quantitative spatial analysis fall under the heading of *volumetrics* and are discussed in more detail below.

Complementing the quantitative component of spatial analysis, in most characterizations there is also a large component of *qualitative analysis*, generally provided by various forms of computer *visualization*. The amounts of information we are dealing with are generally so large that the only way we can obtain an overall appreciation of complex conditions, and the most efficient method of verification, is by appropriate visualization. Visualization in hard-copy paper format is a conventional, time-honored approach to presenting and disseminating geological information. A prerequisite of any computerized approach is the ability

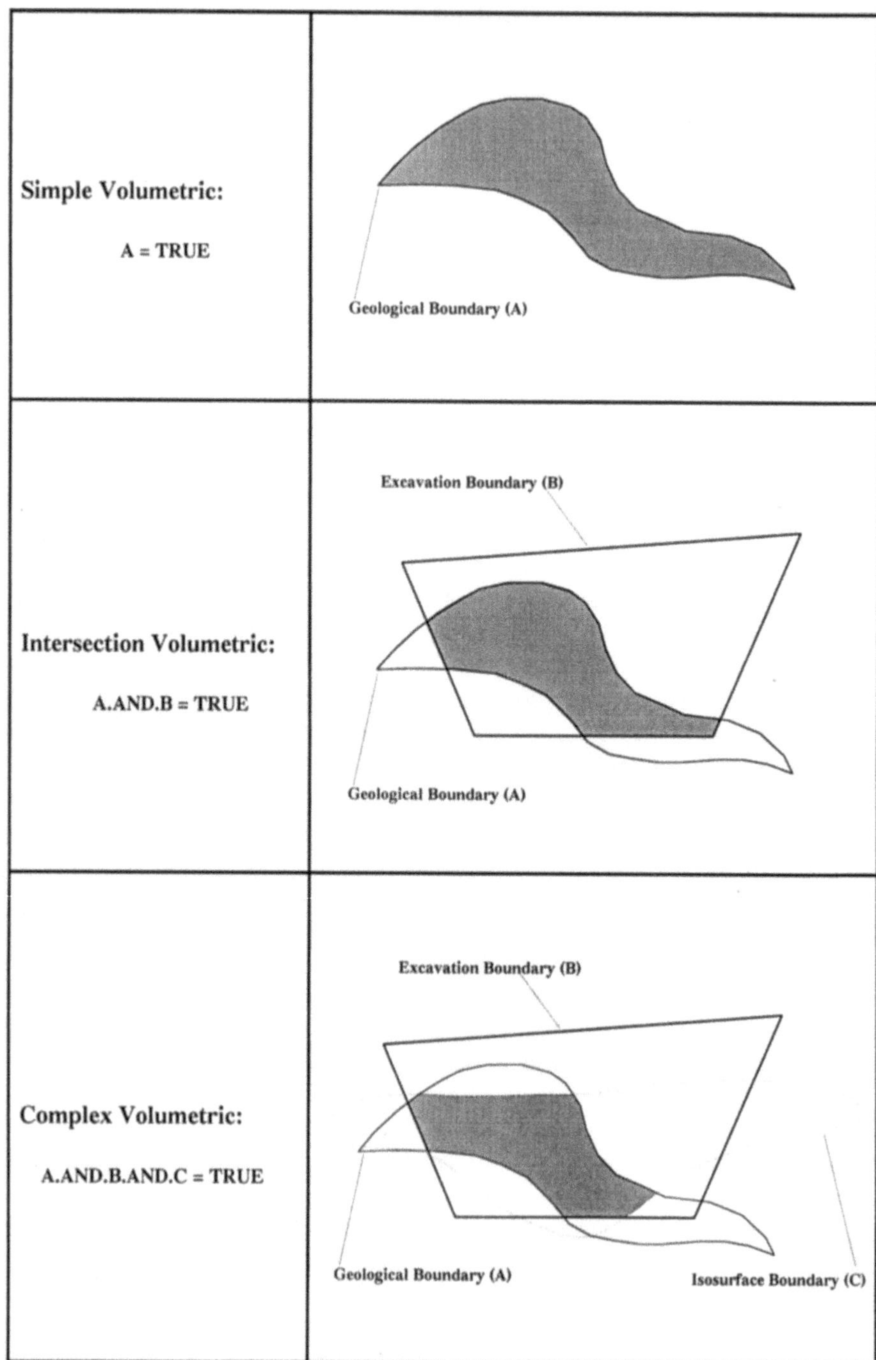

Fig. 9.1 Volumetric logic for spatial analysis of the data structures created for geological characterization

to produce these conventional 2D section and plan visualizations both on the computer screen and as hard-copy plots or graphic prints. With modern visualization techniques we can actually do a lot more than this. We have already seen many examples of computer visualization in the illustrations in this text, and there are many more in the characterization examples presented in Part II. This chapter closes with a brief discussion of basic computer visualization capabilities.

In many cases, spatial analysis with volumetrics and visualization tools represents the end objective of characterization. The subsequent step is frequently a data transfer of the characterization results to an external computer technology for planning or simulation purposes, as discussed at the end of Chapter 4. In certain cases, the inherent capabilities of the techniques discussed in previous chapters lend themselves to extended application in these areas. For example, the vector-based volume modeling technique can be applied just as efficiently to interactive excavation design and planning as to interactive geological interpretation. This area of application is really outside the scope of geological characterization. However, since so many of our characterizations are performed with a view to making holes of one sort or another in the subsurface, we will briefly explore this extended application at the end of the chapter.

9.2 Simple Geological Volumes and Intersections

At the simplest level of quantitative spatial analysis we merely need to know the volume of a particular lithological unit, e.g. the volume of a mineral reef deposit say, or the total volumes of sand, clay and soft and hard rock in a geotechnical investigation. Irrespective of the spatial complexity of our representation of the relevant geological characteristics, this is readily achieved by the volumetrics technique of integrating planar intersections, as presented in Chapter 6. The spatial interrogation logic for this operation is represented by the schematic for a *simple volumetric* presented in Fig. 9.1. The geological boundary in the schematic may be made up of the intersection of one or more volume model component boundaries with the integration plane. The precision of the analytical results is verified by repeating the analysis for several integration intervals.

More frequently, we need to know the partial volumes of the particular lithological or mineralogical units intersected (contained) by an excavation volume, in other words, the volumes within the excavation that are associated with different values of a particular geological characteristic. In this case, the spatial interrogation logic is represented by the schematic for an *intersection volumetric* in Fig. 9.1. On each integration plane we need to determine the common area of intersection of the two boundaries we are analyzing. Figure 9.2b is a visualization of the results of a volumetric intersection between a mine stoping excavation and a volume model representation of four different ore types within the mineral deposit. The diagonal gaps through the intersection are the result of barren (non-ore) intrusions in the deposit. This ability to visualize the analytical results provides an

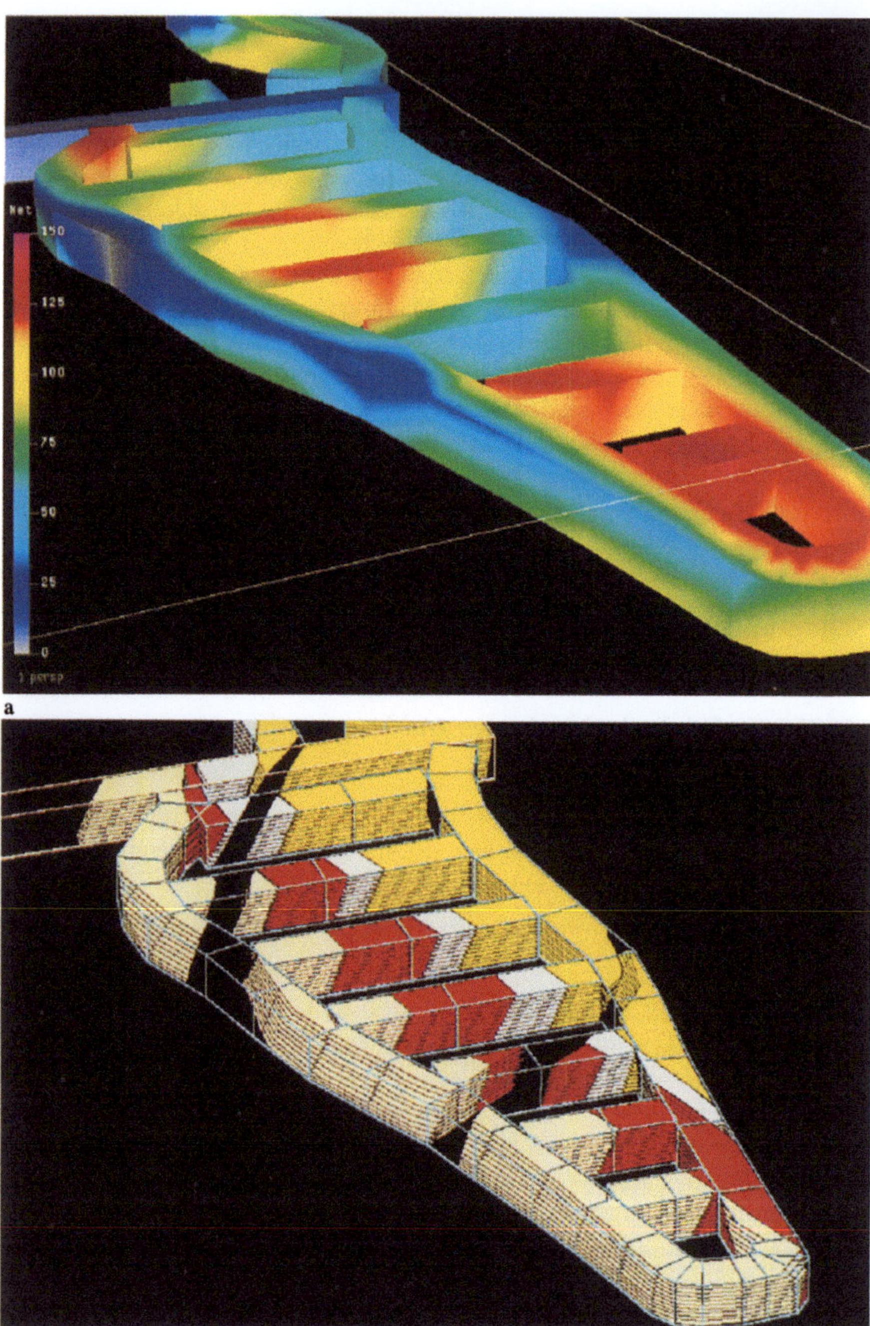

Fig. 9.2   **a** Geometry of an underground mine stoping excavation superimposed with a color mapping of ore grade variation;   **b** visualization of a volumetric intersection of the stope geometry with a volume model representation of mineralogical ore types

efficient and useful visual verification tool. The volumetric analysis results are of course reported in text format; the application examples in Part II provide several examples.

9.3 Volumes Defined by an Isosurface of a Variable

More frequently still, we need to know the partial volumes within all geological units in which the value of a variable exceeds a stipulated threshold. The spatial interrogation logic in this case is similar to the previous example; the excavation boundary is replaced by an isosurface boundary for the threshold value of the variable. This is represented in the volumetric integration process by the contour of the threshold value on each of the integration planes. The isosurface itself is determined by interpolation between the predicted variable values at the grid-cell centroids in a 3D grid data structure.

Figure 9.3b shows a vertical section through the predicted spatial variation of a variable, determined from samples at the monitoring well locations shown in Fig. 9.3a. We cannot read the values at the scale of the figure, however, the grid `cells with values greater than the threshold are highlighted for clarity. In this example, the results represent the characterization of a groundwater plume contaminated with VOCs (volatile organic compounds). The isosurface defines the limits, and hence the volume, of the plume (cf. Fig. 9.4). However, this is not the only result we need in practice. The average variable value (VOC concentration in this example) within the isosurface (plume) is a frequent characterization requirement. In many cases, the average variable value in the material outside the isosurface (less than the threshold) is also of importance. Furthermore, for future investigation planning, or risk assessment purposes, it may also be important to know the volume of material in which the variable of interest is still undefined. These results are determined by accumulating grid-cell and partial grid-cell volumes' and performing a volume weighted averaging of the variable value for each of the volumes of interest.

The greatest level of complexity we are likely to experience is represented by a spatial interrogation of the intersection of three independent boundaries: geological, isosurface and excavation. This is summarized by the schematic for a *complex volumetric* in Fig. 9.1. Within each excavation boundary we need to know the volumes and average variable values of material with a value above and below a threshold, and undefined volumes, for each of the partial geological volumes intersected. Figure 9.2a is a visualization of the mapping of an ore grade variable onto the surface of a mine stoping excavation, essentially this represents an intersection of the excavation with the spatial variation of ore grade, while Fig. 9.2b represents the intersection with geological units (in this case mineralogical ore types).

With the spatial interrogation capabilities discussed above we can cover most of the spatial analysis requirements we are likely to meet in practice.

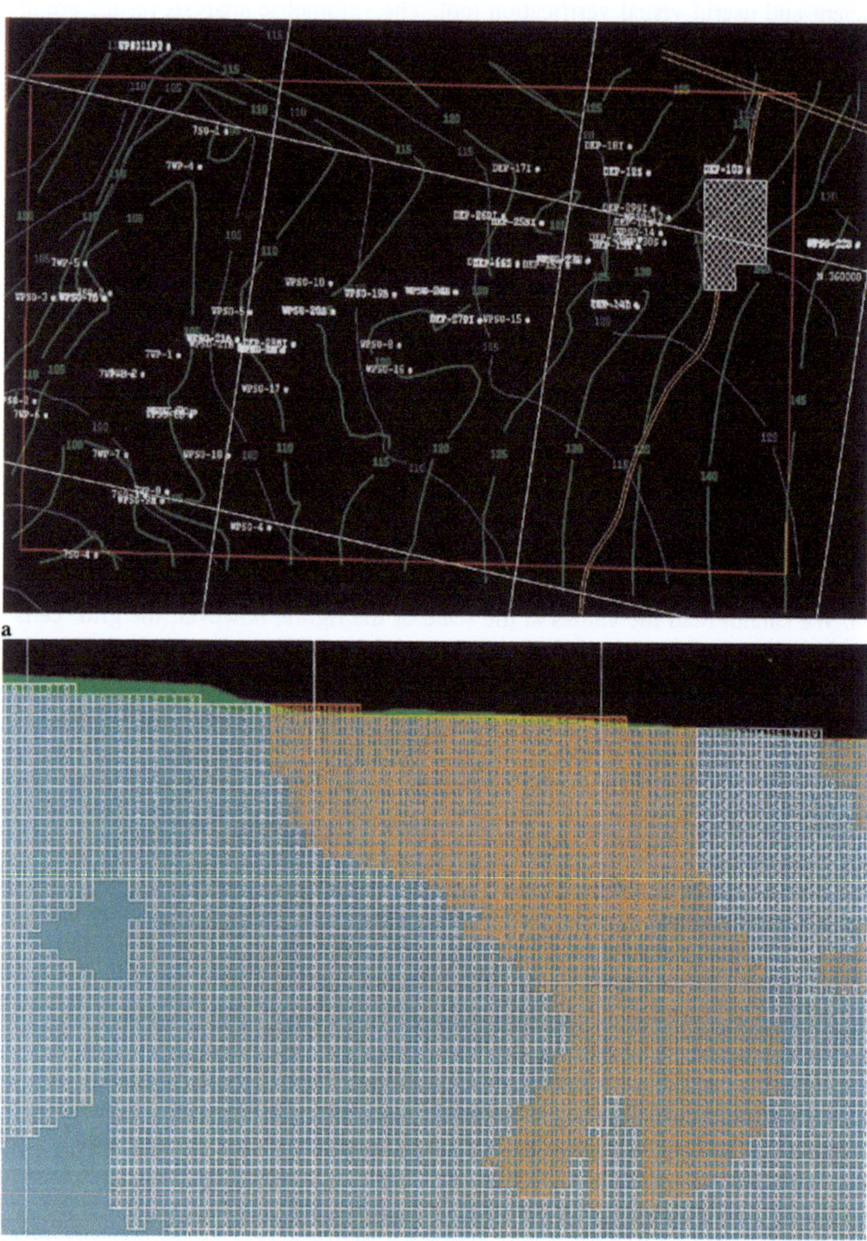

Fig. 9.3 a Site plan of a contaminated groundwater investigation - ground contours (*green*), water table contours (*blue*), well locations and contaminant source; **b** vertical section through a grid model representation of predicted VOC contamination in the groundwater (with vertical scale distortion); grid cells with values exceeding the action level are colored orange; the contaminant plume boundary is interpolated from these results (cf. Fig. 9.4).

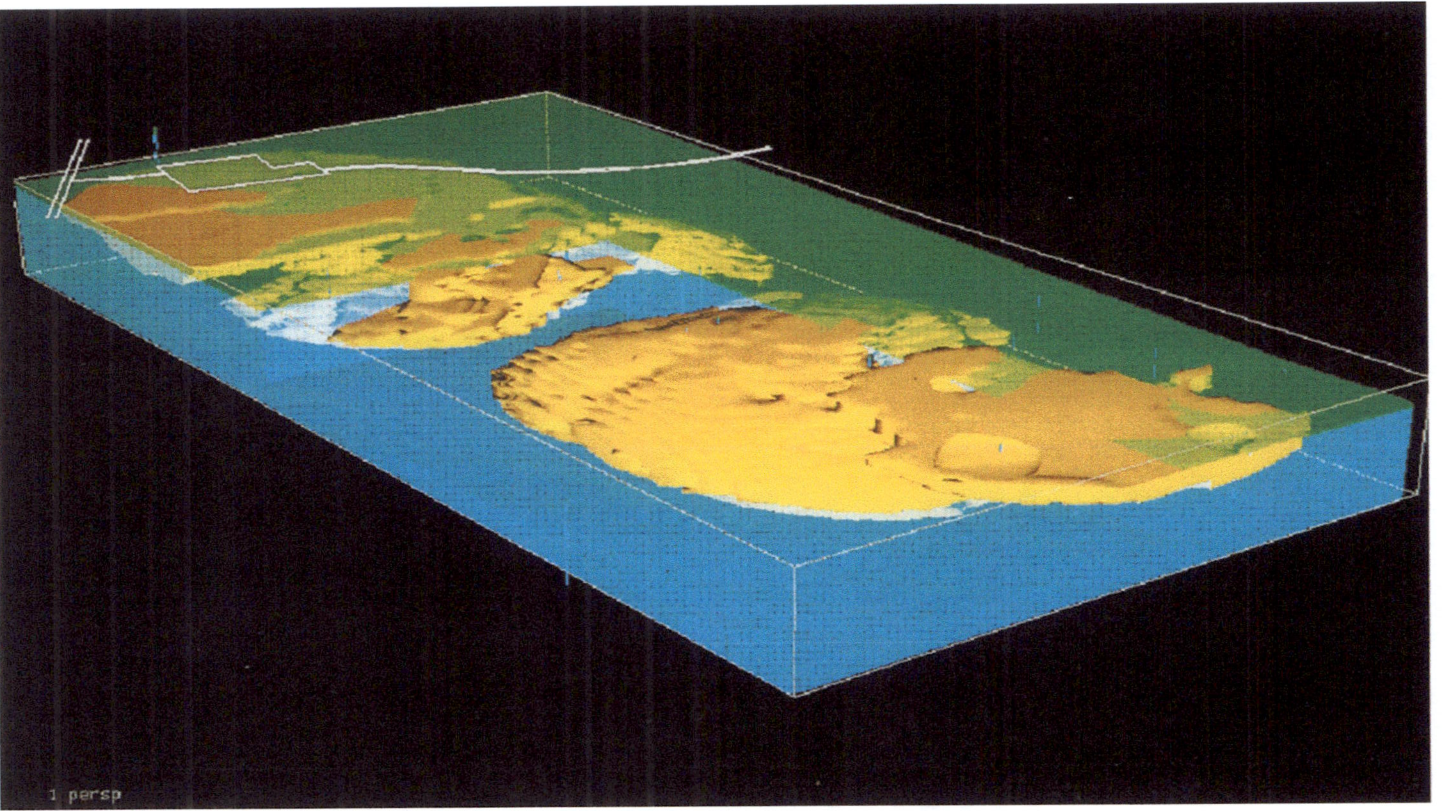

Fig. 9.4 Visualization of the VOC contaminated groundwater plume reported in Fig. 9.3, with a cut-away, semi-transparent model of the saturated and unsaturated zones superimposed

These capabilities are achieved by appropriate application of the simple volumetric integration techniques presented in Chapter 6. In the next section we discuss techniques for extending our spatial analysis capabilities even further.

9.4 Manipulation of Spatial Variations

A distinct advantage of the grid model data structure is the facility with which the variables stored within it can be manipulated and combined with each other in logical and/or arithmetic expressions to derive new variables with their own spatial variation. This may sound suspiciously like fiddling the books, but it is an extremely useful and viable feature of spatial analysis.

In many geological characterizations we are concerned with two or more variables with different spatial variations, yet the volume we need to analyze may depend on a combination of all of them. This is best illustrated by an example. In a mining context we may be concerned with the spatial variations of several minerals within an ore deposit, each of which must be predicted independently. Each of these minerals contributes to the viability (profitability) of the ore deposit as an economically mineable proposition. What we require in order to perform a meaningful spatial analysis of our characterization of the deposit is an *equivalent variable* which reflects the overall profitability of the ore per unit volume. For example

$$\text{Net\$} = \text{Cu\$} + \text{Zn\$} \quad = \text{net profitability of ore per unit volume}$$

$$\text{where} \quad \text{Cu\$} = 10.2 * \text{Cu\_grade}$$
$$\text{Zn\$} = 4.8 * \text{Zn\_grade} (\text{Zn\_grade} \geq 2\%)$$
$$\text{Zn\$} = 0 (\text{Zn\_grade} < 2\%).$$

In this expression Cu_grade and Zn_grade represent the predicted spatial variations for the metals copper and zinc, respectively. The interim variable Cu$ is derived by simply multiplying the Cu_grade values (wherever they are defined) by a factor of 10.2. Derivation of Zn$ requires two steps: first, using a logical expression to derive a new spatial variation which is equal to Zn_grade wherever the latter is equal to or greater than 2% and zero elsewhere, then multiplying this new spatial variation by a factor of 4.8. The equivalent variable Net$, the end product of the manipulation, is obtained by adding the spatial variations of Cu$ and Zn$. Net$ now represents the spatial variation of the net profitability of the ore per unit volume, or, if we include densities in the expression, the net profitability per ton. In its mining context, Net$ is a much more meaningful variable for spatial analysis and planning purposes since it can be related directly to the profitability of the proposed mining operation. The spatial variations visualized in different ways in Figs. 9.2a and 9.5 have been obtained in this way; the most profitable regions of the mining operation are immediately identifiable.

Similar manipulations may be performed in an environmental context. Chapter 11 reports the characterization of a site where the soil is contaminated by a total of five contaminants (variables), each of which has its own independent spatial variation and action level (threshold value). The objective of the characterization is to determine the total volume of soil that is contaminated by one or more of the contaminants which is to be remediated by excavation and backfilling with clean material. To achieve this we require the spatial variation of an equivalent variable that represents maximum contamination from any source. This is obtained from an expression of the form

$$C\_max = MAX\left(\frac{C_1}{L_1} + \frac{C_2}{L_2} + \ldots\ldots + \frac{C_5}{L_5} \right)$$

$$\text{where} \quad C_i = \text{concentration of contaminant i}$$
$$L_i = \text{action level for contaminant i.}$$

The equivalent variable C_max now represents the spatial variation of maximum contamination in terms of ratios of the action levels. The total volume of contaminated soil is obtained by performing an intersection volumetric analysis with the isosurface for C_max = 1.

When manipulating spatial variations in this fashion we should bear in mind that the prediction techniques discussed in Chapter 8 result in independent spatial variations of a variable for each value of the controlling geological characteristic. Derivation of a single spatial variation for the variable throughout the characterization region is a frequent requirement prior to manipulations of the type discussed above. In order to obtain a representative value of the variable in grid cells that have two or more characteristic intersections, we employ a *volume-weighted averaging algorithm* of the form

$$v^* = \frac{\sum v_i * V_i}{\sum V_i}$$

$$\text{where} \quad v_i \text{ is the variable value for characteristic intersection i}$$
$$V_i \text{ is the volume of characteristic intersection i.}$$

Manipulation also provides us with a vehicle for applying *back-transformations* to spatial variations that have been geostatistically predicted from transformed sample values. For example, spatial variations that include a logarithmic transformation must be back-transformed by applying anti-logarithms to the predicted values prior to any other corrections or manipulations that may be required (see Chapter 8).

Finally, manipulation is the vehicle through which we access the geostatistical measures of uncertainty that are stored in the 3D grid data structure with the predicted variable values. We discuss this aspect of spatial analysis in Chapter 10 (Sect. 10.4).

9.5 Visualization Techniques

There are several objectives to visualization; all involve displaying information in a way that is appropriate (meaningful? realistic? presentable?) to the task at hand. One objective is to provide a background display that acts as a visual/spatial reference for interaction with the computer, for example, during interactive geological interpretation. Another is to provide a vehicle for rapid verification of large amounts of information. Attempting to verify the integrity of a 3D interpretation of complex geology by wading through reams of coordinate values that define its geometry would be foolish and time-wasting, and error-prone in itself. We can achieve the same objective far more efficiently by stepping through the interpretation on viewplane sections and/or rotating a 3D perspective visualization of it. A third objective involves the dissemination, presentation or communication of our characterization results, frequently requiring a transfer of the computer visualization to paper.

For visualization to be effective in all characterization contexts we obviously require an extensive library of visualization options. At the grass-roots level we need options for (1) selecting any combination of the available data structures and (2) selecting any viewplane orientation or perspective orientation for data transformation and projection (or intersection). We should be able to control the display *thickness* to limit the depth of information displayed. The display scales should be independently variable, in three dimensions where appropriate. We should be able to apply display formats selectively to different data structures in a single visualization, ranging from a simple display of values, to line drawings, contours, histograms, surfaces, solids, color mapping, texturing, semi-transparencies, cut-aways and light sources. Further, we always need coordinate references to locate the visualization in 3D space.

Visualization is essentially a qualitative form of spatial analysis, although it can also be used to a limited extent in a quantitative sense, provided we have the means of tracking (coordinating in three dimensions) any point on the display in real coordinate terms, or interrogating any point in terms of characteristic or variable values.

There are many examples of visualization scattered throughout this text, including examples of most of the options referred to above. There is no necessity to present any more here. However, Fig. 9.5, a visualization from the characterization of a mineral ore deposit, is a particularly effective example of a variety of data structures displayed in different ways. The spatial variation of ore grade is color-mapped onto a series of transverse sections, or cut-outs, through the deposit. Superimposed on these is a high-grade isosurface displayed as an irregular solid, the geometry of an exploration adit, also displayed as a solid, and

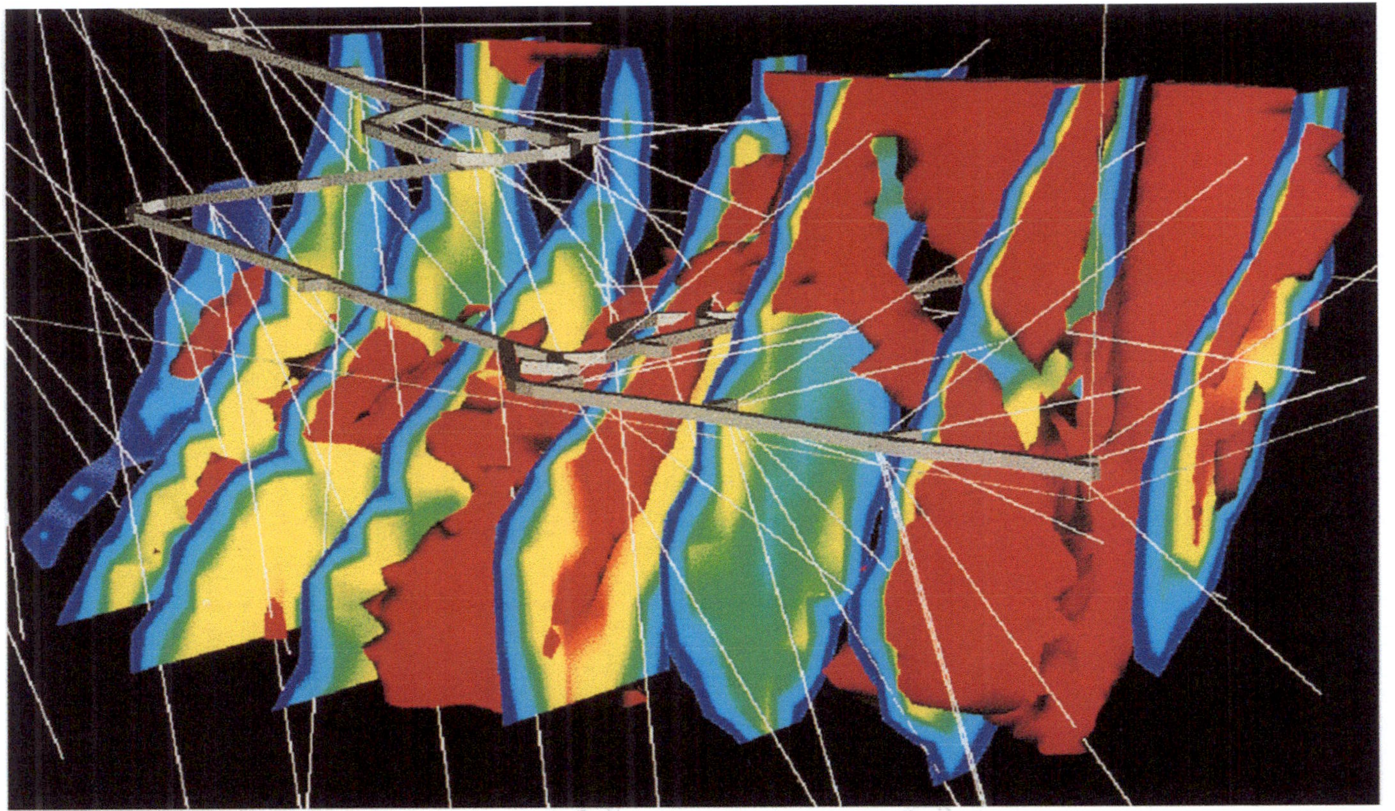

Fig. 9.5   View of sectional visualizations of color-coded mineral grades through a mineral ore deposit, with a high-grade ore isosurface volume, exploration adit and exploration drilling superimposed

exploration borehole traces displayed as line geometries.

Finally, the ability to interactively cut sections through a characterization at any desired orientation and immediately generate a viewplane section visualization through all available information is something that most of us have always looked forward to. This is a feature of the integrated computer approach to characterization that invariably appeals to geologists, made possible by the volume modeling techniques discussed in Chapter 6.

9.6 Excavation Design Applications

The end objective of many geological characterizations is to serve as an information base for the design of surface or underground excavations. This is invariably the case for applications in the mining and tunneling sectors, and frequently the case in environmental remediation applications as well. The vector-based volume modeling techniques discussed throughout this text lend themselves equally well to the representation of excavation geometries and volumes. The geometry of an excavation, whether highly irregular (a mine stoping excavation, or an open pit mining profile) or prismatic (a tunnel, ramp or shaft), can be represented by sets of volume model components in the same way as an irregular geological unit. The advantages of this are obvious; any excavation represented in this way is immediately compatible with the spatial analysis techniques for volumetrics and visualization discussed earlier.

The feasibility of this area of application is dependent to a large extent on the provision of *excavation design tools* that are both compatible with excavation design methods and provide an interactive interface between the designer and the volume model representation. The variety of design methods developed by the mining sector to handle different mining situations is large, resulting in a correspondingly wide variety of excavation shapes. These range from irregular underground stopes that follow the limits of an ore deposit, to strip-mining of coal deposits, to prismatic shapes for tunnel excavation, to large open pit excavations for surface mines. All of these must be definable in terms of excavation components (units) that suit the planning and scheduling requirements.

In practice, irregular excavations are designed in much the same way as the volume modeling approach to interactive geological interpretation discussed in Chapter 6. This involves an interactive definition of excavation profiles on vertical and horizontal sections and subsequent extension of these profiles in the third dimension. Regular, prismatic excavations are more easily defined by a cross section, a center-line geometry and computer generation of suitably shaped volume model components. Other excavation methods have their own specialized design methodologies. The advantage in all cases is the ability to perform all design operations with visual and spatial reference to the results of the geological characterization, to immediately perform a volumetric spatial analysis, and to iteratively refine and optimize the excavation design. The result is a 3D CAD capability, tailored to the special requirements of excavation design, which is fully integrated with the geological characterization process.

A few examples of excavation representations are presented, particularly in Chapter 13. However, limited space and the specific objectives of the text preclude a detailed discussion.

10 Uncertainty, Sampling Control and Risk Assessment

10.1 The Implications of Geostatistical Uncertainty

We have discussed the uncertainty associated with spatial prediction at various stages throughout the text, and in Chapter 5 we covered its derivation in more detail. The first half of this chapter is devoted to a discussion of the implications of uncertainty to spatial prediction, what it means in real terms, and how it can be applied to sampling control and risk assessment. The latter half of the chapter attempts to deal with some of the more intractable sources of uncertainty, including those resulting from site investigation and geological interpretation.

The geostatistically predicted value of a variable at any point is the *most likely value* at the point, based on the neighboring samples and the selected prediction model and parameters. An implicit assumption in geostatistical theory is that the *possible variation in value* at the point has a statistically normal distribution (cf. Fig. 10.1a). The measure of uncertainty that is produced by the prediction process, or *standard error*, is a statistical measure of this possible variation, and is in turn dependent on the neighboring sample values and the prediction model. As discussed in Chapter 5, a different predicted value at the point can theoretically be obtained from each of the samples in its neighborhood. Geostatistical theory ensures that the predicted value (based on all of the samples) has minimal variance, and hence minimal standard error. Thus, in Fig. 10.1a the mean μ (or average) of the normal distribution is equal to the predicted (most likely) value at the point, and the shape of the probability density curve is dictated by the value of the standard error σ. If the standard error is small, then the statistically possible variation in value at the point is correspondingly small; if it is large, then the probability density curve is wide and relatively flat and the possible variation is large.

Every point (grid-cell centroid) for which we predict a value of the variable has its own unique predicted value and standard error, and therefore its own unique possible variation in value. If we have used volume kriging for prediction (see Chap. 8) then the standard error at a grid-cell centroid is unique to the volume of that particular cell. For this reason, the standard error values cannot be combined, manipulated or averaged with each other in any way for spatial analysis purposes, each is the result of a unique normal distribution of values. For example, we

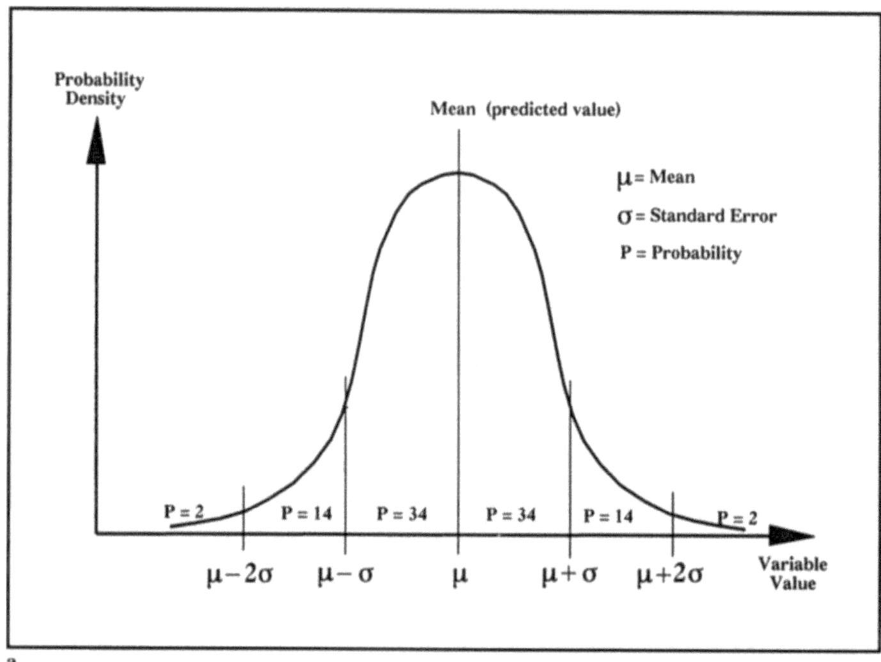

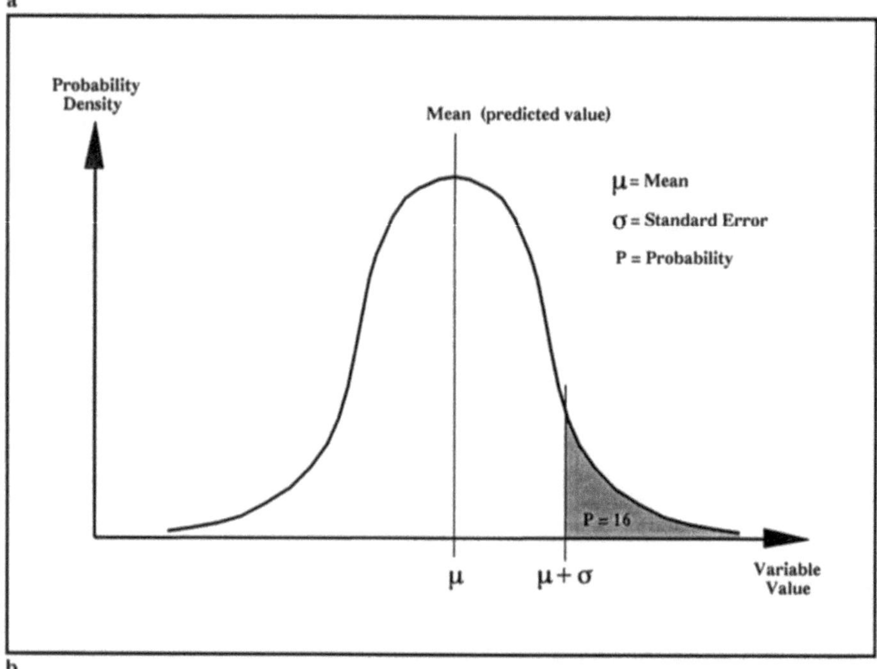

Fig. 10.1   a The normal distribution, standard error intervals and associated probability values; **b** the probability (*P=16%*) of a variable value being greater than the mean plus one standard error

cannot obtain the standard error for an irregular volume by averaging the standard errors of the grid cells contained by the volume. In Chapter 16 we discuss a new geostatistical technique that eliminates this deficiency. Any necessary manipulation of the standard errors (as discussed in Chap. 9) must therefore be performed on the individual values at the grid-cell centroids, prior to subsequent spatial analysis in terms of volumetrics or visualization.

The concept of a normal distribution of the possible variation in value at a point is inherent to geostatistical theory. We are implicitly accepting that we do not have sufficient information to define the *true value* of a variable at a particular point. Instead, we settle for the statistically most likely value and a measure of its possible variation. The reason why we sometimes have to apply data transforms to the sample values (see Chaps. 5 and 8) is to ensure compliance with the geostatistical assumption of a normal distribution of values. If it is necessary to combine the standard errors in any way with the predicted values, then this must be performed before a back-transformation to real values is applied. For similar reasons, if we are combining independent spatial variations within adjacent geological volumes in order to achieve a global variation, then the standard error variation must be applied beforehand.

Knowing that a predicted value has a normal distribution of possible variation in value is all very well, but how do we relate this to uncertainty? To achieve this we make use of the statistical properties of the normal distribution with respect to probability. The integral of the probability density curve over a particular value interval of the variable (the area under the curve in Fig. 10.1a) is equivalent to the probability that the true value of the variable actually falls into this interval. Obviously, the probability that the true value falls somewhere within the total extent of the curve is 100%, and the probability that it is either greater than (or less than) the predicted value is 50%. More appropriately, the probability that it is within one standard error on either side of the predicted value is 68%; and the probability that it is within two standard errors on either side is 96%. This is potentially useful information, but in practice we generally need to know the probability that the true value exceeds (or is less than) a *threshold value*. In a mining context, the threshold is typically an *economical grade cut-off value*; in an environmental remediation context, it is likely to be a *regulatory action level*. Thus, in Fig. 10.1b, the probability that the true value exceeds the predicted value plus one standard error is 16%, and this reduces to 2% for the predicted value plus two standard errors. These probabilities are intrinsic properties of the normal distribution. Values for intervals of the standard error are shown in Fig. 10.1a, and any statistical reference provides tables of probabilities for intermediate fractions of the standard error.

Thus, if we add one standard error to the predicted value at a point then we can say that the *statistical certainty* (probability) of the true value being less than this value is 84%; alternatively, the *statistical uncertainty* (risk) of it being greater than this value is 16%. The standard error has its own spatial variation of values and these are stored in the 3D grid data structure together with the predicted values. Thus, by extension, if we add (or subtract) one times the spatial variation of standard error to the predicted spatial variation of the value of a variable then the resulting (combined) spatial variation represents the *best case scenario (or worst case, depending on the context)* with a statistical certainty of 84%. We can then apply spatial analysis techniques to these results as described in Chapter 9.

This is an extremely useful concept since it allows us to relate probability (or risk) directly to the volumes determined by spatial analysis. However, as we shall see, its application in practice does have some limitations. We discuss the concept in the context of risk assessment later in the chapter and first explore the application of uncertainty in a sampling control context.

10.2 Sampling Considerations and Optimum Sample Locations

The spatial variation of uncertainty represented by the standard error values can be analyzed with any of the spatial analysis techniques discussed in Chapter 9. For example, we can determine the location and extent of volumes within which the uncertainty exceeds a stipulated threshold, or we can visualize contours or color-mapped displays of uncertainty on a horizontal or vertical viewplane section. The volumes or areas of highest uncertainty (relatively speaking) represent those regions from which we know least about the real conditions. Logically, in the interests of achieving (within a limited investigation budget) a characterization that is as representative as possible of the real conditions, the emphasis in any further sampling should be focused on these regions. This ensures the greatest return in terms of reduced uncertainty and increased precision. Obviously, we require a first round of sampling on which to base any analysis of uncertainty. However many investigations are performed in stages and this geostatistical approach to sampling control provides a logical methodology for selecting the optimum locations for each round of sampling.

In many characterizations, particularly in an environmental remediation context, it is more important to predict the location of the action level isosurface than the overall spatial variation of the contaminant. If the cost of remediation is dependent solely on the contaminated volume, and basically independent of the concentration levels within the volume, then we are wasting our site investigation budget if we continue to sample in regions that have already been established as having a high level of contamination, in other words, known *hot spots*. Instead, we should be focusing our sampling efforts on those regions that have both a relatively high level of uncertainty and predicted contaminant concentrations close to the action level. As a result, we achieve the most precise definition possible of the contaminated volume within the constraints of our investigation budget, and minimize the associated uncertainty and risk of residual contamination.

This approach is best illustrated by its application to a characterization study of soil contamination. Figure 10.2a displays sample locations and contours of the standard errors associated with the predicted spatial variation of copper concentrations on a horizontal section through the contaminated site. The areas of

a

b

Fig. 10.2 **a** Sample locations and contours of the uncertainty associated with prediction of copper contamination - shown *red* above the median value of uncertainty; **b** areas of high uncertainty (*hatched red*) overlaid on the isosurface boundary for the contaminant action level (*hatched yellow*)

greatest uncertainty are defined by highlighting the contours wherever they exceed the median standard error value. Some of these areas correlate with a low density of samples, which we would expect. Others correlate with relatively high sample density, implying that the spatial variability itself is relatively high. In Fig. 10.2b the areas of high uncertainty (hatched) are overlaid on the action level isosurface boundary that defines the contaminated volume. From this display we identify areas with both high uncertainty and proximity to the contamination boundary. Further infill sampling in these areas will have optimum benefit with respect to reduced uncertainty and more precise definition of the contaminated volume. The last point of note in this example is that the spatial variation of standard error is quite different and independent of the variation of the copper concentrations, in fact, much of the low uncertainty correlates with relatively high concentrations.

In practice, the standard error values may be relatively large in comparison to the predicted values; in extreme cases, they may even be greater than the predicted values. This is typically the case whenever we have too few samples with a high degree of spatial variability. In itself this is a strong indication that further sampling is required. Despite the magnitude of the standard error values obtained on these occasions, they are still a true reflection of the relative spatial variation of uncertainty, and using them for sample control purposes is still an acceptable practice.

In Chapter 5, we touched on another aspect of geostatistical sample control, i.e. the average sample spacing should be between three and five times less than the measured range of influence obtained from semi-variogram analysis. To rephrase this, in theory, we should continue sampling *until* the average sample spacing is between three and five times less than the range of influence. If we do not achieve this, then we are unlikely to obtain any meaningful measure of spatial variability. If the sample spacing is closer than this, then we are probably wasting our investigation budget. In practice, we may occasionally determine that the spatial variability is not measurable before we reach this level of investigation. However, by applying this consideration together with a spatial analysis of uncertainty we generally achieve a limited but useful degree of control over investigation sampling, and a means of measuring the incremental benefits obtained by iterative rounds of sampling.

10.3 Risk Assessment Based on Uncertainty

The use of geostatistical uncertainty in a risk assessment context is primarily concerned with the derivation of volumes from spatial variations that are associated with varying degrees of probability or risk, as introduced at the beginning of this chapter. These are obtained by adding (or subtracting) multiples of the spatial variation of standard error to the predicted values. The degree of associated probability is directly linked to the value of the selected multiplying factor.

Consider the application of the concept in an environmental remediation context. Suppose we use the spatial analysis techniques discussed in Chapter 9 to determine the volume of contaminated material contained by the isosurface for the action level of a contaminant variable, based on the predicted spatial variation. The result is the *statistically most likely contaminated volume*. In reality, there is a 50% probability (or risk) of residual contamination occurring outside of this volume. This does not imply that we have only identified 50% of the true contaminated volume; the residual contamination may represent only a small fraction of the total. On the other hand, if we add one times the spatial variation of standard error to the predicted values and repeat the spatial analysis for the same action level, then the result is a larger contaminated volume. On the negative side, this represents a higher remediation cost; on the positive side, the probability (risk) of residual contamination is reduced to 16%. By using this approach we can relate reduced environmental risk to remediation cost; alternatively, we can derive a remediation plan that results in an acceptable, pre-determined level of risk.

In a mining context we generally approach risk assessment from the opposite direction. The volume of ore contained by an isosurface of the ore grade cut-off, obtained from the predicted spatial variation of grade, is the most likely volume of ore based on the available samples. This volume and the average grade within it have a 50% probability or risk associated with them; with equal probability the true values of both results could be either larger or smaller. This result is illustrated by the *predicted case* results in Fig. 10.3. By subtracting one times the spatial variation of standard error from the predicted values, we obtain a *worst case* spatial variation of grade. Reanalyzing this spatial variation for the same isosurface value results in a smaller volume of ore and a smaller average grade (cf. Fig. 10.3), but the associated probability has increased to 84% and there is correspondingly less risk attached. This represents the worst case mining scenario, a mine plan based on this result would be considered conservative. On the positive side we know that our chances of the conditions being *at least this good* are 84%, and there is a good chance they will be better. At the opposite extreme, we add the standard error variation instead of subtracting it and repeat the analysis. This time the result is a larger volume of ore with a probability of only 16% and a risk of 84%. This represents the *best case* scenario (cf. Fig. 10.3) and any resulting mine plan would be considered a high-risk venture. In each case the ore volume and its average grade represent potential revenue and the corresponding mine plan represents the cost of achieving that revenue. By plotting graphs of revenue and cost against risk, we can (hopefully) arrive at an acceptable investment proposition that balances risk and cost with financial reward. Ideally, we should ensure that the mining venture is a financially break-even proposition for the worst case scenario, there is then a good chance of a substantial return on investment.

The differences in volumes and average grades for the three cases presented in Fig. 10.3 are quite substantial and probably larger than most of us would expect. They are based on exploration results from a mining venture that is about to reach the production stage at the time of writing. The two extreme cases in Fig. 10.3 illustrate the *possible variation in conditions*, based on the available information.

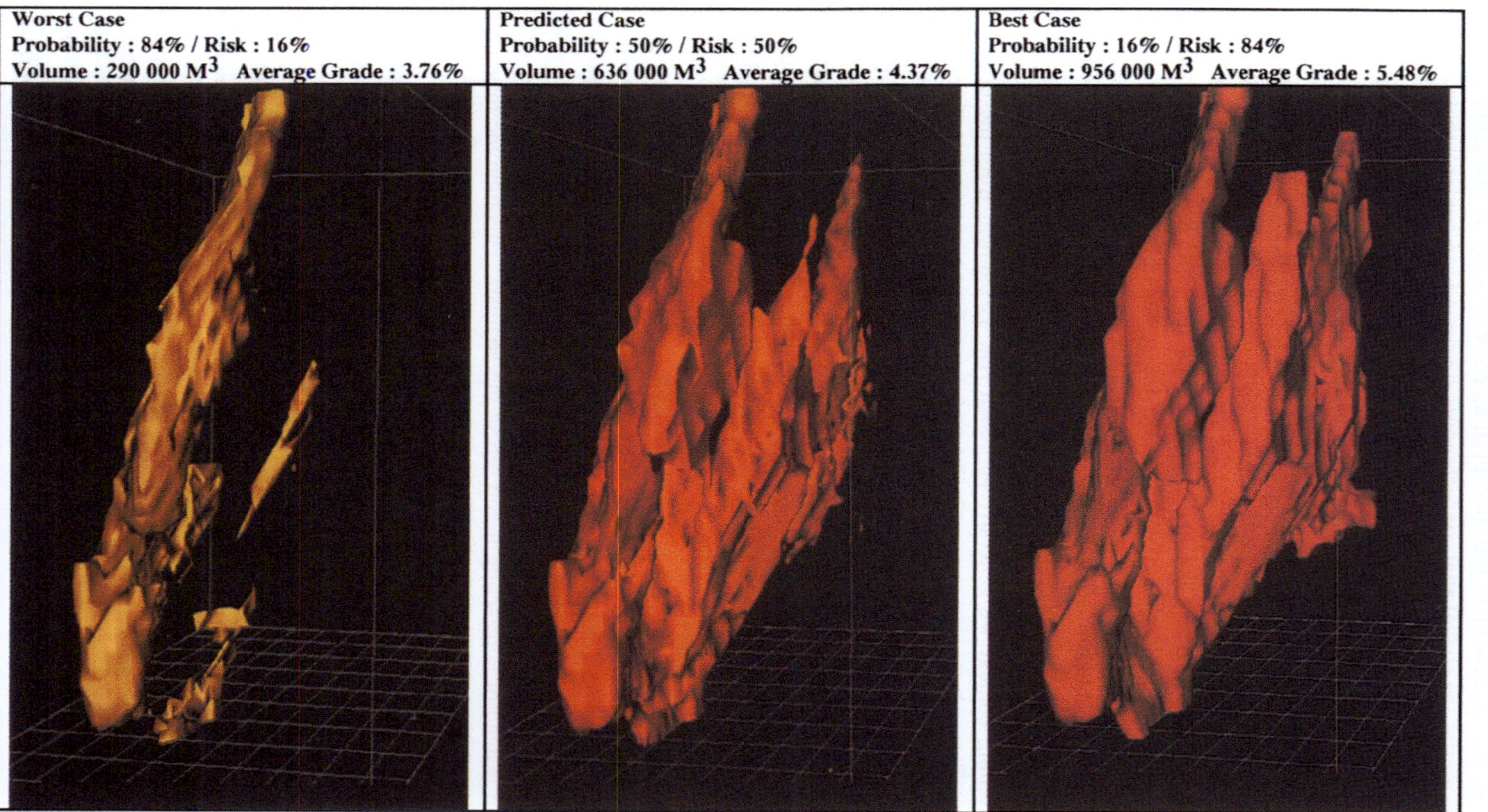

Fig. 10.3   Views of the *2% ore grade isosurface* for the worst case, predicted and best case spatial variations of copper grade for one ore type within a mineral deposit; the corresponding probabilities, ore volumes and average grades are noted for each case

In practice, this approach to risk assessment is not applied as frequently as it should be. In many cases because the standard error values are so large that they are viewed as being unrepresentative, and in some cases embarrassing. In reality this is invariably because we do not have sufficient samples to match the spatial variability of the variable we are interested in. The high degree of possible variation that results from an analysis of the type described above is merely a reflection of this fact. A general attitude of ignoring the implications of uncertainty is particularly prevalent in the mining sector, surprisingly so when one considers that geostatistical prediction evolved within this sector. This is possibly due to investor relations aspects; mining is an investment-driven industry and promoters like to present their mining ventures in the rosiest possible light. Uncertainty in any form is generally considered detrimental to investor relations. There are probably a significant number of failed mining ventures that would never have been put into operation had they been subjected to a proper assessment of risk. This will most likely only happen when investors begin to demand a more thorough assessment.

In an environmental remediation context we are generally more concerned with risk and more aware of the implications of uncertainty, possibly because the bottom line impacts upon our health rather than our pocket book. In Chapter 16, we discuss a new approach to geostatistical prediction that allows the direct prediction of values and uncertainties for irregular volumes. This removes many of the current restrictions on the manipulation and analysis of uncertainty variations that are tied to specific points and volumes.

10.4 The Application of Probability Isosurfaces

We discussed the concept of indicator transforms, and their application to direct estimation (prediction) of probability in Chapter 8. In many ways this provides a more direct approach to risk assessment than the techniques discussed in the previous section. The disadvantages are mostly associated with the inability of the indicator transform approach to predict real values of a variable. It is primarily applicable to those cases that are not directly concerned with the spatial variation of a variable and are focused instead on predicting volumes above or below a stipulated threshold. For this reason the approach is generally most applicable within the environmental sector.

The approach is focused on the direct prediction of the spatial variation of the *probability of a variable exceeding (or being less than) a stipulated threshold*. It eliminates the necessity to manipulate the standard errors with the predicted values in order to achieve a spatial variation with a known probability attached to it. However, the results of prediction with an indicator transform are specific to the particular threshold used in the transformation. They are essentially meaningless in the context of a different threshold value. If we are concerned with several thresholds, then the transformation and prediction processes must be repeated for each one since the spatial variations of probability may be significantly different.

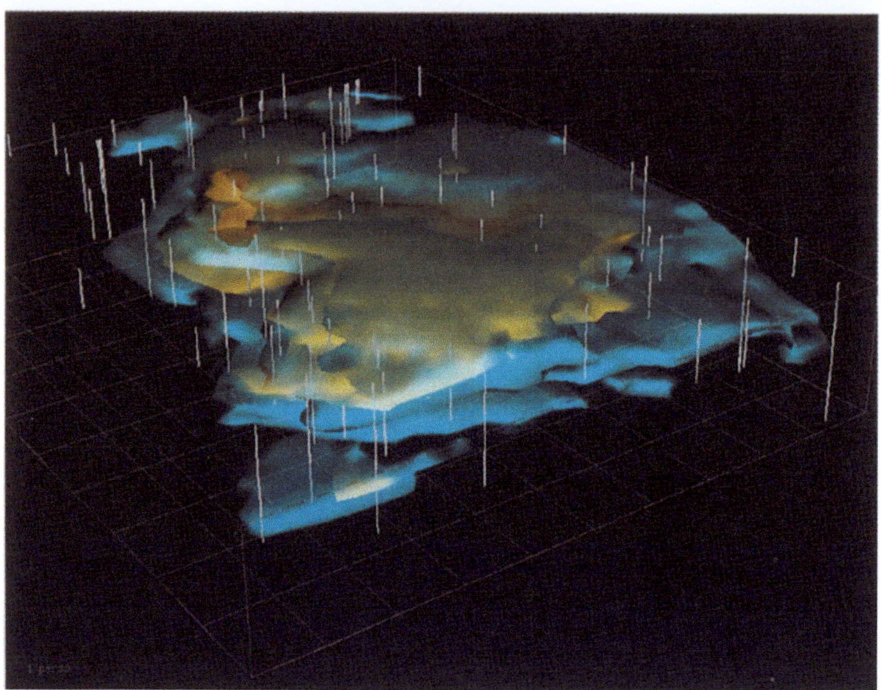

a

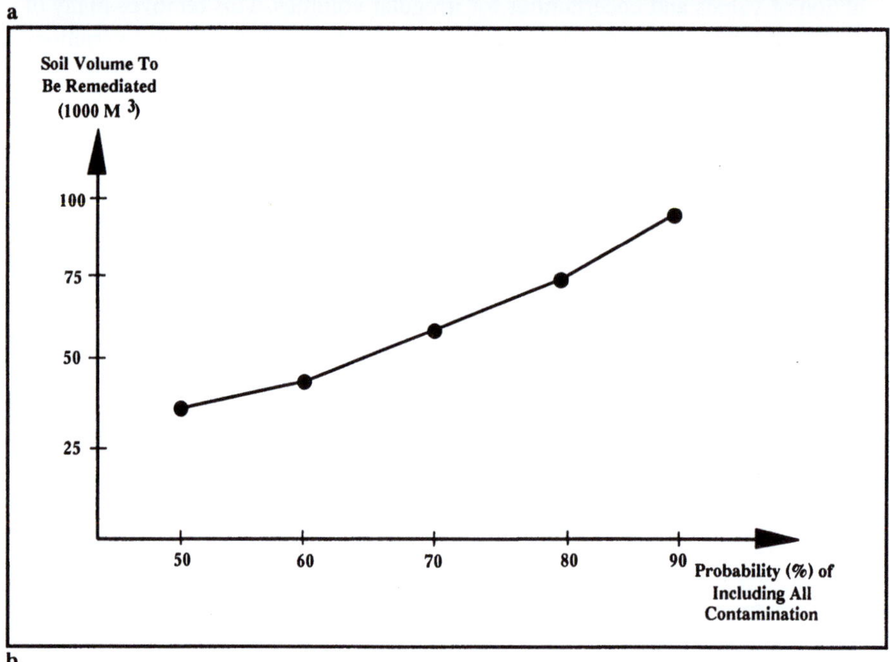

b

Fig. 10.4 **a** Two isosurfaces, one in semi-transparent format, representing increasing levels of the probability of no residual contamination; **b** graph of remediation (isosurface) volume vs. probability of including all contamination

This is a disadvantage that is not shared by the previously discussed approach to risk assessment. On the other hand, an advantage of the probability estimation approach is that we can analyze the resulting spatial variation of probability for a series of probability isosurfaces. This is directly compatible with risk assessment requirements since a relationship between risk and volume (and hence cost or value) is immediately available.

As before, this approach is best demonstrated by an example. Figure 10.4 illustrates an extension of the indicator transform example of soil contamination first presented in Chapter 8. Figure 10.4a shows a perspective view of two isosurfaces for different probabilities, the larger is shown semi-transparent. The indicator transform criterion in this case is for copper contamination less than an action level of 100 mg/g. The smallest isosurface in Fig. 10.4a represents a low probability (50%) of meeting this criterion, in other words, a high risk (50%) of residual contamination outside the isosurface volume. The largest isosurface represents a high probability of meeting the criterion (90%) and a correspondingly low risk (10%) of residual contamination. For each level of probability we perform a volumetric spatial analysis of the corresponding isosurface. Figure 10.4b shows a graph of the contaminated volume to be remediated plotted against the probability of including all contamination. By converting the contaminated volumes to equivalent remediation costs we achieve a direct relationship between probability (or risk) and remediation cost.

As in the previous example, the variation in results is substantial and cost (volume) increases dramatically with decreasing risk. Nonetheless, these results are based on a relatively intensive site investigation, reported in more detail in Chapter 12.

10.5 Geostatistical Prediction of Geological Surfaces

So far, all discussion of geostatistical prediction has focused exclusively on its application to predicting the spatial variations of geological variables. To a limited extent geostatistics can also be usefully applied to the prediction of geological surfaces. In fact, this area of application has a distinct advantage in that it is one of the few ways in which we can quantify the uncertainty associated with observations of geological characteristics.

The application is by nature limited to two-dimensional prediction since the third dimension of the surface we are interested in is the subject of the prediction exercise. This, in turn, limits the application to surfaces that are approximately parallel to the prediction plane, i.e. to relatively simple, layered stratigraphy, or topography, or fault surfaces. If necessary, we can orient the prediction plane appropriately to deal with inclined surfaces. The source information is typically provided by observations of borehole intersections with a particular geological unit. The source data structures must be transformed to the prediction plane

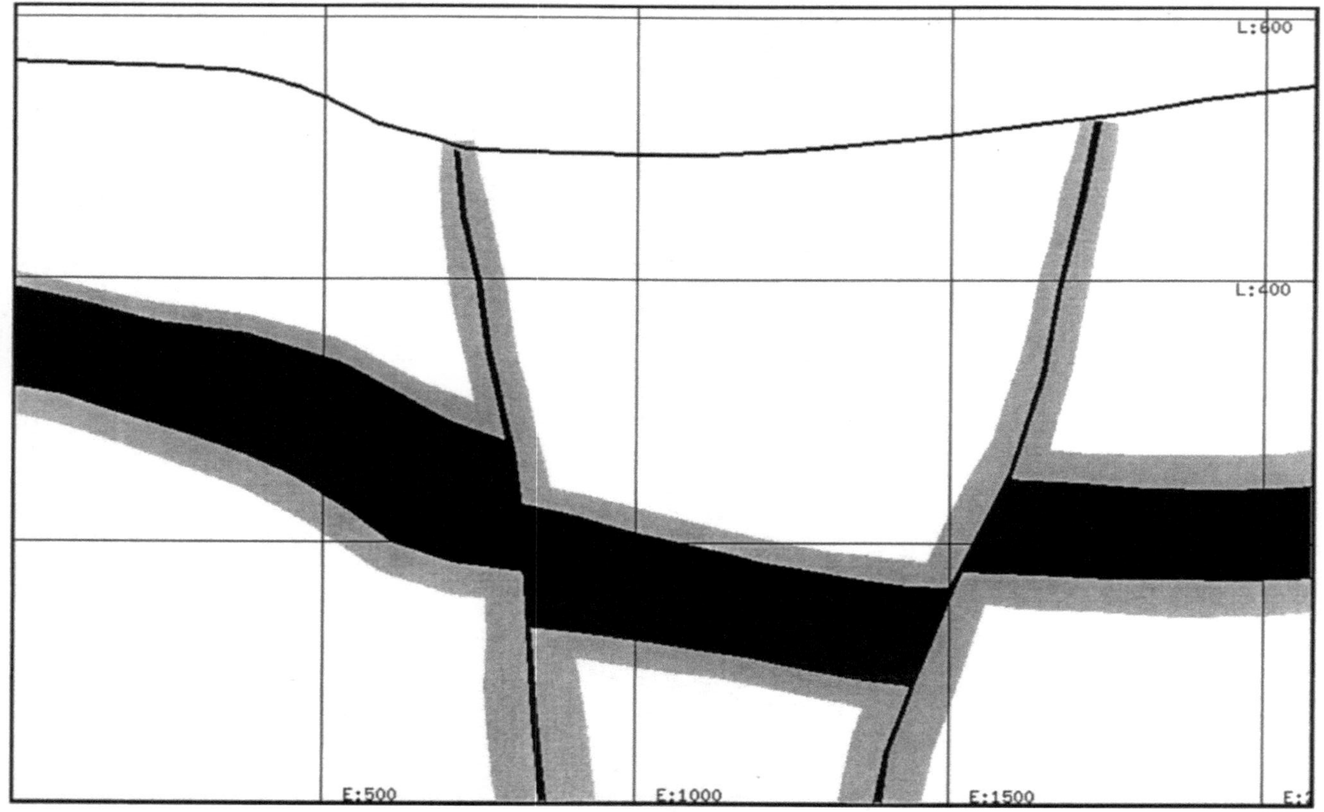

Fig. 10.5   Viewplane section through a volume model representatiom of topography, fault surfaces and faulted strata (*dark gray*); fault and strata surfaces have been determined by geostatistical prediction; the extent of the resulting locational uncertainty is *light gray*

orientation and then reduced to Z offsets normal to the plane, as discussed in Chapter 7. These offset values are the equivalent of sample values in the prediction process. Our objective is to predict the spatial variation of a geological surface and its associated uncertainty based on the isolated offset values.

This approach requires the normal steps involved in geostatistical prediction, starting with a semi-variogram analysis of the offset values and selection of an appropriate prediction model, as discussed in chapter 5. Geostatistical prediction results in a 2D grid representation, in the prediction plane, of the spatial variation of surface offsets. This represents the most likely location of the geological surface, based on the available observations and the selected prediction model. The associated spatial variation of standard error is added or subtracted as discussed previously to derive upper and lower bound limits for the surface with a stipulated degree of probability (or confidence). Using the surface-based approach to volume modeling discussed in Chapter 7, these spatial variations are transformed into a series of representations of the surface that describe its possible variation. This approach has particular application to any form of excavation design (including tunnel alignment) that is concerned with avoiding potentially unstable geological strata or features.

Figure 10.5 shows a viewplane intersection with several surfaces represented in this way. In each case the line intersections (dark) represent the predicted locations of the surfaces and the wide intersections (gray) represent statistical upper and lower bounds. The latter are defined by one standard error on either side of each surface. We therefore have 68% confidence that the surfaces are located somewhere between these bounds. Alternatively, if we are only concerned with our excavation being above (or below) a surface, then we can have 84% confidence in this as long as our design does not intersect the upper (or lower) bound.

The approach can be extended to predicting the spatial variation of the thickness of a layer in addition to its offset. For example, in surface mining of coal seams the economic mineability of a seam is frequently dictated to a large extent by the ratio of overburden thickness to seam thickness. In this scenario we can readily develop volume model representations of the most likely case (predicted overburden and coal), the best case (minimum overburden and maximum coal) and the worst case (maximum overburden and minimum coal) for a coal deposit. In each case the predicted thicknesses of the two layers are appropriately adjusted by the associated standard error variations and then mapped to the topographical surface.

10.6 Other Sources of Geological Uncertainty

In addition to the uncertainty associated with spatial prediction, there are two other sources of uncertainty that we should at least be constantly aware of throughout any characterization, although we may be unable to do anything about them. The first concerns the potential for investigation errors in sample and observation values and their locations. The second concerns the uncertainty inherent to any interpreted information. The uncertainties arising from both of these sources are cumulative with the prediction uncertainty.

There is generally little we can do about the potential for investigation error in sample and observation values without comprehensive research into each of the wide variety of investigative techniques in common use. There is a general scarcity of information in this regard; if the research has been done, then the results are by no means freely available.

Information on the potential for locational errors is almost as scarce. However, a limited amount of valuable information has been published. For example, the uncertainty in the location of a borehole sample, or its intersection with a geological feature, should be based on a knowledge of the errors inherent to borehole survey measurements and the depth of the intersection. Sides (1992) has quantified the locational errors that occurred on an underground mining project, by comparing observed and interpreted locations of geological features with the actual (surveyed) locations determined during mining excavation. Results in close proximity to borehole traces indicate a *vertical accuracy* (mean error) of approximately 2 m and a *horizontal accuracy* of approximately 3 m for drilling depths of between 400 m and 500 m. The associated *precision* (standard error) is roughly double the accuracy in each direction. Based on these results, we can predict a probable locational error to depth ratio in the vicinity of a borehole trace of approximately 1:200. This is attributable primarily to errors in survey measurement since little interpretation is involved. If we wish to attach a measure of confidence, then we are forced to use precision instead of accuracy; with 68% confidence, the ratio of potential locational error to depth increases to approximately 1:100.

The same mining study indicates that the accuracy and precision of interpreted feature locations between borehole sections are approximately double the values closer to the borehole traces. The incremental increase in locational error is attributable primarily to errors of interpretation. However, the magnitude of this component of error is obviously dependent on geological complexity, as well as interpretational expertise, intuition and experience and its quantification in the general case remains a largely subjective exercise. Nonetheless, this is a valuable study. If nothing else, it is educational in terms of the potential magnitude of locational errors. For example, the study indicates that the potential error in location of an interpreted geological feature at a depth of 100 m and 50 m away from the nearest borehole trace is of the order of 1.5 m in any direction. This is probably larger than most of us would expect. Chapter 12 includes an attempt to apply these results in the context of a regional characterization.

As noted above, the potential for error and uncertainty resulting from interpretation of geological features is largely subjective. There is no way of incorporating it into a computerized approach unless we are prepared to quantify it ourselves during the interpretation process. This would essentially require a subjective estimate of the possible variation (in three dimensions) of each geological feature of concern, however, most of us are probably a long way off from attempting this. In the mean time, we should not forget that any geological

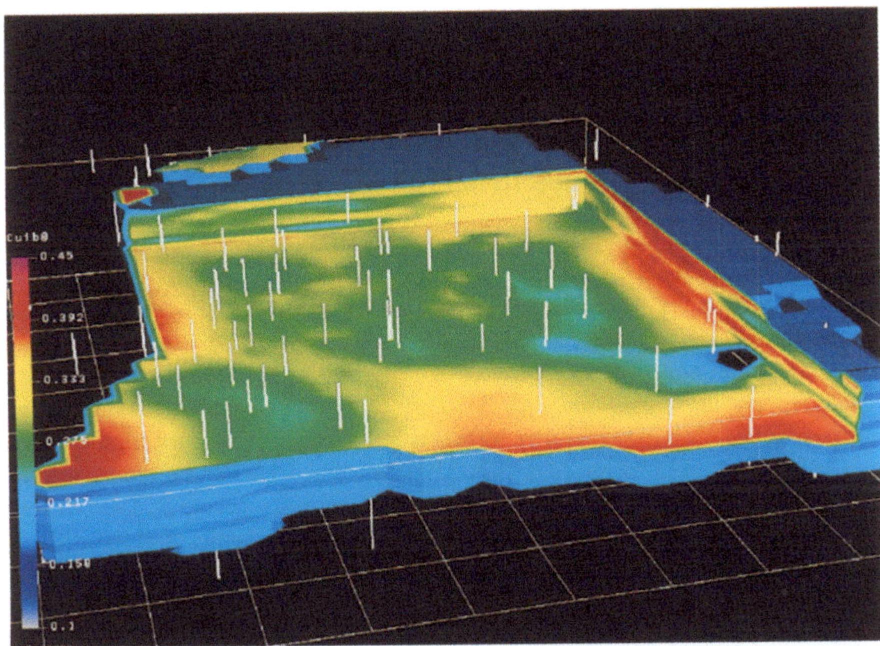

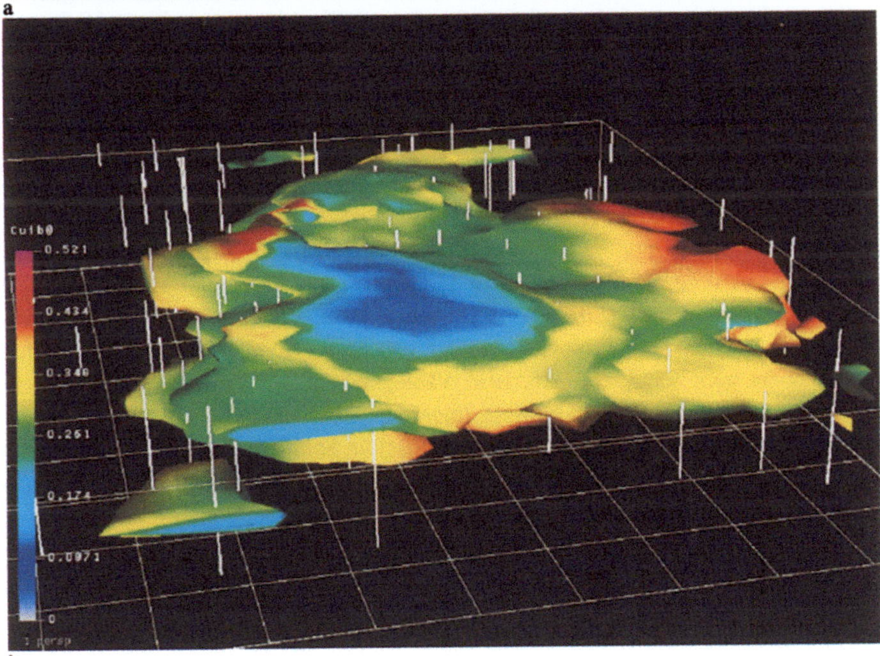

Fig. 10.6 **a** Color-mapped visualization of the spatial variation of uncertainty associated with prediction of soil contamination, cut-away to expose internal variation; **b** the spatial variation of contamination uncertainty mapped onto the action-level isosurface of a contaminated volume

interpretation is just that, an *interpretation*, and represents only one of a range (variation) of possible interpretations that fit the available information.

10.7 Visualization of Uncertainty

Computer visualization capabilities are somewhat limited with respect to uncertainty. This is because (1) uncertainty essentially represents another dimension and (2) we invariably want to display uncertainty together with the information with which it is associated. We can display any 2D information very effectively on viewplane sections. Display of 3D information as perspective transformations on a 2D computer screen is less effective for interaction but still meaningful and useful for presentation. The time dimension can be represented as motion in either of the above. The only way we can accommodate another dimension such as uncertainty is to overlay it on the associated information in an appropriate format. This means that we either have to be able to see through the representation of uncertainty, or we have to be able to discern the shape or form that it refers to beneath it. Visualization of uncertainty on its own is generally only useful in the context of determining future sampling locations; Figs. 10.2a and 10.6a present examples.

Thus, on a 2D viewplane section, contours of uncertainty can be overlaid on a color-mapped representation of the variation of a variable. In contrast, overlaying two sets of contours generally results in an almost unreadable display, even if they are in different colors. In 3D perspective views we can use color-mapping, as in Fig. 10.6b, or semi-transparency, as in Fig. 10.4a, to represent uncertainty. Both solutions are only moderately effective. When it comes to representing the uncertainty associated with geological boundaries, the capabilities, in 2D or 3D, are even more limited. Gradational (fuzzy) color variation between adjacent geological units is unacceptable because it can easily be misread as a gradational variation in geological characteristic value, whereas what we are trying to represent is a hard boundary with an uncertain location. Contours of uncertainty overlaid on geological information are a possible solution (see Chapter 12). The remaining alternative is to create several geological models that represent the range of possible conditions, however, this is generally expensive.

Until computers can provide a few more visualization options to represent multi-dimensional information, we remain somewhat limited in this respect.

Part II. Applications in the Geosciences

11 Subsurface Soil Contamination Assessment

11.1 Project Overview

The majority of contaminated site characterizations are still performed with conventional GIS and CAD tools, essentially working in two dimensions. Few characterizations employ geostatistical analysis or prediction techniques. The following study of subsurface soil contamination illustrates the advantages of using an integrated 3D approach. As a result, both environmental risk and remediation cost are substantially reduced.

Some of the results presented here will be familiar, they have been used to illustrate various points in the text of Part I. They are repeated here for the sake of completeness.

This study involves a characterization of subsurface soil contamination based on information from environmental site investigation of the Pacific Place Project, Vancouver, Canada. The study was performed with the assistance and approval of the British Columbia Ministry of Environment and the environmental consulting group responsible for investigation and remediation of the site.

The site is an 80 ha waterfront property on the north shore of False Creek, close to the downtown core of Vancouver. In the early 1900s much of the site was reclaimed from False Creek by imported fill material, some of which was contaminated prior to placement. It was subsequently subjected to mixed industrial usage for many years, resulting in further significant contamination over parts of the site from a variety of sources. For investigation purposes the site is divided into land parcels according to historical records of industrial usage. The extent of this characterization measures approximately 200 x 200 m. It has been subjected to several phases of subsurface investigation and is known to be heavily contaminated, principally as a result of former usage by a coal gasification utility and an electroplating operation. It forms part of a larger area which is now re-zoned for high density residential and commercial usage.

The objective of the study is to determine the location, extent and volume of unacceptable contamination from a range of contaminants, and the total volumes of soil requiring either containment, remediation or removal. The features of the site include relatively simple, shallow, layered soil geology that requires little interpretation. However, this is combined with complex spatial variations of contaminants from several industrial sources.

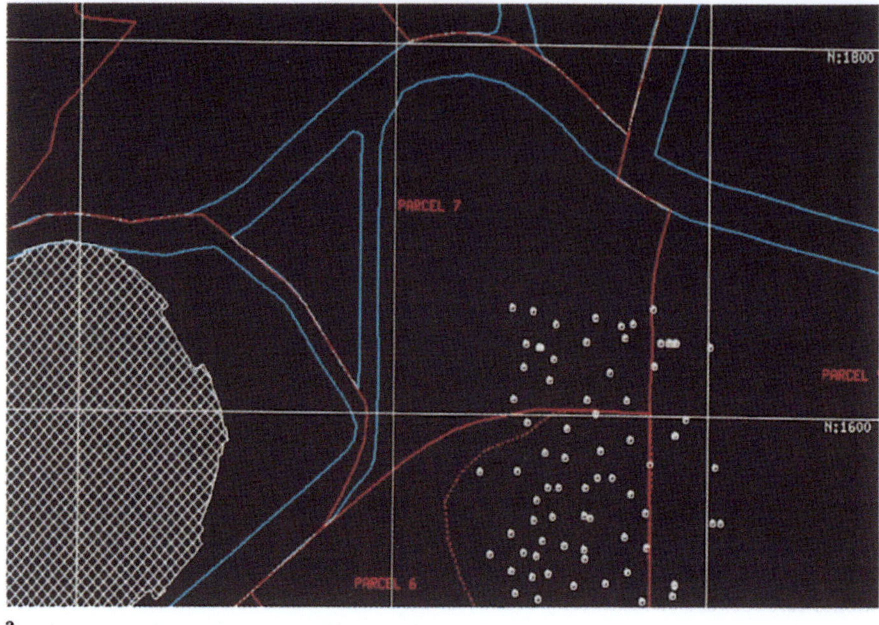

a

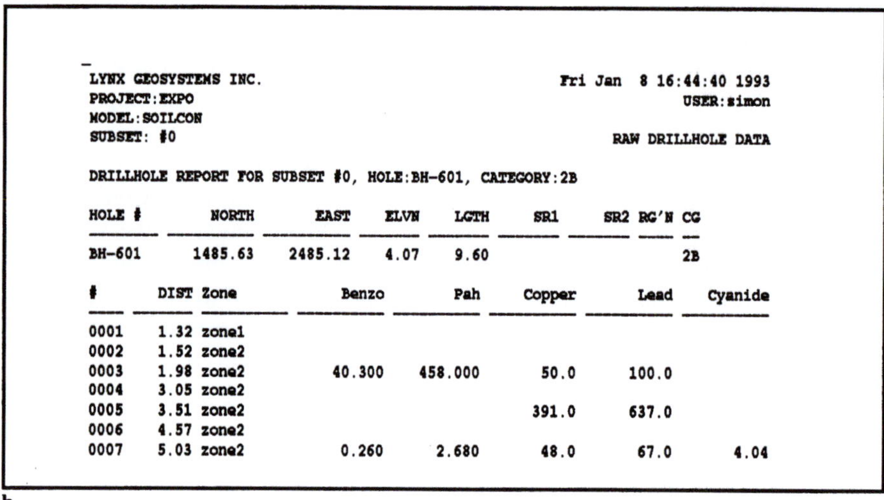

```
  LYNX GEOSYSTEMS INC.                              Fri Jan  8 16:44:40 1993
  PROJECT:EXPO                                               USER:simon
  MODEL:SOILCON
  SUBSET: #0                                        RAW DRILLHOLE DATA

  DRILLHOLE REPORT FOR SUBSET #0, HOLE:BH-601, CATEGORY:2B

  HOLE #        NORTH      EAST    ELVN    LGTH     SR1      SR2 RG'N CG
  ________   ________   ________  ______  ______  ______  ______  ___ __

  BH-601      1485.63   2485.12    4.07    9.60                      2B

  #      DIST Zone        Benzo        Pah     Copper      Lead    Cyanide
  ____   ____ ____      ________    ________  ________   ________  ________

  0001   1.32 zone1
  0002   1.52 zone2
  0003   1.98 zone2        40.300    458.000      50.0      100.0
  0004   3.05 zone2
  0005   3.51 zone2                              391.0      637.0
  0006   4.57 zone2
  0007   5.03 zone2         0.260      2.680      48.0       67.0      4.04
```

b

Fig. 11.1   **a** Plan of the contaminated site with borehole, well and test-pit locations superimposed;   **b** site investigation information in hole data format containing soil strata intersections and sample analysis results of contaminant concentrations

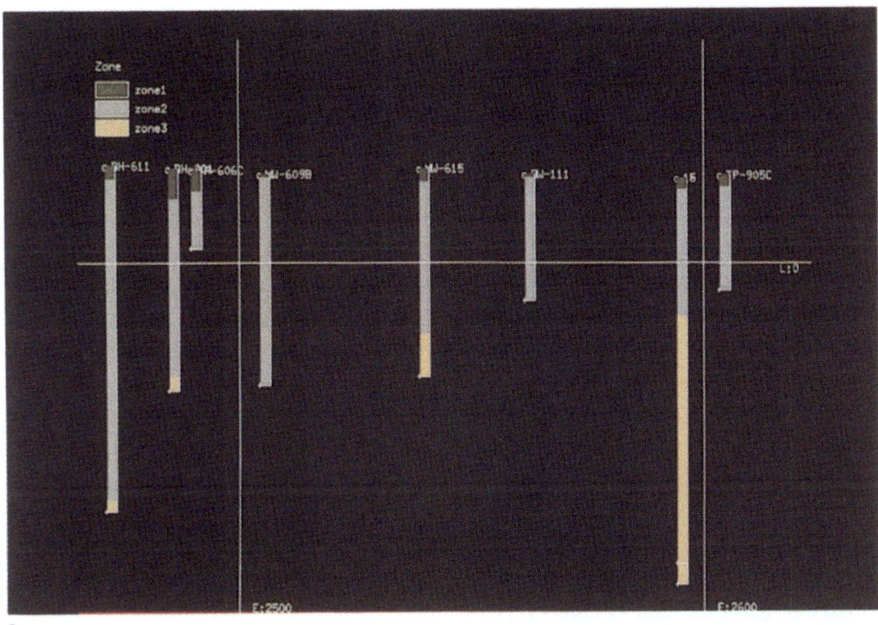

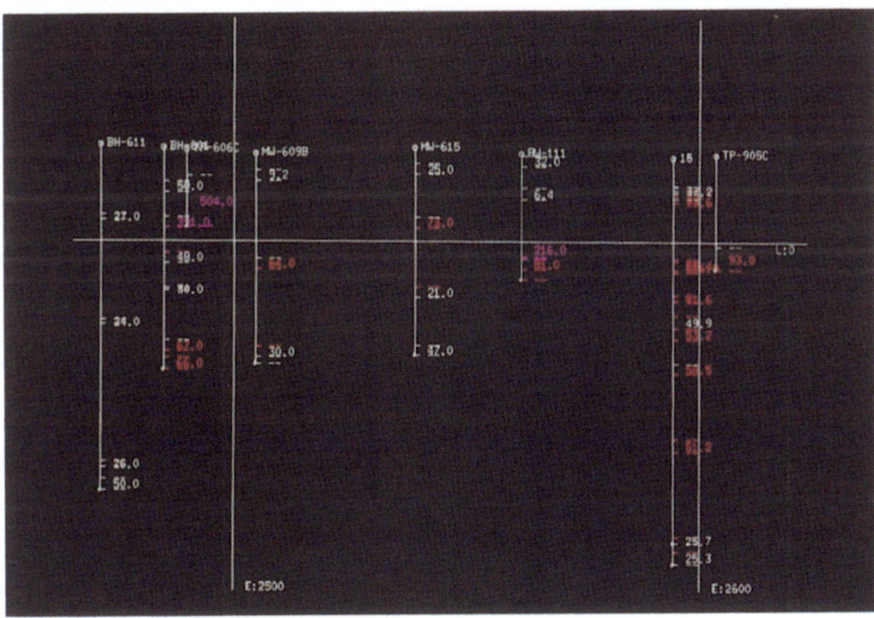

Fig. 11.2 **a** Vertical viewplane section with color-coded borehole intersections of soil strata; **b** sample values of copper concentrations plotted against borehole traces

11.2 Site Investigation Records

Information available to the study includes site plans, parcel boundaries and investigation records for 97 test pits, boreholes and monitoring wells (cf. Fig. 11.1a). The latter are already available in a computer database and are imported directly in ASCII file format and stored as hole data structures. The site plans are digitized, referenced to appropriate elevations and stored as map data structures.

The hole data includes ground surface coordinates and elevations, soil strata intersections, sample locations and sample results of contaminant concentrations (Fig. 11.1b). Sample results are available for five contaminants from approximately 700 sample locations. The five contaminants of concern to the study include benzo(A)pyrene and total PAH (polyaromatic hydrocarbons) which are believed to originate from the coal gasification utility, and cyanide, copper and lead which probably originate from electroplating operations. The contaminant concentrations are the *variables* of our subsurface characterization. The pertinent regulatory limits, or remediation *action levels*, for contaminant concentrations in high-density residential areas are the Level C Standards published by the BC Ministry of Environment (Table 11.1).

The soil strata classification, contained in the hole data intersections and designated by soil zone, is the only *characteristic* available (Fig. 11.1a). Three soil strata are significant to the study; a thin, recently placed layer of clean fill, designated as *zone1* material; a layer of original reclamation fill between 6 and 9 m thick which contains most of the contamination and is designated *zone2* material; and a thick natural base layer of dense silty clay, designated *zone3* material, which is relatively impervious.

11.3 Information Review and Evaluation

Visualization of the investigative information on a series of vertical viewplanes through the characterization region provides us with an initial appreciation of subsurface conditions (see Fig. 11.2a,b). These confirm that the soil geology is a simple layered formation, that contamination is widespread throughout the area and depth of the site, and that spatial variation differs significantly between the different contaminants.

Statistical analysis of sample subpopulations from each of the soil strata confirm that the uppermost zone1 material is relatively free of contamination and that the lower zone3 material contains only a small number of contaminated samples. In contrast, the intermediate zone2 fill material is shown to be heavily contaminated (see Table 11.1).

For two of the contaminants, benzo(A)pyrene and total PAH, the *averages* of the sample values exceed the action level, and for all contaminants the statistical

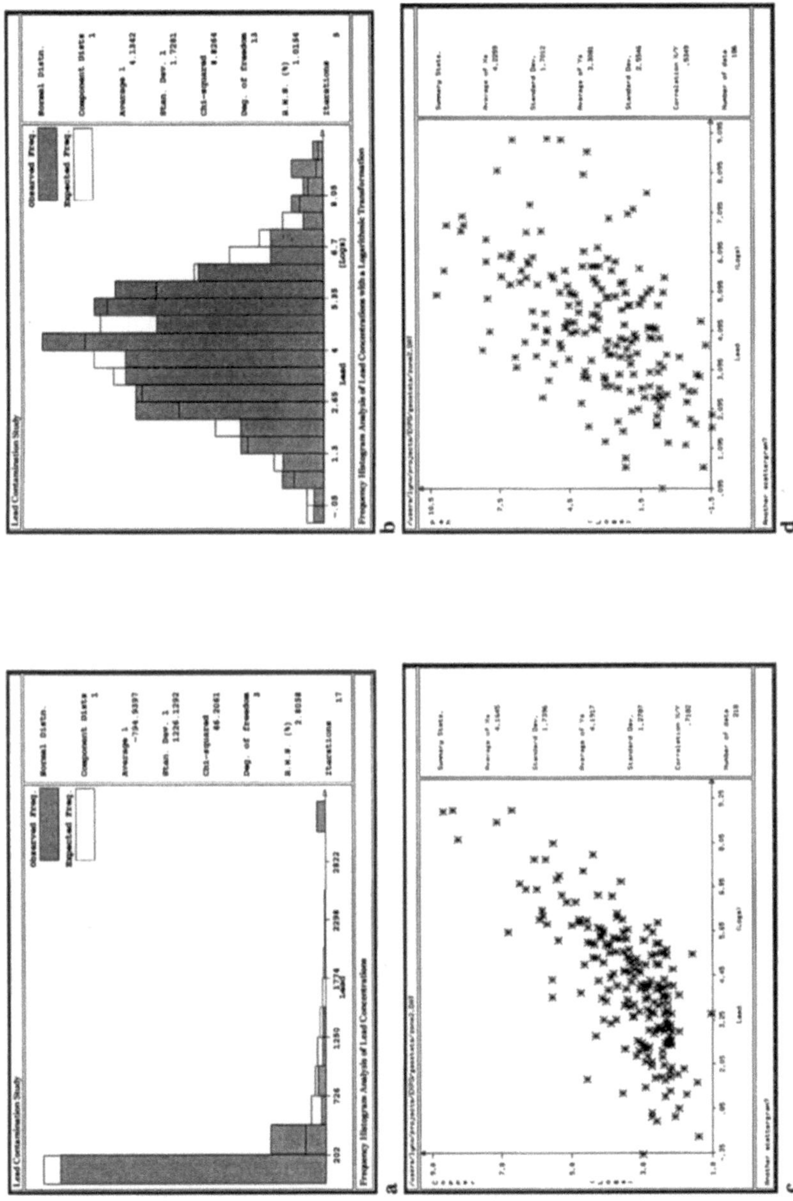

Fig. 11.3 a Frequency histogram of lead concentrations; **b** histogram of lead concentrations with a log data transform; **c** correlation analysis of lead and copper concentrations; **d** correlation of lead and total PAH concentrations

standard deviations (standard errors) are unacceptably high. Frequency histogram analyses show that the samples exhibit a log-normal distribution of values for all of the contaminants (Fig. 11.3a,b), which is typical to many cases of soil contamination.

Statistical correlation analyses show that copper, lead and cyanide samples are well correlated (cf. Fig. 11.3a), as are the second group of benzo(A)pyrene and total PAH samples. This indicates that the samples of the contaminants that comprise each group exhibit similar spatial variability, whereas analysis of lead and total PAH samples, for example, shows little cross-correlation between the two groups (Fig. 11.3b). This tends to confirm that we are dealing with two groups of contaminants, either originating from different sources and/or exhibiting different migratory behavior within the subsurface. The observed complexity in the spatial variations (both vertical and horizontal) of the contaminants dictates that geostatistical semi-variogram analysis of the spatial variability of the samples is a critical step in our characterization.

Contaminant statistics			
Geology : ZONE2 **Samples : 553**			
Contaminant	**Level C standard (micro-g/g)**	**Sample average (micro-g/g)**	**Sample standard deviation**
Benzo(A)pyrene	10	45	212
Total PAH	200	842	3642
Cyanide	500	128	517
Copper	500	293	1285
Lead	1500	331	1843

Table 11.1   Level C regulatory limits (remediation action levels) and statistical analysis results for contaminant samples occurring in zone2 fill material

11.4 Spatial Variability Analysis of Contaminant Samples

The fact that all contaminant samples display a logarithmic distribution of values indicates that a similar transformation is required for semi-variogram analysis.

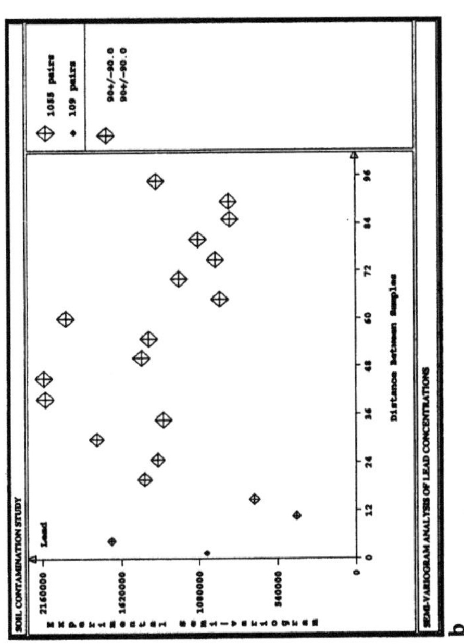

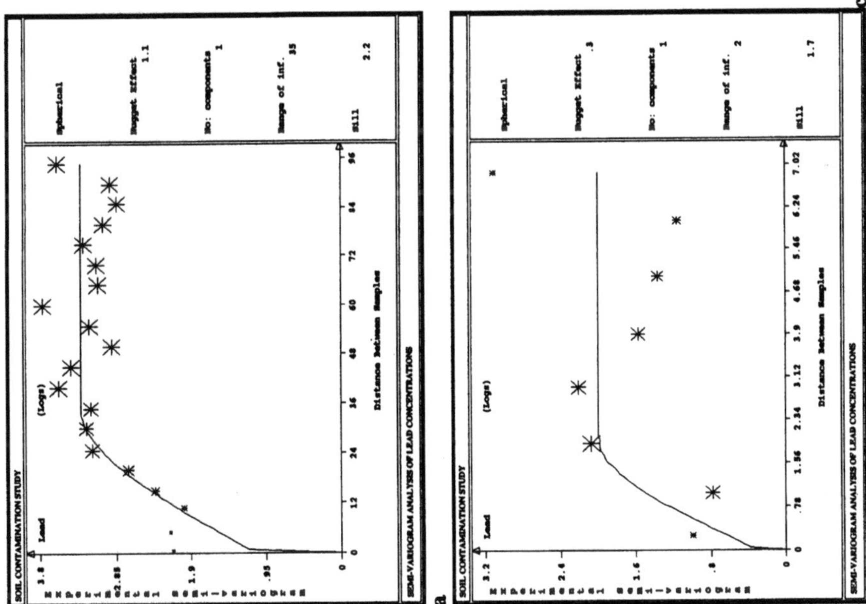

Fig. 11.4 **a** Semi-variogram results and prediction model for the horizontal variability of lead concentrations with a log transform; **b** semi-variogram results for lead concentrations without transformation or directional control; **c** semi-variogram results and prediction model for the vertical variability of lead concentrations with a log transform

Visual inspection of viewplane sections through the site indicates that a significant degree of anisotropy may also be present in the spatial variability of the samples. Figure 11.4b shows the semi-variogram results for lead samples in zone2 material without data transformation and with no directional control. It is difficult to discern any trend at all in these results. In contrast, Fig. 11.4a shows the results for the same samples with a logarithmic data transformation and horizontal directional control; in this case, only sample pairs with an orientation within 10° of the horizontal plane are included. These results exhibit a quite noticeable trend between sample variability and distance.

The differences between these two sets of results demonstrate the importance, in terms of achieving a meaningful result, of selecting the appropriate data transformation and degree of directional control for semi-variogram analysis. The results of Fig. 11.4a represent the spatial variability of lead samples in approximately horizontal directions. They indicate that the horizontal *range of influence*, or distance within which sample variability is directly related to distance, is approximately 35 m. Copper and cyanide samples exhibit similar spatial variability to the lead samples, whereas benzo(A)pyrene and total PAH samples exhibit ranges of influence of approximately 50 m.

In contrast, semi-variogram analysis of transformed lead samples with vertical directional control indicates that the range of influence in the vertical direction is of the order of 2 m only (cf Fig. 11.4c), although the trend is not as obvious in this case. This confirms our suspicion that there is significant anisotropy present in the spatial variability of the lead samples. The ratio of the vertical to horizontal ranges of influence is a measure of the degree of anisotropy present. Similar anisotropy is evident in the spatial variability of the other contaminants. We can only speculate that a possible cause of this anisotropy is that the zone2 fill material was originally placed and compacted in sublayers, resulting in anisotropic permeability conditions within the layer.

Since reasonably good semi-variogram results are obtained for all contaminants in this manner, geostatistical kriging is chosen as the appropriate *prediction technique*. We therefore need to derive suitable semi-variogram relationships from our results, which can later be used as *prediction models* to obtain the spatial variations of the contaminants. These are obtained by interactively fitting *spherical* semi-variogram models to the measured spatial variability for each contaminant, using the empirical approach discussed in Chapter 8. The prediction models for lead contamination with a logarithmic transformation are shown overlaid on the semi-variogram results in Fig. 11.4a,c. In each case, these models reflect the strong degree of anisotropy present in the spatial variability of the samples. The models are cross-validated against the samples (cf. Chap. 8) to obtain a measure of their *goodness*, i.e. the degree to which they represent the measured sample variability.

11.5 Representation of Soil Geology

At this stage, since both statistical and geostatistical analyses have confirmed that contaminant variation is significantly different in the three soil strata, we need to develop a model of soil geology that can be used to apply spatial control to our subsequent prediction of contaminants. Because we have sufficient investigative information available, and because the geology is a relatively simple, layered formation, we can derive the model directly from the zone characteristic that defines strata intersections in the hole data structures (see Figs. 11.1b and 11.2a).

Using the surface-based approach discussed in Chapter 7, intersection elevations are reduced from this information and used to develop triangulated surfaces for each soil strata horizon. A set of triangulated prism-shaped volume model components is then generated between each pair of surfaces to represent the volume of the contained soil stratum. Each set is assigned a code that designates its characteristic and controls its display color. The resulting model of soil geology, with a three-dimensional cut-away to expose internal irregularity, can be seen in Fig. 11.7.

11.6 Geostatistical Prediction of Contamination

Our first step in the spatial prediction of contamination is to define a 3D grid data structure with a geometry that encompasses the site to an appropriate depth. This provides a data structure in which to store the predicted contamination values. Grid-cell dimensions of 5 x 5 m in plan and 1 m vertically are chosen. These are based on average sample spacings of approximately 15 m horizontally and 3 m vertically, and semi-variogram ranges of 35 to 50 m horizontally and 2 m vertically. In general, the cell dimensions should be between two and five times less than the minimum of these in each direction. The chosen dimensions result in prediction of values at 5 m centers horizontally and 1 m vertically, throughout the region of characterization.

Our second step is to intersect the grid model with the volume model of soil geology. This involves intersecting each grid cell with each of the sets of components representing the different soil strata. The results of this intersection, consisting of soil codes and volumes of intersection, are stored in the grid data structure.

These results provide a means of identifying which grid cells intersect which soil strata. We can now predict contaminant concentrations throughout the volume of the zone2 material (industrial fill layer), using the prediction model and sample subpopulation appropriate to the contaminant and the material, and only predicting values for those cells which intersect the zone2 material. We also predict a measure of the uncertainty associated with each predicted value. In similar fashion,we predict the contaminants in the upper zone1 material and the lower

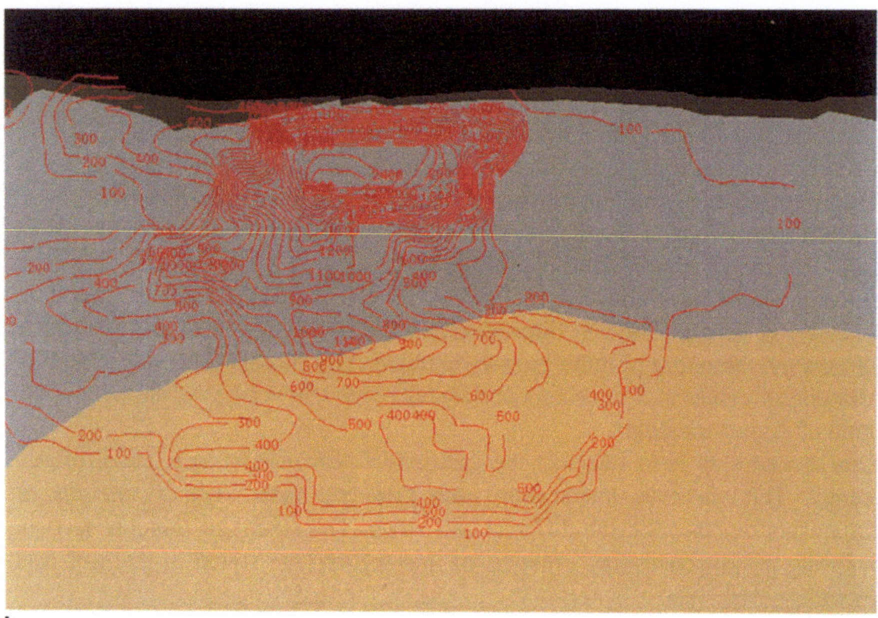

Fig. 11.5 a Combined values of predicted copper contamination for all soil strata, shown *red* where they exceed the specified threshold (action level); **b** the same values displayed in contour format

zone3 material, in each case using the appropriate prediction model and sample subpopulation.

The importance of being able to constrain the prediction process in this way requires emphasis. Without geological control, unacceptable contamination would be falsely predicted within the overlying zone1 material and the underlying zone3 material, whereas we know from statistical analysis that the unacceptable contamination occurs primarily in the intermediate stratum of zone2 material, and that the adjacent strata are relatively uncontaminated. These false predictions would distort our final evaluation of contaminated volumes.

Some grid cells intersect several soil strata and therefore contain a corresponding number of predicted values. We combine the predicted contamination for all three strata by determining volume-weighted average values for these multiple intersection cells at the soil strata boundaries (as discussed in Chap. 9).

The predicted spatial variation of each of the contaminants can be viewed on any viewplane through the site in a variety of display formats. For example, we can request that values above a specified threshold be highlighted in a different color (cf. Fig. 11.5a). Or, instead of values, we can request a contour display of contaminant concentrations, with contours above the threshold highlighted as before (cf. Fig. 11.5b). These results confirm that most contamination occurs in the zone2 material, while relatively minor amounts of contamination have penetrated the zone3 material (natural base layer). These contours of predicted values form the basis of our determination of contaminated volumes. A contour for the regulatory limit of contaminant concentration, extended into 3D, becomes the isosurface which delineates unacceptable contamination. We discuss this aspect in more detail later.

11.7 Geostatistical Control of Sampling

One of the uses to which we can put the uncertainty values is a verification of sampling adequacy. For example, suppose we reorient our viewplane to a horizontal (plan) orientation through the middle of the zone2 material. We can then display contours of the uncertainty (standard error) values, with the areas of greatest uncertainty highlighted (cf. Fig. 11.6a). The sample locations are overlaid on the uncertainty contours and show that uncertainty is generally, but not always, higher where sample density is low (which we would expect). Some of the areas of high uncertainty also contain high predicted values of contaminant concentrations and therefore are included in our final determination of contaminated volumes anyway. Others contain low concentrations and are therefore of little concern.

The areas of greatest concern to us are those which have both high uncertainty and predicted values close to the regulatory limit or action level. It is questionable whether these areas should be remediated or not. Obviously, any further sampling should be concentrated in these areas to obtain maximum benefit in terms of

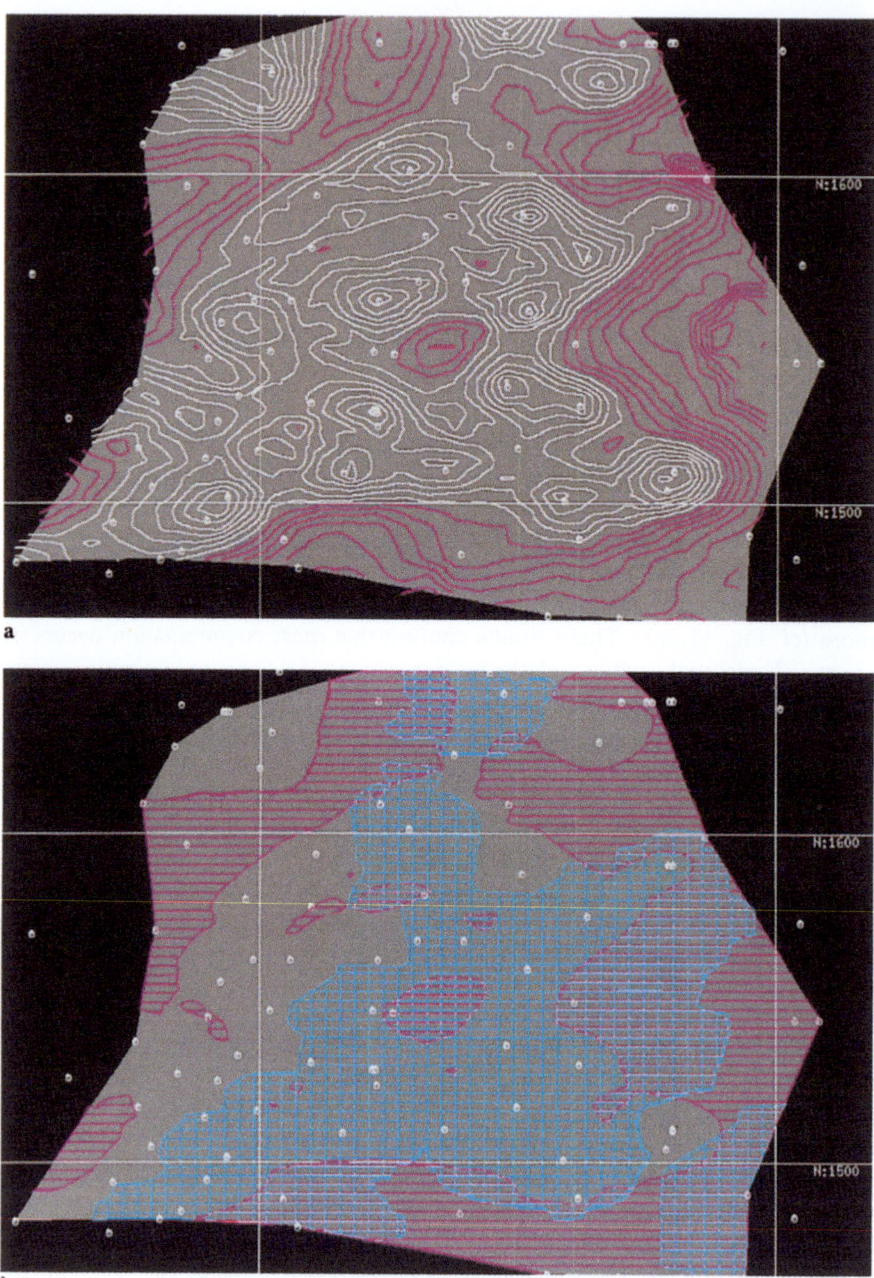

Fig. 11.6   **a** Contours of uncertainty for benzo(A)pyrene contamination on a plan section through the industrial fill layer, shown *magenta* above the median value, with sample locations overlayed; **b** areas of high uncertainty (*hatched magenta*) overlaid on the isosurface boundary for the contaminant action level (*hatched blue*)

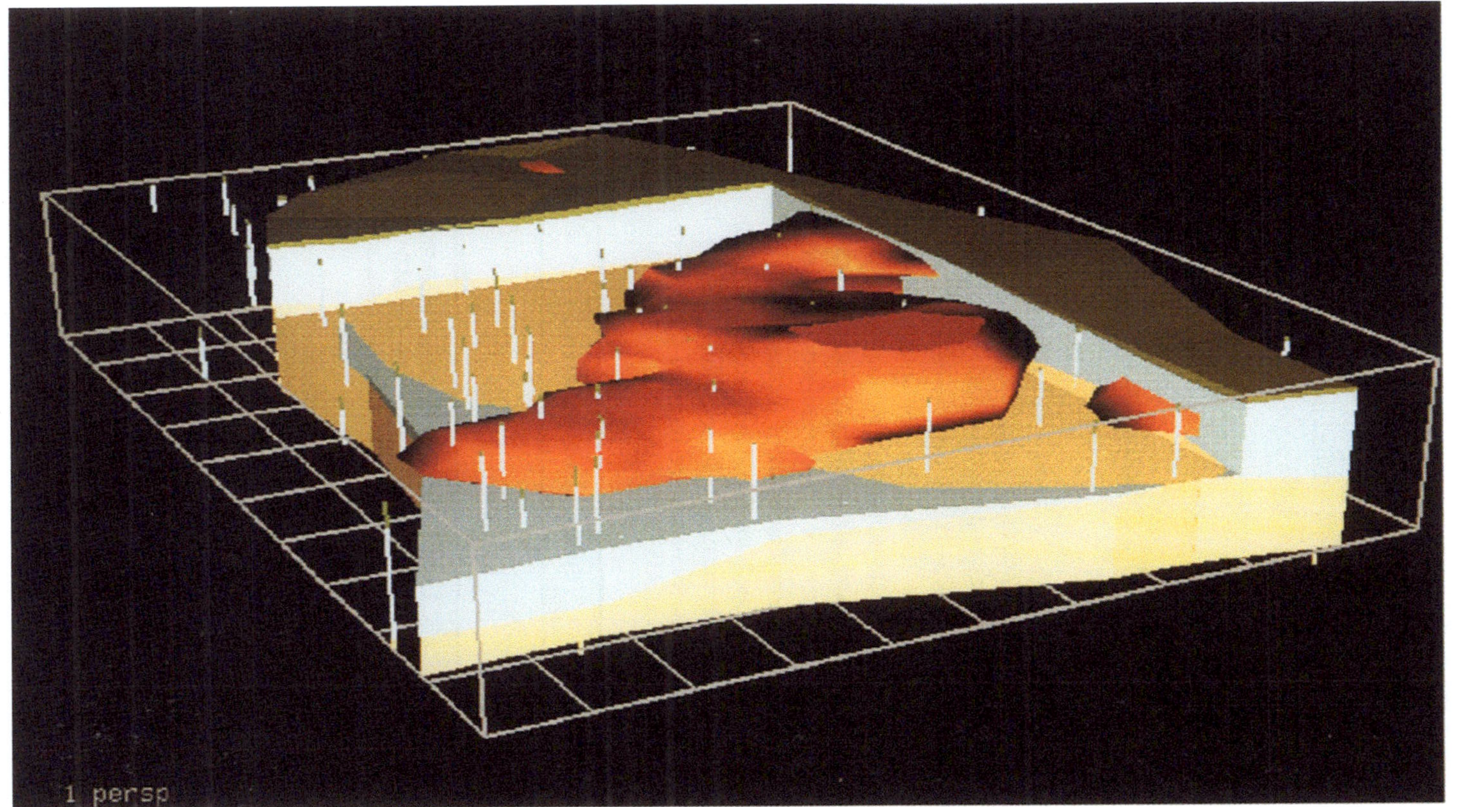

Fig. 11.7   3D cut-away section through the volume model of soil geology, exposing the contaminated volume for benzo(A)pyrene concentrations, with borehole and well sampling locations superimposed

defining the contaminated volumes with greater certainty. We can identify these questionable areas by superimposing contours of predicted contamination on the uncertainty contours (cf. Fig. 11.6b).

Geostatistical kriging thus provides us with a limited, but logical, degree of control over sampling and investigation costs. This contrasts with the common practices of *blanket* sampling on a gridded basis, or concentrated sampling around predetermined *hot spots*. There is little point in over-sampling hot spots if they are to be remediated anyway!

11.8 Spatial Analysis of Contaminated Volumes

Our final objective is to quantify the volumes containing unacceptable contamination and to visualize them in their correct locations so that appropriate remediation plans can be designed around them.

Contaminated soil is defined as any material with a concentration exceeding the regulatory limit or action level. The contaminated volume for each of the contaminants is therefore determined by applying spatial volumetric analysis to the predicted grid-cell values and interpolating the isosurface with a value equal to the limit, as described in Chapter 9. The combined predicted contamination for all three strata show that minimal amounts of contamination are present in the layers above and below the zone2 material and must also be included in the final determination. The contaminated volumes which exceed the limit, and their average contaminant concentration, are reported for each of the contaminants in Table 11.2.

Contaminated volumes report			
Geology : ALL Code : 0			
Contaminant	Level C standard (micro-g/g)	Volume exceeding level C (M3)	Volume average (micro-g/g)
Benzo(A)pyrene	10	61,600	99
Total PAH	200	58,300	1864
Cyanide	500	12,500	898
Copper	500	16,200	1912
Lead	1500	7,300	2467

Table 11.2   Contaminated volumes report for each of the contaminants studied

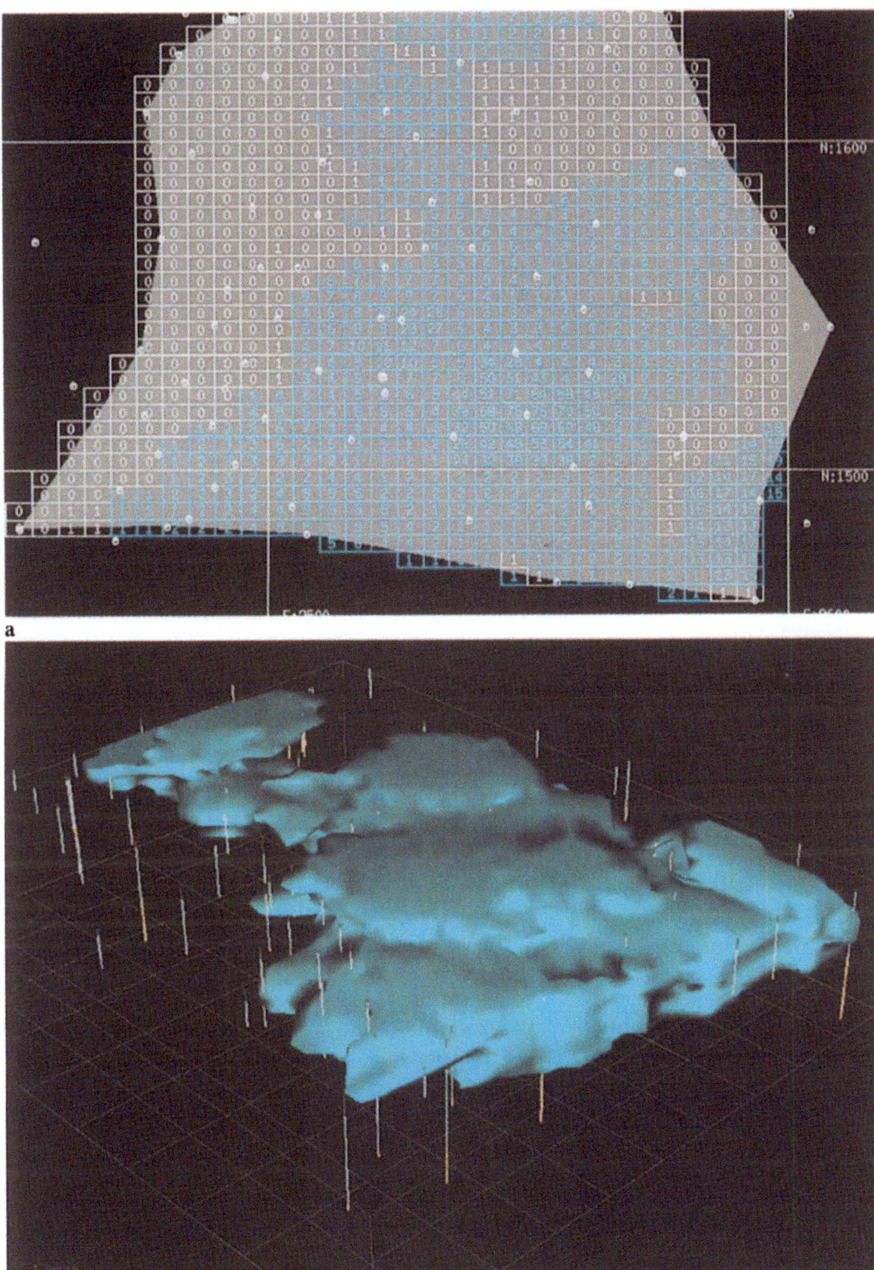

Fig. 11.8 **a** Variation of "total" contamination on a plan section through the industrial fill layer, shown *blue* above a regulatory limit ratio of one; **b** 3D visualization of the "total" contaminated volume, with borehole and well locations superimposed

For presentation purposes the volume of benzo(A)pyrene contamination is shown superimposed on a cut-away 3D perspective view of site geology in Fig. 11.7. Borehole and well locations are also shown.

These results are obtained by independent analyses of the spatial variations of each of the contaminants. The resulting volumes have a significant degree of spatial overlap, i.e. some soil volumes contain contamination from several sources. However, the remediation technique for much of the site involves excavation and removal. We therefore need to know the *total contaminated volume* from any source. To achieve this we use grid model manipulation techniques (see Chap. 9), firstly to *normalize* the predicted values by dividing them by the appropriate regulatory limit for each contaminant, and secondly to compare the normalized values of the five contaminants for each cell and extract the maximum. The result represents the spatial variation of a new variable which reflects the maximum degree of contamination from any source.

Contaminated volumes report			
Geology : ALL **Code : 0**			
Contaminant	**Level C ratio**	**Volume exceeding level C (M3)**	**Volume average ratio**
Total	1.0	73,800	9.6

Table 11.3   Contaminated volumes report for total contamination from any source, expressed as level C ratios

We now repeat the volumetric analysis, this time for an isosurface value of one since the results are normalized with respect to the regulatory limits. The spatial variation of the new variable is illustrated on a plan section through the site in Fig. 11.8a. The predicted values cannot be read at this scale, however, the contaminated area on the section is identified by highlighting the values wherever they exceed a value of one. The final analytical results of contaminated volumes, reported in Table 11.3 and illustrated in Fig. 11.8b, can be used for remediation planning purposes. In this case, the *minimum* excavation which encompasses the *total* contaminated volume is required. A significant feature of the result of our assessment is that the plan extent of the contaminated volume varies significantly with depth. Considerable savings in remediation costs are achieved by designing an excavation that accounts for this variation.

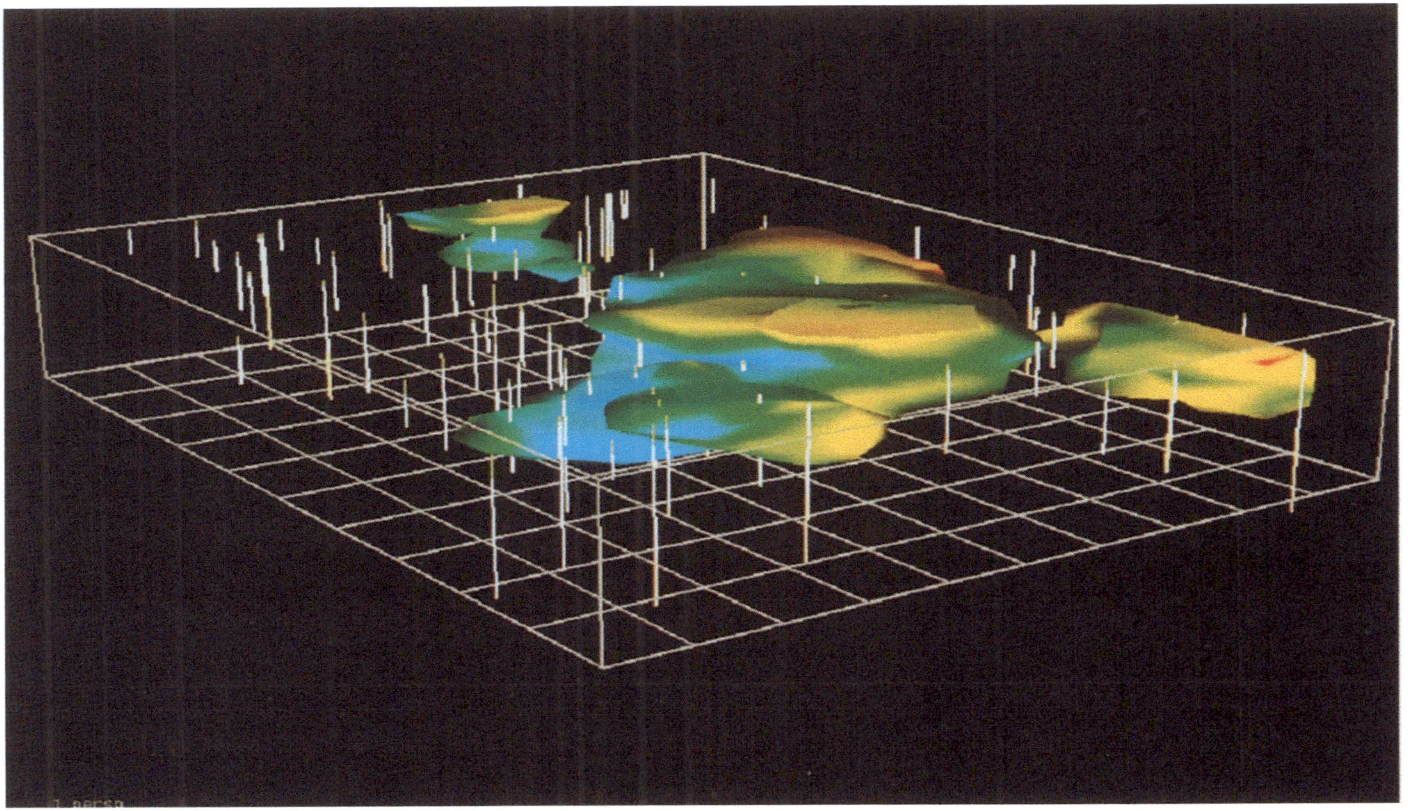

Fig. 11.9   3D visualization of a contaminated volume with a color-coded mapping of prediction uncertainty overlaid on the action level isosurface for the contaminant

11.9 Remediation Risk Assessment

The volumes determined above are the *probable*, or *most likely*, contaminated volumes, based on the available sample information and appropriate prediction models. However, there is obviously a risk that some unacceptable contamination might remain following remediation. Geostatistical kriging provides an avenue for quantifying this risk and equating it to the cost of remediation. The vehicle for risk assessment is the geostatistical measure of uncertainty, the *standard error* (or standard deviation), that is stored with the predicted contaminant variations in the 3D grid data structure. Provided that our semi-variogram prediction models are representative of the measured spatial variability of the samples, then we are justified in using the measure of uncertainty in a risk assessment context, as discussed in Chapter 10.

Figure 11.9 shows a color-coded mapping of the spatial variation of uncertainty onto the isosurface that defines a contaminated volume. Using the grid manipulation techniques discussed earlier, we add a multiple of the uncertainty values to the predicted contaminant values. We then repeat our volumetrics analyses for the regulatory limit isosurface to obtain a *worst case* scenario of soil contamination. For example, in the case of benzo(A)pyrene concentrations the resulting contaminated volume is effectively doubled by adding one standard error to the predicted values (cf. Fig. 11.10). However, the risk of leaving unacceptable concentrations of benzo(A)pyrene in the soil is reduced to 16%. Further, if we add a multiple of two standard deviations, then the risk reduces to 2%. By comparing the contaminated volumes for each risk scenario, we effectively relate risk to remediation cost, and integrate the concept of risk into our remediation planning.

11.10 Project Summary

This completes our discussion of subsurface soil contamination. In summary, the integrated approach to computerized characterization provides a logical and justifiable approach to *qualification* and *quantification* of all of the aspects of subsurface contamination considered in the study. The technical features of the study can be summarized as follows:

· Geostatistical prediction of contamination is a significant improvement on conventional methods in that (1) it is three-dimensional, (2) it employs prediction models based on actual sample variability, and (3) it is spatially constrained by the geology of the site.

· The subsequent determination of contaminated volumes is significantly more precise than conventional methods. In the study, the final volume requiring

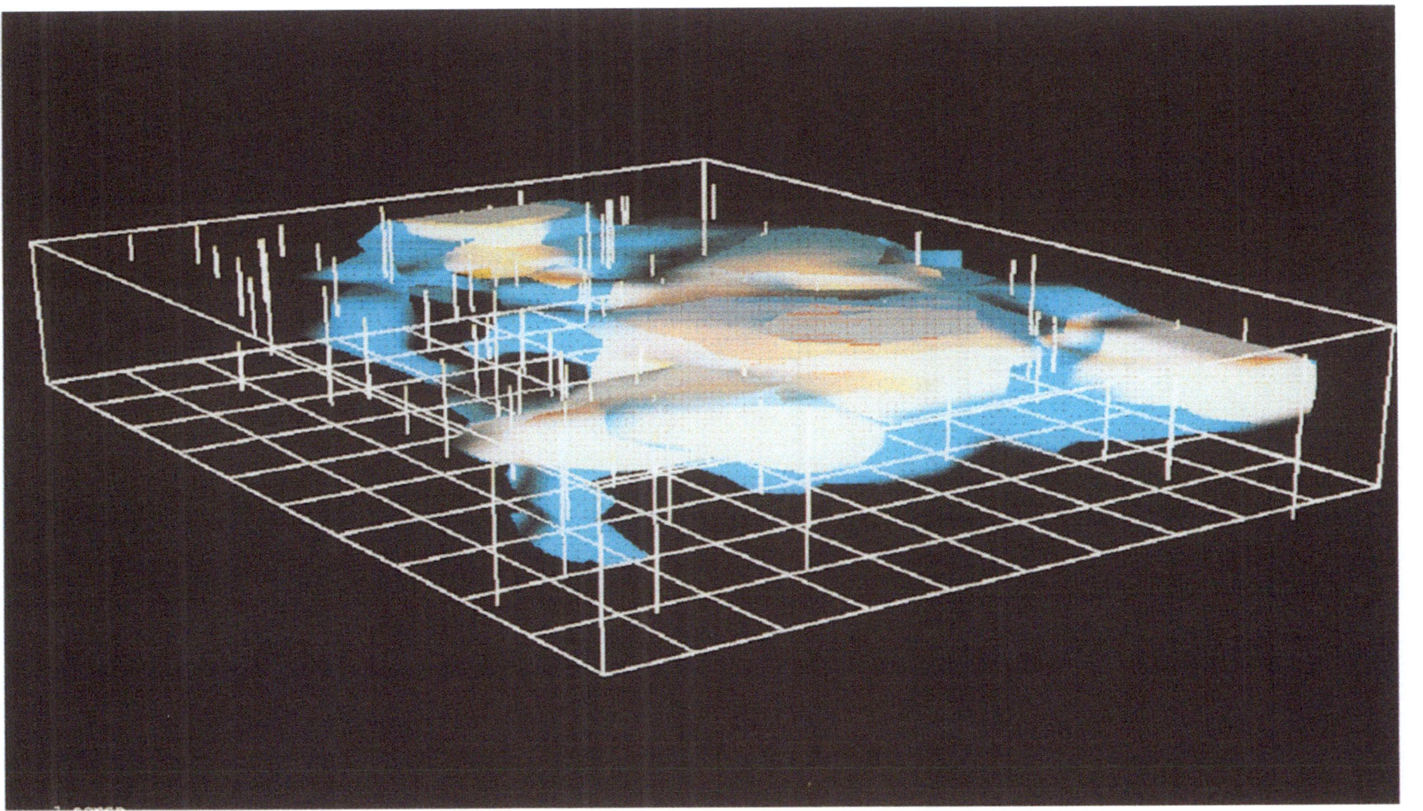

Fig. 11.10   The worst case contaminated volume (semi-transparent) overlaid on the most likely (predicted) contaminated volume; the former is obtained by manipulation of the spatial variation of uncertainty with that of the predicted values

remediation is approximately 30% less than that determined by a parallel, conventional study.

· Geostatistical kriging provides an avenue for quantifying uncertainty and relating risk to remediation cost. Kriging also provides a logical approach to controlling sampling and investigation costs.

· All information either used or generated in the study is precisely located in three-dimensional, coordinated space and is therefore immediately available and relevant to the remediation planning process.

The principal procedural advantage of this approach to contaminated site characterization is the efficiency and rapidity with which meaningful results can be obtained. Once the investigative source information is available in suitable data structures, the entire characterization and assessment are completed within a 1-week period. This highlights the advantages of using an integrated approach to the characterization process. A secondary benefit is provided by use of the enhanced visualization capabilities as the primary communication medium for nontechnical audiences, in this case, property owners, investors and the general public.

12 Hazardous Waste Site Characterization and Underground Repository Design

12.1 Project Overview

This study involves a regional-scale geological characterization for a proposed underground high-level radioactive waste repository at Yucca Mountain, Nevada. Investigation of the site by the US Department of Energy, US Geological Survey and various contractors has been in progress for more than a decade, resulting in large volumes of information from a variety of investigative sources. Since this is the first time that geoscience modeling technology is to be applied to a regional characterization of this scale and level of detail, there is a requirement to evaluate the technology with respect to project requirements. Our study is performed solely for the purpose of demonstrating the suitability of the integrated computerized approach for this task. It covers only part of the repository site and is based on a limited selection of the available public-domain information and published documentation.

Our immediate objective is to demonstrate how computer-assisted characterization can be used to develop a detailed 3D interpretation and model of site geology that includes stratigraphy, lithology and structure. Longer term project objectives include using similar models to control prediction of the spatial variations of geomechanical and hydrogeological properties within the different strata, leading to engineering design of the proposed pilot repository and, ultimately, to final evaluation of the suitability of the repository site.

Public concern regarding the environmental risk associated with the proposed repository has generated corresponding concern within the project regarding the precision of interpretation and modeling, and the uncertainty (or confidence) attached to the results. Public and regulatory scrutiny of the project have generated a requirement for dissemination of information; visualization and presentation capabilities are therefore primary considerations. Our study emphasizes the capabilities of the integrated approach with respect to all of these requirements and includes an attempt to model the spatial variation of uncertainty associated with geological interpretation.

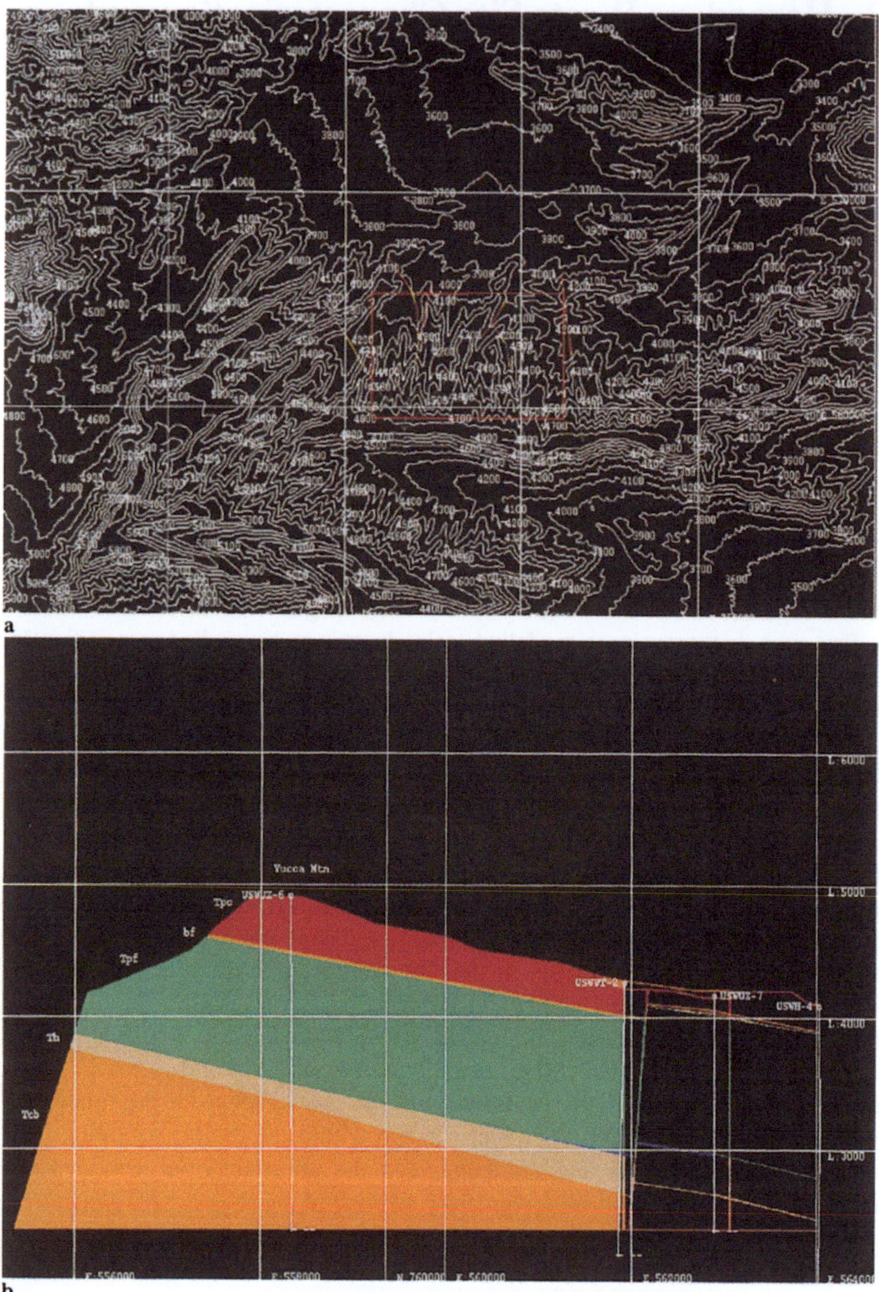

Fig. 12.1 **a** Regional contour map of site topography with the rectangular extent of the study area (*red*) superimposed at lower center (north to left); **b** preliminary geological interpretation of stratigraphy on a vertical section between boreholes USWUZ-6 and USWWT-2; off-section geology is shown in outline only

12.2 Site Geology and Source Information

Yucca Mountain is underlain by a sequence of silicic volcanic rocks, varying from 3000 to 10 000 ft in thickness with a dip (inclination relative to horizontal) between 5° and 10° to the east. These rocks consist mainly of welded and nonwelded ash-flow and ash-fall tuffs resulting from volcanic activity. They are underlain by deep-seated granitic basement rock. The ash-flow units are generally dense, welded material with widely varying degrees of fracturing; whereas the ash-fall units consist of nonwelded, bedded tuff that is considerably more porous but generally containing fewer fractures. The variation of geomechanical, thermal and hydraulic properties varies widely between the different geological units. The proposed repository is to be sited in the Topopah Springs (Tpf) member (cf. Fig. 12.1b), an ash-flow unit comprising moderate to densely welded, devitrified tuff. It is the most extensive unit in the region, with a thickness of approximately 1100 ft at the repository location, thinning towards the south. The water table at the proposed site is deep, approximately 2500 ft below ground surface and 700 ft below the proposed repository elevation.

The development of geological structure in the region has been complex. Two overlapping phases of tectonic activity have been identified; older extensional faulting associated with silicic volcanism, and more recent basin-and-range faulting. The latter caused the crust to fragment into basins and ranges oriented along northerly trends. Yucca Mountain is a series of north-trending structural blocks that have been tilted eastward along steeply west-dipping faults and transected by local faulting. The proposed repository is located in a relatively unfaulted region of one of the larger structural blocks. Vertical fault displacements in the region vary between 15 to 100 ft in the vicinity of the repository location, to 700 ft towards the south. Precise location of faults is of particularly importance to the characterization since there is some evidence of relatively recent seismic-induced movement.

The information available as the starting point for our study includes contour mapping of surface topography, fault mapping, borehole logs of lithological intersections, and preliminary geological interpretations on vertical sections between boreholes. The regional stratigraphy of the area is well established and this knowledge is applied to the interpretation process. The density of borehole exploration in the region is sparse, due to public concerns and resulting restrictions on exploration. The principal characteristic of our characterization is *lithology*, in the form of both borehole intersections and preliminary sectional interpretations. The variable data, including samples of geomechanical properties, hydraulic parameters and thermal properties, that will ultimately be required for a complete characterization is unavailable at the time.

The topography of the study area is clipped from a regional contour map of topography which has previously been entered from a digitizing table and stored as a map data structure (cf. Fig. 12.1a). Fault mapping in the vicinity of the study area is also extracted from regional maps and stored as map data (cf. Fig. 12.2b).

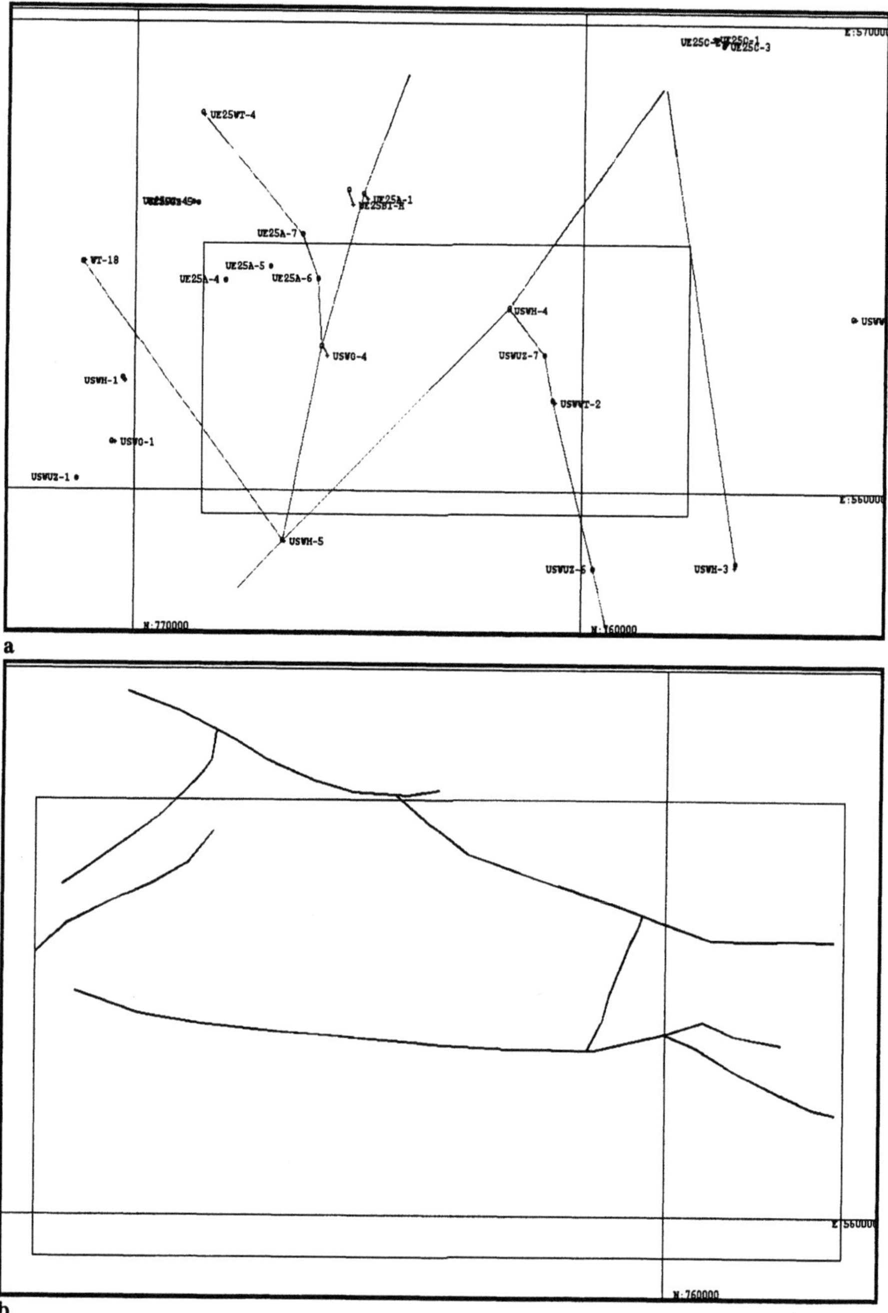

Fig. 12.2 **a** Plan section of borehole locations and preliminary geological interpretations on vertical sections in the vicinity of the study area (north to left); **b** mapping of primary fault traces in the vicinity of the study area

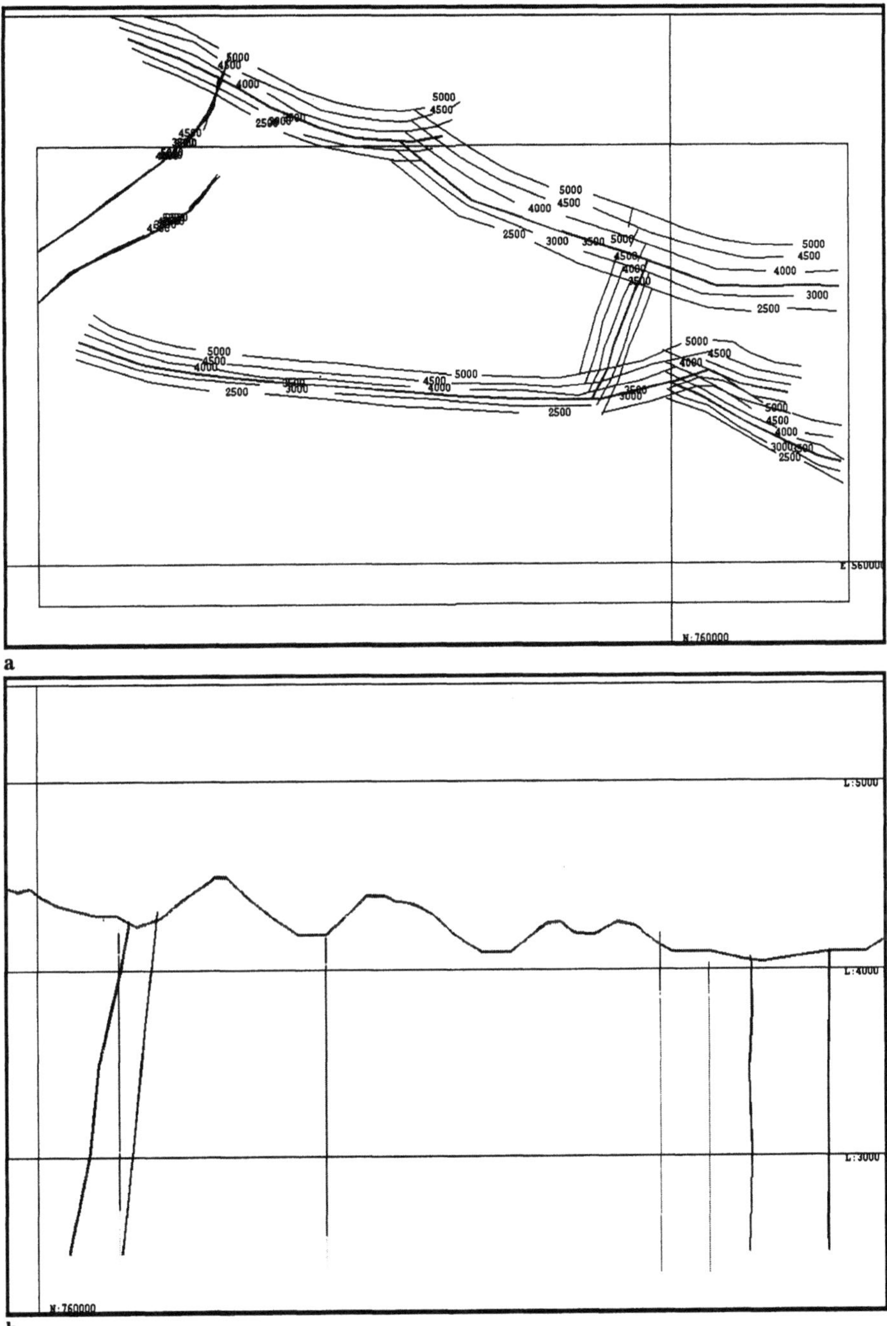

Fig. 12.3 **a** Plan view of contours used for generation of inclined fault surfaces; **b** vertical viewplane section trending southwest/northeast through the study area which intersects topography, fault surfaces and preliminary geological interpretations

The fault traces shown in the figure are projected to an elevation of 3500 ft. We also know the approximate dip angles (vertical inclinations) and displacements of the faults from surface investigation and borehole intersections. Borehole logs are entered and stored in the hole data structure. The preliminary geological interpretations between boreholes, resulting from manual interpretations on paper sections, are digitized on vertical viewplanes and stored as volume model fence sections of nominal thickness (cf. Figs. 12.2a and 12.4).

The proposed underground geometry for a pilot repository facility and a ramp access design are defined using the volume modeling excavation design tools and stored as volume data structures (see Fig. 12.10a).

12.3 Information Review and Evaluation

The topography, fault structure, regional stratigraphy and borehole intersections are established; these form the basis of our geological characterization. The primary objective is to interpret the vertical and horizontal extent of each of the geological members throughout the study area, resulting in a full three-dimensional interpretation of geology and fault structures. The preliminary sectional interpretations between boreholes are used as a guide only and are modified as inconsistencies become apparent during detailed interpretation.

The first stage of interpretation involves development of surfaces which represent topography and fault structure. This establishes a structural framework which can be used to apply spatial control to the interpretation of stratigraphy. The second stage involves interpretation of the individual geological members within this framework, in accordance with the established stratigraphic sequence and the recorded borehole intersections. The primary purpose of information review and evaluation in this case is to establish interpretation and modeling procedures which are appropriate to the available information and the objectives of our characterization.

12.4 Representation of Geological Structure

We model the topographical surface using the surface-based approach discussed in Chapter 7, first generating a triangulated network that represents the topology of the surface. This network honors the topology by ensuring that none of the triangles cross any of the contours. We then assign a nominal thickness to the surface and generate a set of prism shaped volume model components which represent topography. These operations are performed from a plan (horizontal) viewplane that is approximately parallel to the surface.

Fig. 12.4   Preliminary geological interpretations on fence sections between boreholes, with irregular fault surface representations superimposed

To model the fault discontinuities we first generate a set of contours for each fault which correspond to the mapped fault trace and the observed inclinations (cf. Fig. 12.3a). We then employ a similar process to that used for topography, orienting our viewplane approximately parallel to each fault surface. This is accomplished by starting from the plan viewplane shown in Fig. 12.3a, interactively cutting a vertical viewplane section which is approximately parallel to the strike (plan orientation) of the fault, and then inclining the viewplane parallel to the dip of the fault. The triangulated network that represents the topology of the fault is truncated by the topographical surface prior to generation of volume model components, thus ensuring that the upper extremities of the fault surface match topography (cf. Fig. 12.3b).

These surfaces are now available for spatial reference and control of our interpretation of stratigraphy. For example, a viewplane at any orientation through the study area displays intersections with topography and fault surfaces, as well as intersections with the preliminary geological interpretations (Fig. 12.5). Alternatively, we can superimpose the fault surfaces on the geological interpretations in a three-dimensional visualization (Fig. 12.4). From the latter it is apparent that the preliminary geological interpretations do not correlate exactly with the fault representations in some areas.

12.5 Interpretation of Geology

The relatively thick Tpf ash-flow member that hosts the proposed repository is the first to be interpreted, using the interactive sectional approach discussed in Chapter 6. Initially, two-dimensional interpretations are interactively defined on a series of north/south (east-facing) vertical sections (cf. Fig. 12.5a). These interpretations are parallel to the direction of greatest continuity in structure and stratigraphy, i.e. approximately parallel to the primary basin-and-range fault structure. This new information is added to the background display and used to enhance the final interpretation of the member on a set of transverse east/west (north facing) vertical sections.

Working within one fault block at a time, these secondary interpretations are extended into three-dimensional volume model components and interactively matched on their common planes with each other and the control references (fault surfaces and intersecting interpretations), using the sectional interpretation approach to volume modeling discussed in Chapter 6 (cf. Fig. 12.5b). The result is an almost complete three-dimensional interpretation of the member within each fault block, consisting of sets of vertical, north-facing, slab-like components whose boundaries match the fault surfaces (Fig. 12.6a). However, because this work is performed on parallel east/west sections, some voids remain at the northern and southern extremities adjacent to the faults, particularly where the inclined fault surfaces intersect. These voids in the interpretation are filled with additional

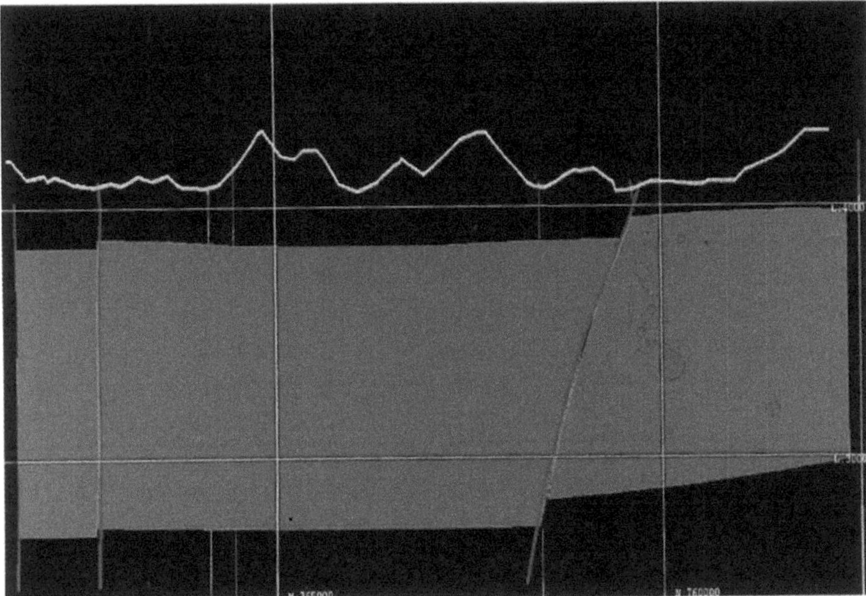

a

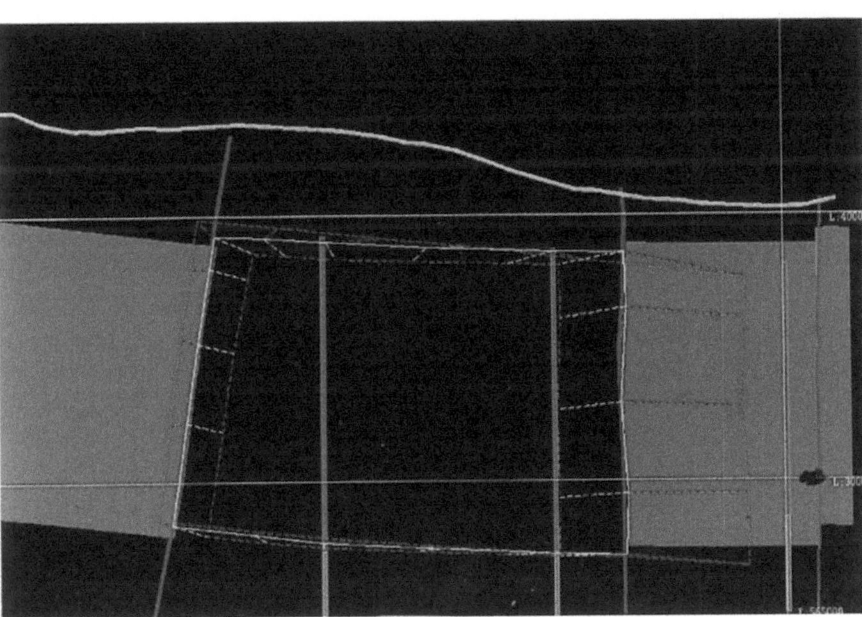

b

Fig. 12.5 **a** Initial interpretation of the Tpf member on one of a set of vertical north/south section through the study area; **b** interactive interpretation on a transverse (east/west) section of the boundaries of a Tpf member volume model component within a fault block

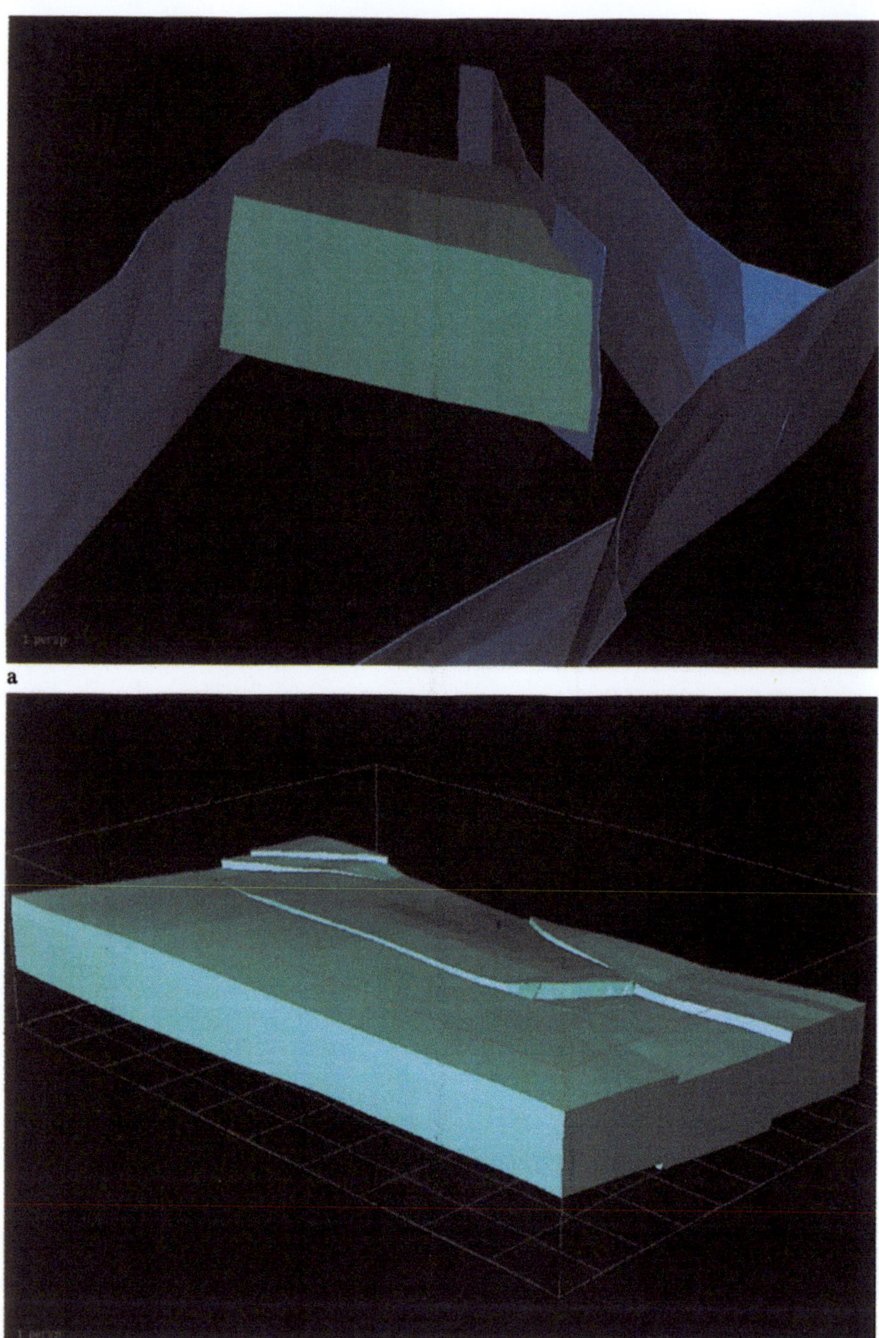

Fig. 12.6   **a** 3D visualization of the Tpf member component interpreted in Fig. 12.5b, bounded by fault surfaces; **b** final volume model representation of the Tpf member comprising individual interpretations within each fault block

appropriately shaped and oriented components that are defined from inclined viewplanes. The final result is a geometrically accurate volume model representation of our interpretation of the Tpf member within each fault block (Fig. 12.6b). The set of components which represent the member is finally assigned a code which identifies their geological type.

In the next step we model the geology above the Tpf member. For demonstration purposes these members are composited into one. No interpretation is required since the upper extent of the composite member is fully defined by topography and the lower extent by the upper surface of our interpretation of the Tpf member. A volume model representation can therefore be obtained by computer generation of prism-shaped components between the two surfaces, using the surface-based approach to modeling discussed in Chapter 7. However, this is complicated by the requirement for the lower extent of the member to honor the fault discontinuities of the Tpf member.

To overcome this we again have to work within one fault block at a time, first generating a set of contours which represent the top of our Tpf member interpretation (cf. Fig. 12.7a); then triangulating both surfaces and clipping them to the outline of the fault block (Fig. 12.7b); and finally generating a volume model representation between the two surfaces (Fig. 12.8a). As before, all of these components are assigned an appropriate geological type code. The final representation of the irregular volume of the upper members honors both topography and the discontinuities of the top of the Tpf member (Fig. 12.9). In the final stage, the member immediately underlying the Tpf member is interpreted interactively in similar fashion to the Tpf member itself, using the sectional approach.

This represents the extent of geological interpretation and modeling for our site characterization study. The end result is a "three-dimensional map" of geology with stratigraphy and structure represented as irregular volumes and surfaces. The model can be viewed in perspective, with cut-aways to expose the interior, or on any arbitrarily oriented two-dimensional section through the site (Fig. 12.9). The geometry of the geological structure and stratigraphy is fully defined in terms of a "real-world" coordinate system, and geological volumes are characterized by their geological type codes. The model thus represents a qualification of subsurface space by geological type. Any point or subvolume within the model can now be interrogated as to its geological type. This feature will later be used to constrain the spatial prediction of variables throughout the volume of each geological member. The model also provides a platform for precise design of the repository excavations, since the intersection of any excavation geometry with the geological members can readily be reported or visualized.

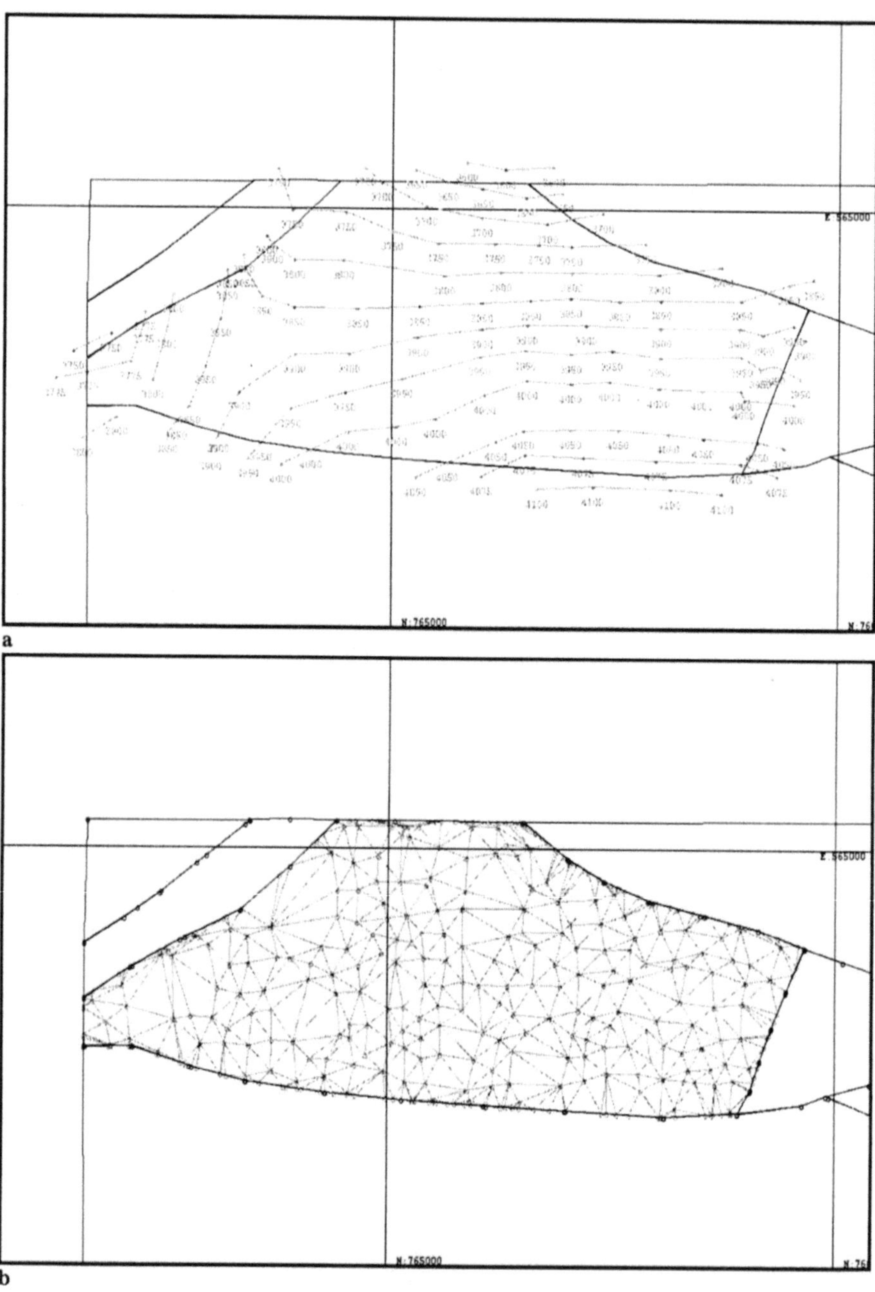

Fig. 12.7   **a** Generation of contours that define the top of the Tpf member within a fault block;   **b** clipped triangle set for generating volume model components of the upper composite member within a fault block

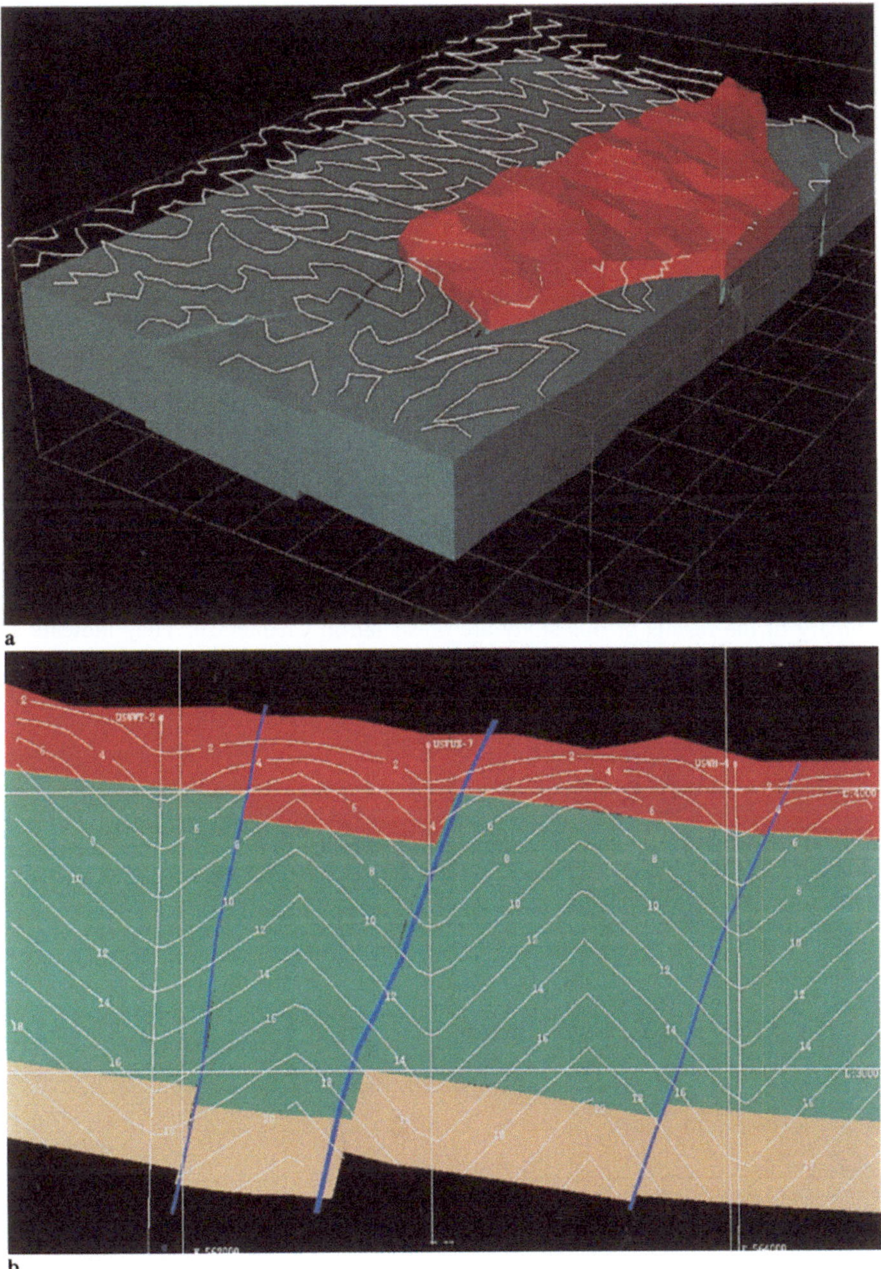

Fig. 12.8 **a** Volume model representation of the upper composite member within a single fault block, generated between surface topography and the top of the Tpf member interpretation; **b** vertical viewplane section through the final geological model, showing contours of possible locational error based on potential borehole errors and distances from borehole traces and the topographical surface

12.6 Modeling Geological Uncertainty

Largely as a result of public concern surrounding the proposed repository, there is a requirement to qualify all information in terms of its associated uncertainty (or precision). This requirement naturally extends to any interpretation and modeling of the site geology. No fixed procedure or convention has yet been approved for this. The following discussion represents an attempt to quantify the *potential geological uncertainty* associated with a site characterization of this nature.

Any attempt to represent geological uncertainty essentially requires us to quantify the *sources* of uncertainty beforehand. These are associated, firstly, with the available information and, secondly, with our interpretation of it. The former is easier to approach on a logical basis. The uncertainty in the observed location of a borehole intersection with a geological member can be based on a knowledge of the errors inherent to borehole survey measurements and the depth of the intersection. Attempts to quantify this type of uncertainty have been made in the mining sector, by comparing observed and interpreted locations of geological features with the actual locations determined during mining excavation. A summary of the results of this study are presented in Chapter 10. They indicate a standard error in locational precision (in any direction) of the order of depth/100 for features close to a borehole trace, and double this for features between borehole sections. The first component refers to potential borehole measurement error, the second to potential interpretational error. However, the magnitude of the second component is largely dependent on geological complexity, which can only be estimated, as well as our interpretational expertise, intuition and experience. Its quantification is a largely subjective exercise. In the case of Yucca Mountain, although geological structure is moderately complex, the stratigraphy is known to be a relatively simple layered formation. We therefore have some justification in assuming a similar standard error to distance ratio of 1:100 for locations away from borehole traces.

This discussion provides some justification for attempting to predict the spatial distribution of possible error in location based on borehole information and its interpretation. In this case, considerable surface mapping of geology has also occurred; for demonstration purposes, we assign a possible locational error of 0.5 ft to points in the topographical surface. Figure 12.8b illustrates application of this approach to a vertical section in the study area that includes three exploratory boreholes. The contours of predicted locational error indicate a maximum possible error of approximately 25 ft, at a depth of 1500 ft between boreholes approximately the same distance apart. In other words, the precision with which we can predict or interpret the location of a particular geological feature at this location is 25 ft in any direction.

This is probably larger than most of us would anticipate and is based to a large extent on supposition. However, we do well to bear in mind the potential magnitude of the uncertainties associated with our interpretations and models of geology.

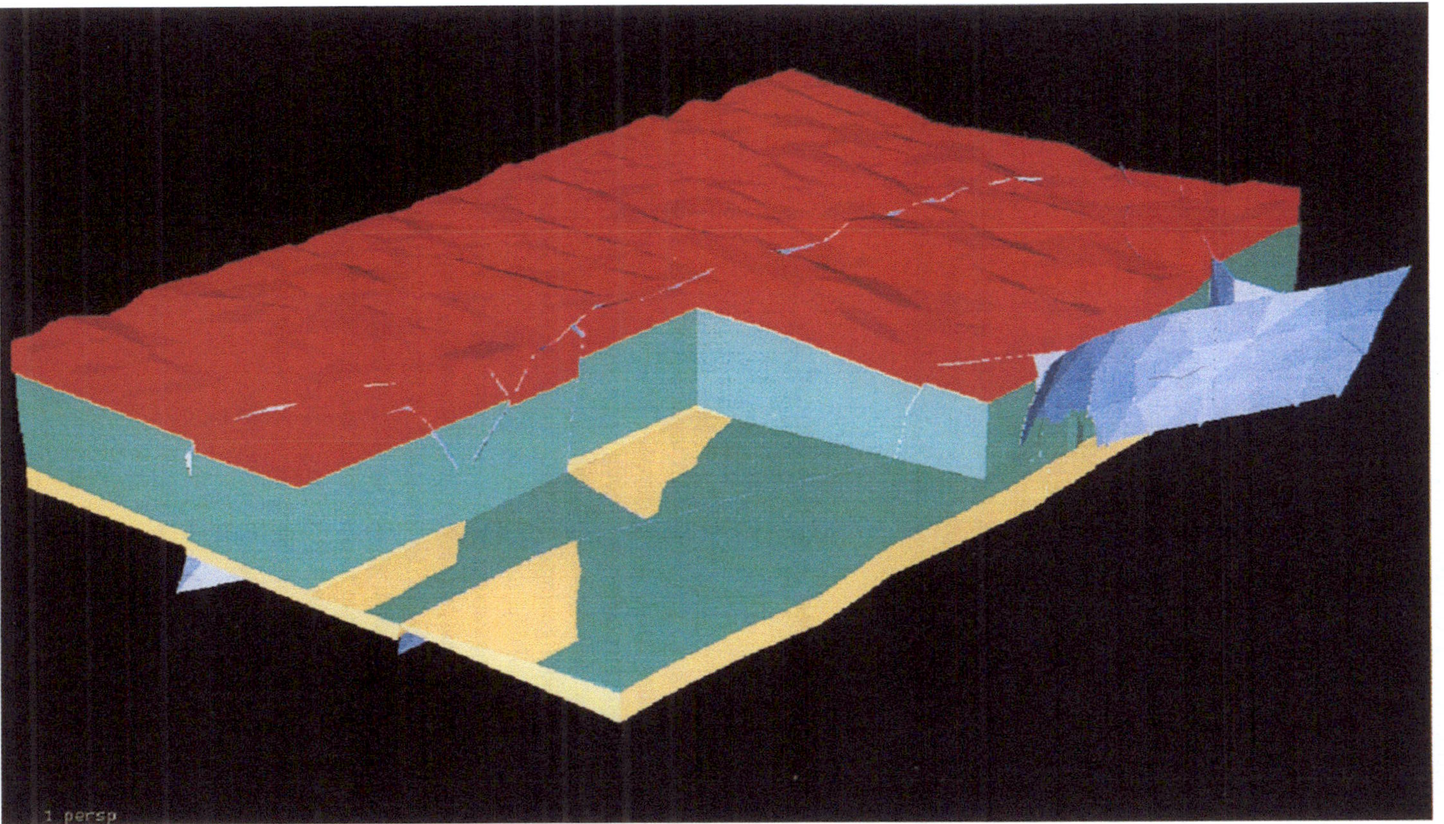

Fig. 12.9   Cut-away view of the final volume model representation of site geology and fault structures; Tpf member is shown green

12.7 Prediction of Geomechanical Properties

The next stage involves prediction of the spatial variation of geomechanical properties, such as fracture density and hardness, and other properties such as porosity, hydraulic conductivity and thermal conductivity. These predictions will be based on borehole samples, geostatistical techniques and spatial control provided by the geological model. The planned prediction technique is *conditional simulation*, a new, stochastic approach to spatial prediction discussed briefly in Chapter 16. Although the prediction technique is different to the geostatistical procedures discussed in Chapter 8, the requirements for geological control and sample selection control are identical. These results will provide a basis for final evaluation of the suitability of the site for an underground repository.

At the time of writing, detailed geological interpretation is still ongoing and the necessary test results on borehole samples of rock properties are not yet available in the public domain. At this time, there are unfortunately no results available for demonstration of the variable prediction stage of the characterization process.

12.8 Engineering Design of the Repository

Figure 12.10a shows a volume model representation of a preliminary conceptualization of the proposed pilot underground facility, with vertical access and ventilation shafts and access ramps. These are defined with the interactive excavation design tools discussed briefly in Chapter 9. A benefit of using the same volume modeling techniques, and the same real-world coordinate system, for representation of both geology and excavation geometry is that we can take full advantage of the available volumetric analysis capabilities. Reports of volumetric intersections between excavations and different lithologies are readily generated by applying the spatial analysis techniques discussed in Chapter 9. Such results can also be reported visually, as demonstrated in Chapter 13. Once a 3D grid model representation of predicted geomechanical properties is available, then similar volumetric intersections can be performed with it as well; these will highlight areas of concern in the underground repository design.

Detailed design of the pilot repository is ongoing at the time of writing, based to a large extent on the techniques discussed here.

12.9 Project Summary

This completes our discussion of site characterization for the proposed high-level waste repository. In summary, computer-assisted geological characterization has

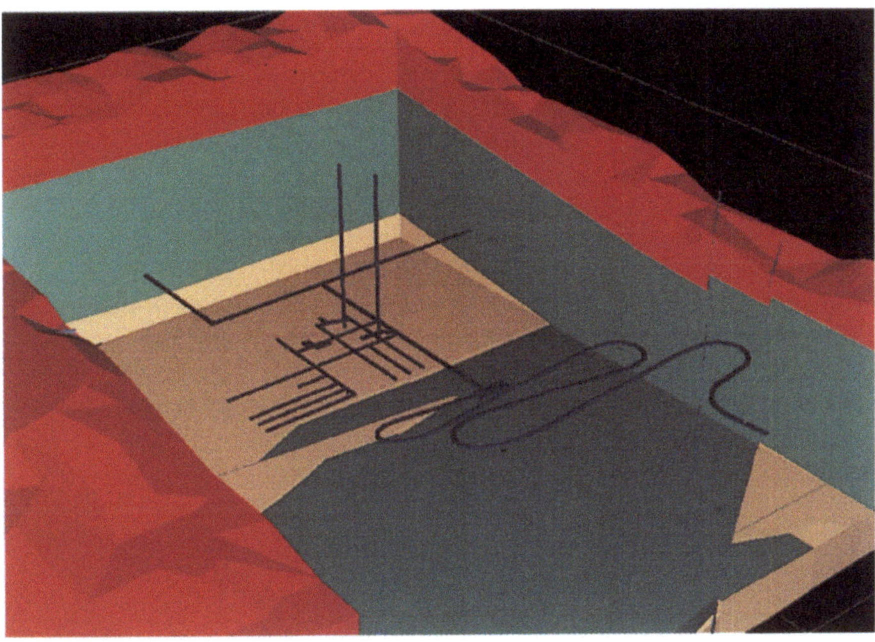

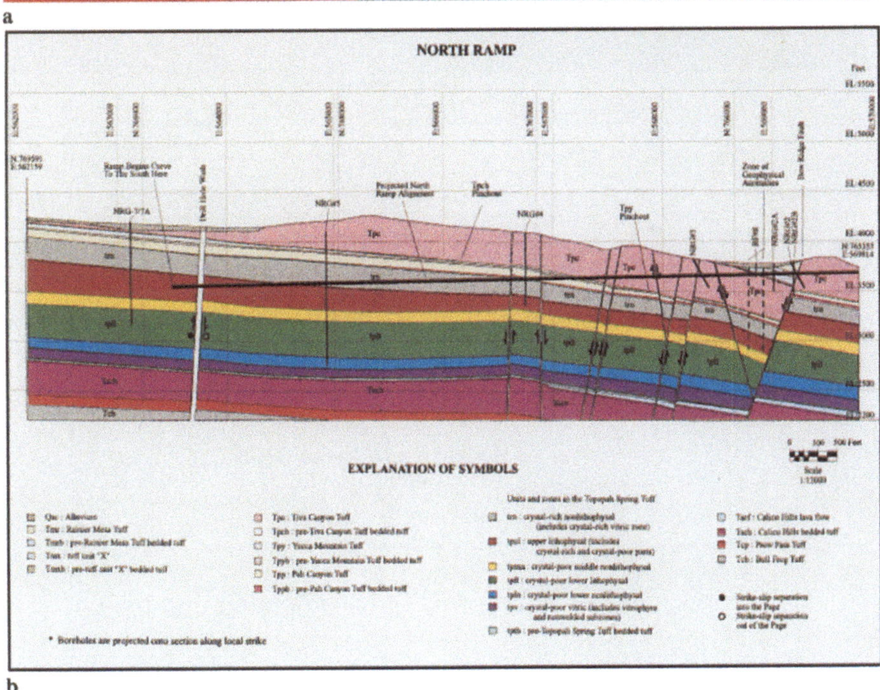

Fig. 12.10 **a** Volume model representation of a conceptual underground pilot repository, vertical shafts and access ramps, superimposed on a cut-away view of site geology; **b** detailed vertical section of site geology generated with 3D volume modeling techniques, reproduced by permission of the US Geological Survey

been shown to provide a workable solution to both interpretation and representation of the geological structure and stratigraphy of the site. These results can subsequently be applied to prediction of geomechanical and other properties and engineering design of the repository.

Use of this approach by the US Geological Survey and other DOE contractors on the Yucca Mountain project has already paid dividends. The Geological Survey is currently involved in interpretation of a considerably more detailed model of geology, using the techniques outlined here (cf. Fig. 12.10b). The three-dimensional modeling and visualization capabilities have assisted in identifying and rectifying inconsistencies in the original fence-section interpretations of site geology. This confirms that extension of the characterization process into three dimensions provides enhanced interpretational capabilities as well as enhanced representation of complex conditions.

13 Ore Deposit Evaluation and Underground Mine Planning

13.1 Project Overview

Many ore deposit characterizations still utilize the traditional block (3D grid) modeling computer approach discussed in Chapter 1. For deposits with little or no geological control, such as disseminated porphyry deposits, the approximations involved are probably acceptable. For deposits with any degree of geological complexity, the approach is generally inadequate, for the reasons discussed earlier. The deposit considered here includes both structural and mineralogical complexity, and its characterization requires a high degree of geological control. This emphasizes the advantages of integrating the vector-based volume modeling approach for geology with the raster-based grid modeling representation of ore grades.

The project covers characterization of a metal ore deposit, its evaluation in terms of grades and tonnage, and a feasibility study of alternative underground mine designs. The potential mine site is located on the Black Sea coast in north eastern Turkey. The deposit consists of a high-grade copper/zinc massive sulfide ore with several mineralogical ore types. A previous evaluation which ignored the mineralogical influences has indicated geological reserves of approximately 11 million tons with mineral grades averaging 4.5% copper and 7.5% zinc.

The deposit is relatively shallow, extending from 50 m below surface to a maximum depth of approximately 250 m. Surface mining of the deposit is prohibited by several factors including ruggedness of the surface terrain and proximity to a river. Despite the high mineral grades, the feasibility of mining the deposit on a profitable basis is influenced by a number of other factors. Those with greatest impact include the geological and mineralogical complexity of the deposit itself, regions of instability in the surrounding host rock that influence underground mine design, and the financial constraints imposed on the project. The viability of the deposit as a potential mining project has already been proven. The objective of this study is to establish the most economical mining configuration. This requires that we perform a detailed characterization of the deposit, a precise volumetric evaluation in terms of grades and tonnage, and a

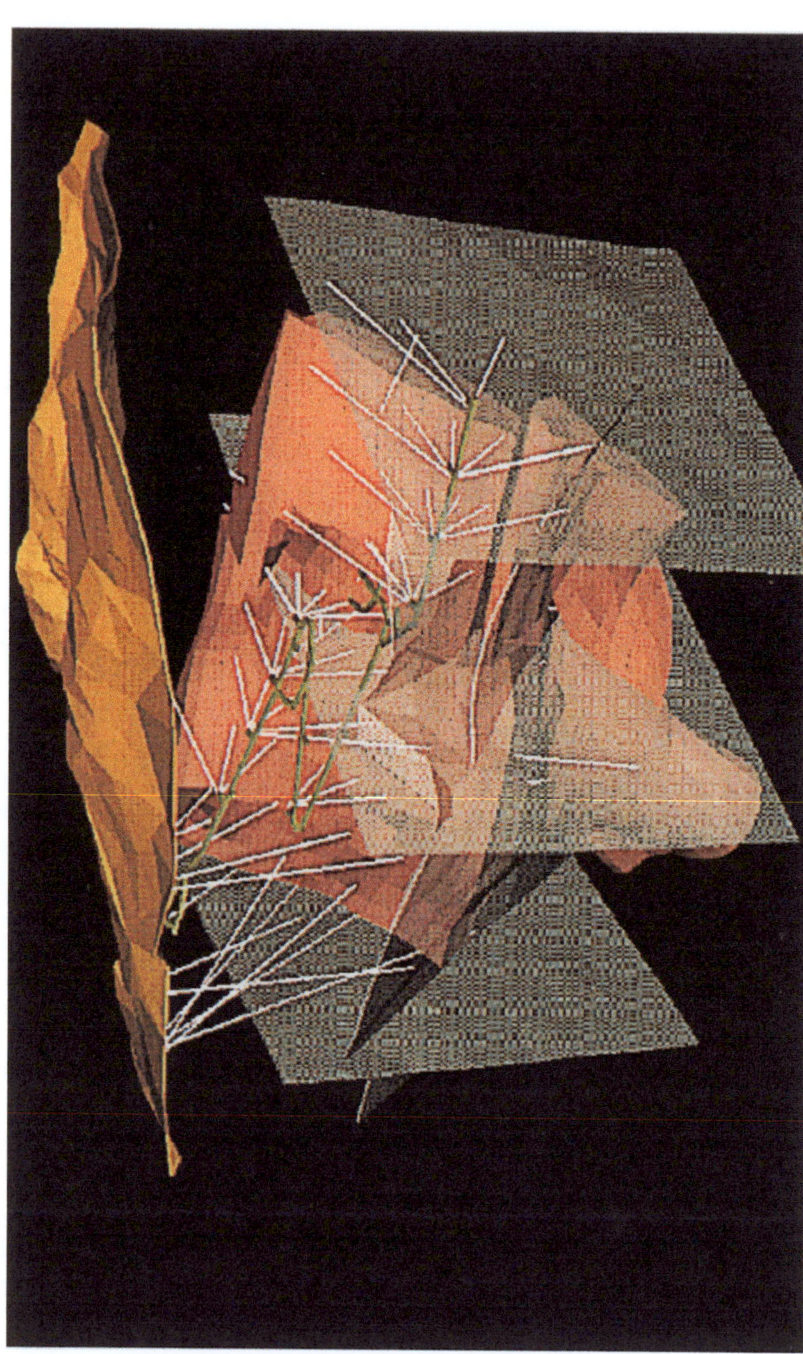

Fig. 13.1 View of the ore deposit model, with transverse and strike faulting (*semi-transparent*), topographical surface, exploration adit and borehole drilling traces superimposed

comparative analysis of two proposed alternatives for underground mine development.

The study itself is subject to strict time constraints and the period available is limited to 3 weeks. Use of the integrated computer approach to characterization is a significant factor in achieving the study objectives within this time frame.

13.2 Geology, Structure and Mineralogy of the Deposit

The deposit is thought to have originated as a submerged volcanogenic massive sulfide *black smoker* deposit on the sea bed. It was deposited on an uneven substrate of acid volcanic flows and pyroclastics that now form the *footwall rock* (beneath the deposit). The deposit was subsequently covered with interbedded ash and tuff layers and capped with a basalt layer. The entire structure was then uplifted, faulted and inclined by tectonic activity, resulting in a dip (vertical inclination) of 75° to the west and a strike (azimuth rotation) of 30° east of north (Figs. 13.1 and 13.2a).

The deposit is bounded and intersected by several distinct faults, including three transverse faults and a strike fault. The north and south transverse faults delineate the *mineable* region of the deposit that is the focus of our study. However the mine plan must accommodate eventual extension of mining activity into the north and south regions of the deposit as well. The strike fault parallels the deposit, and intersects it with an opposing dip of 70° to the east. It appears to have a *scissor displacement*, with indications that it developed contemporaneously with the massive sulfide deposit. The central transverse fault exhibits considerable horizontal and vertical displacement and effectively splits the mineable region of the deposit into two portions. The complexity of the deposit is further compounded by the presence of numerous *dikes* (basalt intrusions) in the southern region.

The contact between ore and waste is distinct with no gradation from high grade ore to waste rock in either the footwall or the hanging-wall (above the deposit). This feature greatly simplifies the mine planning requirement provided that these *hard* geological boundaries can be accurately represented in the computer for geological control of grade prediction.

Mineralogical studies indicate that the deposit is comprised of two primary ore types, characterized as *black ore* and *yellow ore*, each of which requires special treatment in the milling process. This characterization is based on the sphalerite content (respectively greater than or less than 10%) in the ore matrix. Each of the primary ore types is further divided into two subtypes, depending on the concentration and/or smelting treatment required in each case. In summary, the four subtypes of ore to be considered in the characterization include *clastic black ore* (CBO) and *massive black ore* (MBO), with higher and lower concentrations of gold, silver and trace elements, respectively; and *powdery yellow ore* (PYO) and

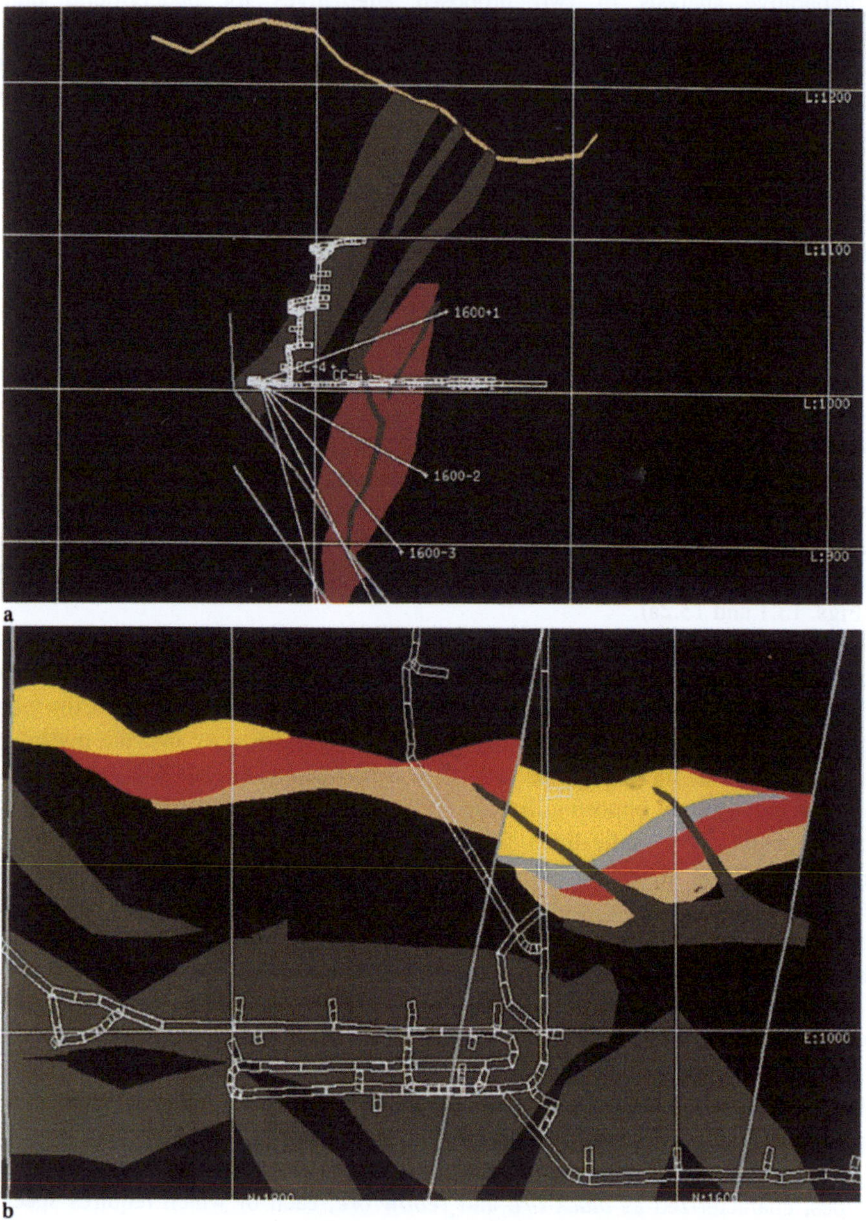

Fig. 13.2   **a** Transverse section through the volume model representations of topography, fault surfaces, ore deposit limits (*brown*), and hanging-wall basalts and dykes (*green*); borehole traces from an exploratory ramp are superimposed; **b** interpretation of mineralogical ore types on a longitudinal plan section showing ore types MBO (*red*), CBO (*light brown*), MYO (*gray*), PYO (*yellow*); dyke intersections (*green*) are at left; hanging wall basalts (*green*) are in the foreground; the exploration ramp is superimposed

massive yellow ore (MYO), distinguished by their rock mass qualities. The complex geometrical relationships of these ore types are illustrated in Fig. 13.2b.

The previous characterization of the deposit ignored the fault structure, the intrusive dikes, and the mineralogical influences within the deposit, resulting in considerable oversimplification. The complex mineralogy was reduced to a zinc-rich/zinc-poor characterization, with a consequent *smearing* of predicted grades across the boundaries of mineralization.

13.3 Project Information Base

The source information available to the study includes a contour map of site topography, exploration drilling logs and mineral assay (sample) results stored in a spreadsheet system, and preliminary, hand-drawn geological interpretations on vertical and horizontal sections. The topographic contours and the geological interpretations are entered via a digitizing table and stored as map data structures at appropriate orientations (see Chap. 4). The drilling logs are imported in ASCII file format and stored as hole data structures. They comprise the bulk of the source information.

Exploration drilling was performed mostly on a set of transverse sections at 40 m spacing. Results are available for both surface drilling and underground drilling from an exploration ramp within the hanging-wall of the deposit. The characteristic information in the hole data includes both lithological and mineralogical classifications, as discussed earlier. The primary variables include assays of copper and zinc ore grades, expressed as percentages of metal content by weight. Secondary variables include gold and silver grades and trace element concentrations.

13.4 Spatial Variability Analysis of Mineral Grade Samples

There is no requirement for statistical analysis in this study since the controlling characteristic (mineralogical ore type) for mineral grade prediction has already been established by previous studies. Preliminary semi-variogram analysis of the mineral assays indicates that ranges of influence of 70 m in the plane of the ore deposit and 35 m normal to its inclination are generally appropriate throughout the deposit (cf. Fig. 13.3). The time constraints of the study unfortunately preclude any detailed analysis of spatial variability and a decision is made to subsequently use a conventional prediction technique (rather than geostatistical kriging) for the mineral grades, with the established ranges of influence as the search volume parameters. This is justified by a judgment that precise geological control of

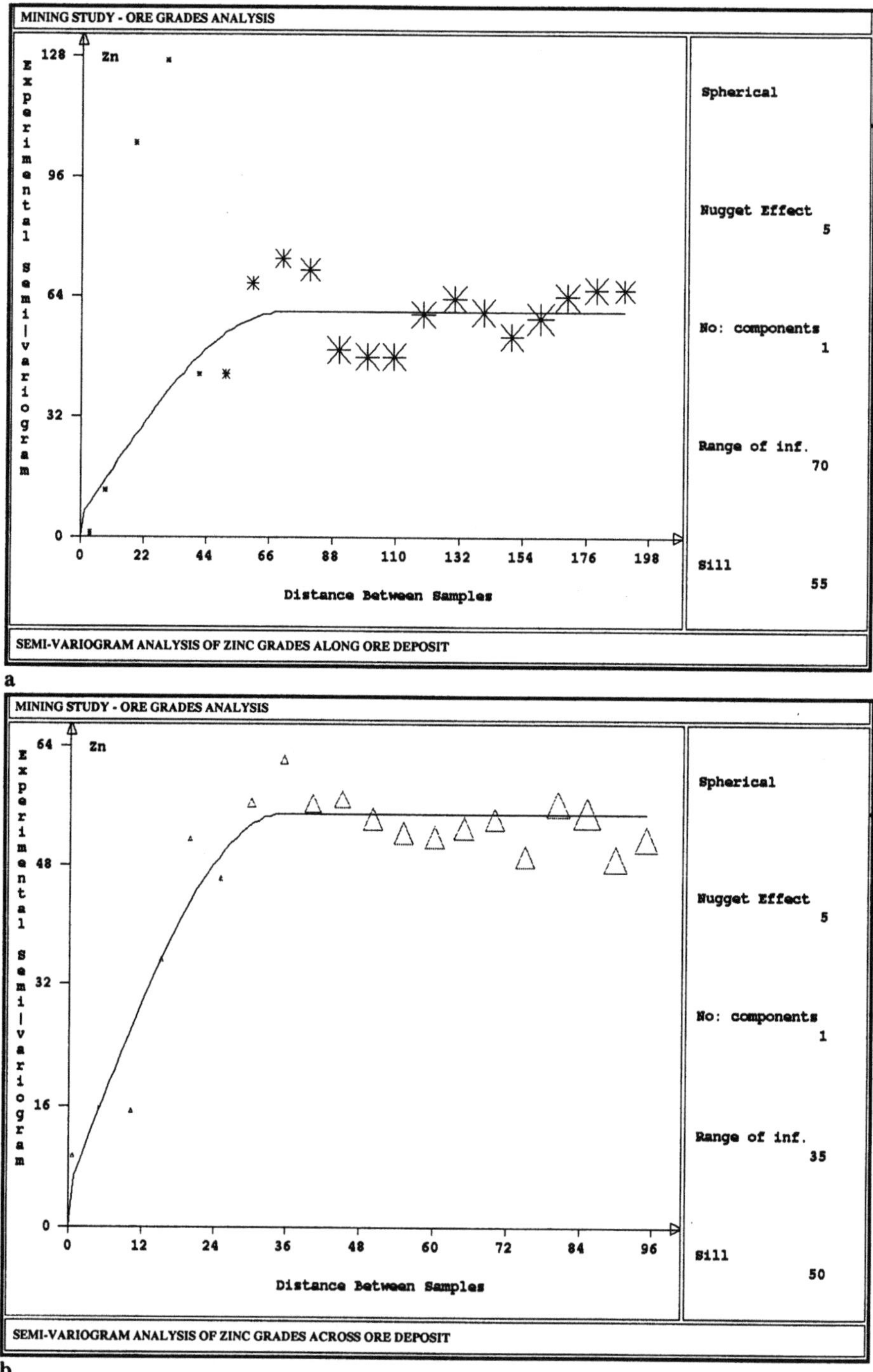

a

b

Fig. 13.3 Semi-variogram analysis of the spatial variability of zinc gradesa in the plane of the ore deposit (range = 70 m) and **b** normal to the deposit plane (range = 35 m)

prediction has a much greater influence (in this case) on mineral grade prediction than the choice of prediction technique itself.

13.5 Geological Interpretation

The digitized sectional interpretations are used as the starting point for a full 3D interpretation of the deposit. The interactive volume modeling approach described in Chapter 6 provides distinct advantages, notably in the reconciliation of vertical to horizontal interpretations and iterative reinterpolation of complex conditions in the vicinity of faults and dikes. It is worth noting that the necessity for reinterpolation is a frequent occurrence whenever interpretation progresses from a conventional 2D approach to an interactive 3D approach. This emphasizes the limitations and difficulties of attempting to interpret and represent complex, irregular volumes in a 2D medium.

The difficulties of interpreting complex geology can be significantly reduced by approaching the task in a systematic manner. The first step is to identify the *controlling features*, in this case the topography, the dikes, the faults, and finally the hanging-wall and footwall contacts of the deposit. The intrusive dikes are the most continuous feature in the region since they result from the most recent event in the geological evolution of the deposit. Once these controlling features have been interpreted and represented appropriately in the 3D context, then the task of interpreting the ore deposit and its mineralized ore zones is considerably simplified. If we proceed in this manner, then our interpretation sequence is essentially the reverse of the geological evolution of the deposit. Interpretation therefore proceeds in three stages. The first includes representation of topography, dikes and fault structures. The second involves interpretation of the deposit limits, defined by hanging-wall and footwall contacts. The final stage involves detailed interpretation of mineralogical ore types within the deposit.

The surface-based approach to volume modeling discussed in Chapter 7 is used to generate a triangulated representation of topography, based on the topographic contours stored as map data. The dikes are interpreted on horizontal sections through the deposit region and then extended and correlated in 3D using the sectional approach to volume modeling discussed in Chapter 6 (cf. Fig. 13.2). The surface-based approach is used again to generate triangulated representations of the fault surfaces, based on borehole intersections with the faults. These are reduced from fault observations stored under the lithology characteristic in the hole data structures.

The primary sources of information for interpretation of the deposit limits are the hole data observations of lithology and mineralogy characteristics. These are displayed in appropriate formats and interpreted initially on a set of transverse viewplane sections that coincide with the original exploration drilling sections. These viewplane sections also display intersections with topography, dikes and

Fig. 13.4 Detailed 3D interpretation of mineralogical ore types within the deposit; *dark green* intrusions in right foreground are basalt dikes

faults, thus providing a much-needed source of spatial control to our interactive interpretation. The hanging-wall and footwall contacts are interpreted first from observations of the lithology characteristic. Between these limits the zones of mineralization that define the four ore types are then interpreted from the mineralogy characteristic. The final interpretation of ore types is performed on a set of longitudinal plan sections at 20 m vertical spacing that intersect the initial transverse vertical interpretations (similar to Fig. 13.2b), resulting in iterative refinement of the latter. These sectional interpretations are then extended vertically and correlated in 3D to obtain the final interpretation, using the interactive volume modeling tools (see Fig. 13.4).

At this stage, our geological model includes full 3D representations of topography, dikes, faults and mineralized ore volumes. The final step involves interpretation of the basalt unit in the deposit hanging-wall (cf. Fig. 13.2a). This information is required for design input in the mine planning step.

13.6 Prediction of Mineral Grades

The first requirement is an appropriate 3D grid data structure in which to store the predicted grades. Grid-cell dimensions of 5 x 5 x 5 m are chosen throughout the study region, based on the average sample spacing of 40 m (between drilling sections) and the 30 m minimum range of influence of sample variability (cf. Fig. 4). The grid is aligned with the ore deposit in plan rotation and its overall dimensions encompass the mineable region of the deposit. This grid is intersected with the volume model representation of the four ore type volumes to provide the necessary information for mineralogical control of ore grade prediction.

For the reasons discussed earlier, the prediction technique in this case is the conventional inverse distance squared sample weighting technique (see Chap. 8). The sample search volume is a 65 x 65 x 30 m ellipsoid at the same inclination as the ore deposit with the short axis normal to the inclination. The mineral grades for each ore type are predicted from sample subpopulations associated with the appropriate mineralogical characteristic. The preliminary results include predicted spatial variations of copper, zinc, gold and silver for each ore type. These are combined by averaging the values at shared mineralogical boundaries by a volume-weighted compositing algorithm (see Chap. 9). The result is a spatial variation of grade for each mineral throughout the mineable region of the deposit that takes full account of the controlling mineralogical characteristic and its spatial discontinuity.

The principal disadvantage of using a conventional prediction technique is that no measure of the uncertainty associated with the predicted grades is obtained. This eliminates any opportunities for risk analysis in the mine planning step.

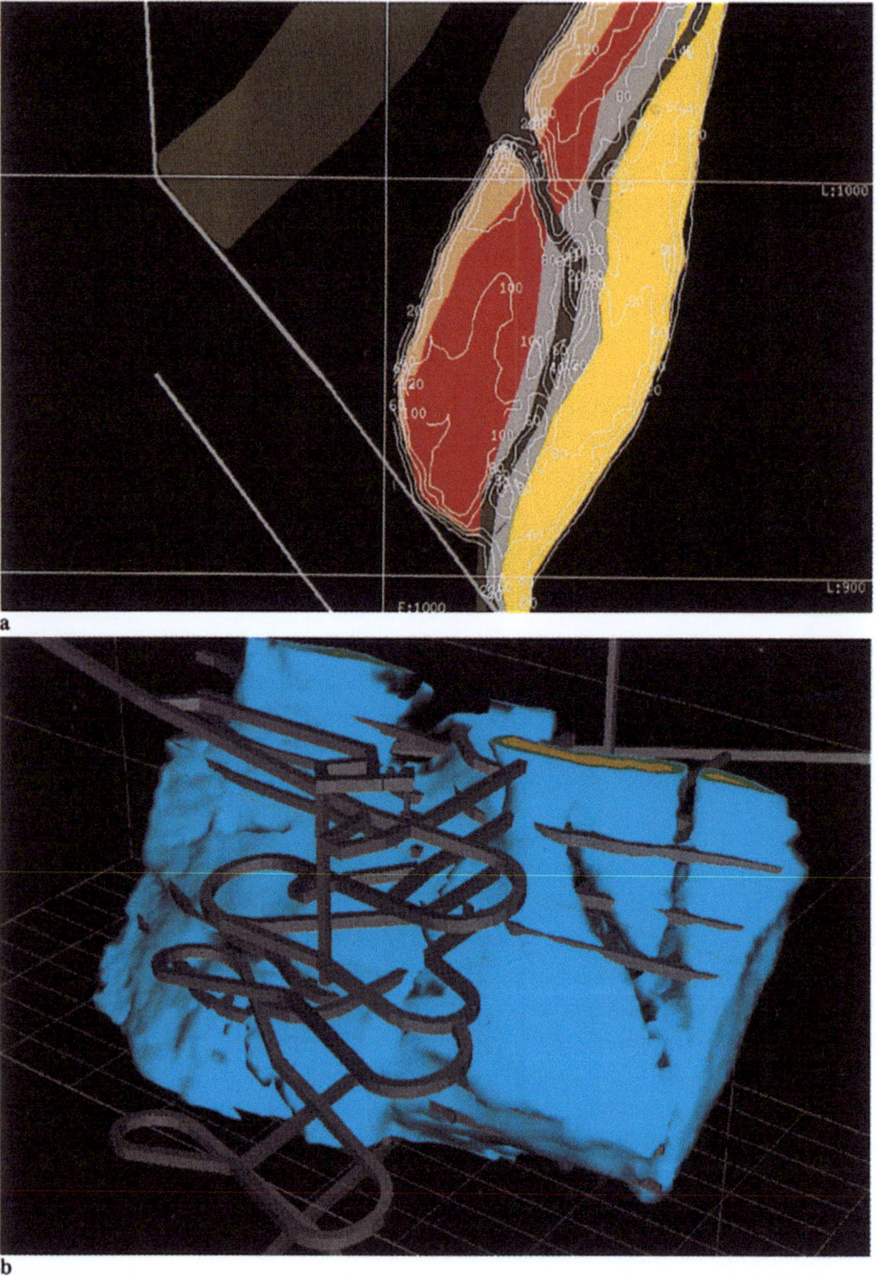

Fig. 13.5 **a** Transverse section through the ore type interpretation with contours of the net smelter return (NSR) variable superimposed; **b** 3D visualization of the volume of economically mineable ore, defined by the isosurface for the NSR cutoff value, superimposed on a volume model representation of mine workings; the effect of the dike intrusions on the ore volume can be clearly seen at the *right*

13.7 Spatial Analysis of Mineral Grades

This is a multi-mineral deposit, and each mineral has a unique spatial variation of grade and a unique economic value. To simplify the processes of deposit evaluation and mine planning we require the spatial variation of a single *equivalent variable* that reflects the total value of the ore, i.e. the combined values of the individual grades. In this case, the requirement is a *net smelter return* (NSR) variable that combines the mineral grades with appropriate dollar factors. We achieve this with a two-stage grid manipulation exercise (see Chap. 9), applying a slightly different algorithm in each case, one for ore with greater than 2% zinc and one for low zinc ore. We now have a continuous spatial measure of the dollar return we can expect from smelting the ore (cf. Fig. 13.5a). The NSR variable also becomes the primary information source for the design of ore excavation limits in the following mine planning step.

Spatial analysis of the deposit now involves a determination of *total geological reserves*, i.e. the total volume and tonnage of ore in the deposit for which the net smelter return variable exceeds an economic threshold or cutoff value. We obtain this result by volumetrically intersecting the 3D isosurface for the economic threshold value of the NSR variable (Fig. 13.5b) with the mineralogical boundaries of the four ore types. The former is obtained from the 3D grid representation of the net smelter return variable; the latter from the volume model representation of the deposit. This integration ensures optimum precision in the volumetric calculations, as discussed in Chapter 9. The results are reported by volume, tonnage and average value for each of the four ore types.

13.8 Ore Excavation Design

The budget constraints imposed on the project dictate an ore excavation design and schedule that yield a rapid cash flow without jeopardizing long-term mining profitability. For this reason, the mine plan focuses on the upper portion of the deposit above the strike fault, at the same time, ensuring future mineability of the lower portion.

The methods of mining are dictated by poor rock conditions in the footwall and hanging-wall rocks which envelop the ore deposit, by the need to maximize ore recovery, by the requirement to differentiate between ore types for processing purposes, and by the requirement to exercise grade control. The mining methods selected are *sublevel retreat* and *drift and fill stoping*. Ore development is initiated on each mining level by driving subdrifts on the footwall contact in the orebody. Cross-cuts are turned from the footwall subdrifts at rightangles to the strike of the deposit and driven to end points not less than 5 m from the hanging wall contact. Ore volumes opened by cross-cuts across strike are extracted by

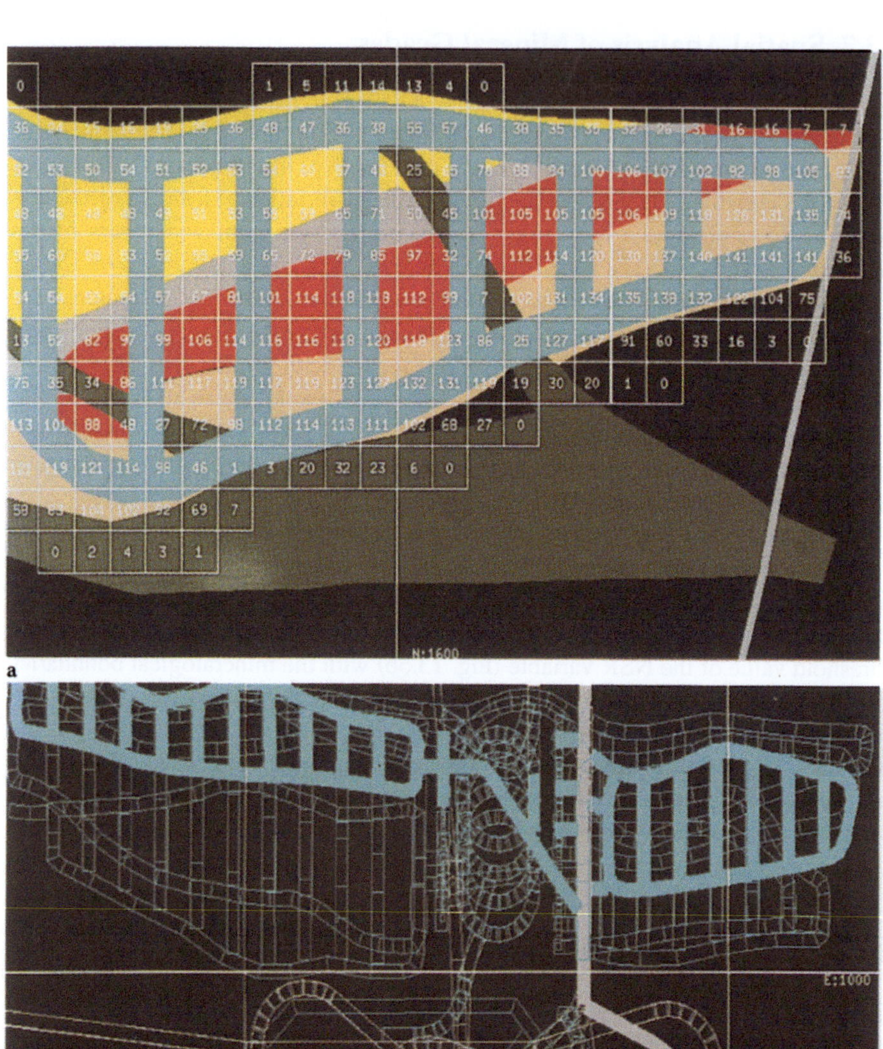

Fig. 13.6   **a** Interactive stope design (*blue*) on a plan section with reference to the deposit limits and the predicted spatial variation of the NSR variable; **b** plan section through the proposed stope excavations and mine development; excavations intersected by the viewplane are *shaded blue*; off-section excavations are *outlined blue*

sublevel retreat mining and subsequently backfilled with cemented rock fill. The barrier pillars left in the ore on the footwall and hanging-wall contacts are subsequently extracted by drift and fill stoping, with the stope cuts being filled tight with cemented rock fill. The use of sublevel retreat stoping and drift and fill stoping in this sequential manner is expected to provide the required ground control, to maximize ore recovery and to differentiate between ore types and grades. Early revenues from mining are enhanced by placing a sublevel access ramp to the stopes in a central *ore pillar*.

The stoping limits are designed on plan sections using the volume model excavation design tools discussed in Chapter 9. The horizontal limits of each stope are determined by reference to the isosurface of economically mineable ore, defined by the NSR variable (cf Fig. 13.6a). Ramp design is also performed in plan, first specifying a ramp cross-section, then interactively defining its center line and grade (slope). Design templates are used to obtain the geometry of curves and spirals.

13.9 Mine Development Design

As discussed earlier, the poor quality of both the hanging-wall and the footwall rocks are the primary constraints on the design of mine development. The footwall consists of highly altered and broken acid volcanics and volcaniclastics; the hanging-wall consists of ash/tuff layers overlain by a relatively competent basalt layer. The hanging-wall is considered the most stable of the two and is therefore chosen as the location of mine access development. As much as possible of the development is located within the basalt layer.

The existing exploration ramp and drifts are modeled as a design reference. Some of these workings have already collapsed where they intersect the hanging-wall ash/tuff layers and the footwall material. Once again, working in a 3D context highlights inconsistencies in the source information; in this case, in the survey data reductions for the exploration workings.

Mine access development includes a haulage decline tunnel, access cross-cuts and a raise (near-vertical shaft) for transporting backfill (Fig. 13.6b). All design is performed using the interactive volume modeling tools (cf. Chap. 9) and with visual reference to the geological model, particularly the hanging-wall basalt representation.

Several long-term options for mine egress development are studied, including a conveyor ramp and a vertical shaft, both located in the footwall and requiring extensive rock support. Short-term mine egress is provided by the haulage decline in the hanging-wall. The conveyor option also requires a ventilation/emergency egress raise in the footwall. The interactive design tools allow rapid design and refinement of these options, and the volumetrics analysis discussed below allows rapid comparison in terms of excavation, support and servicing quantities.

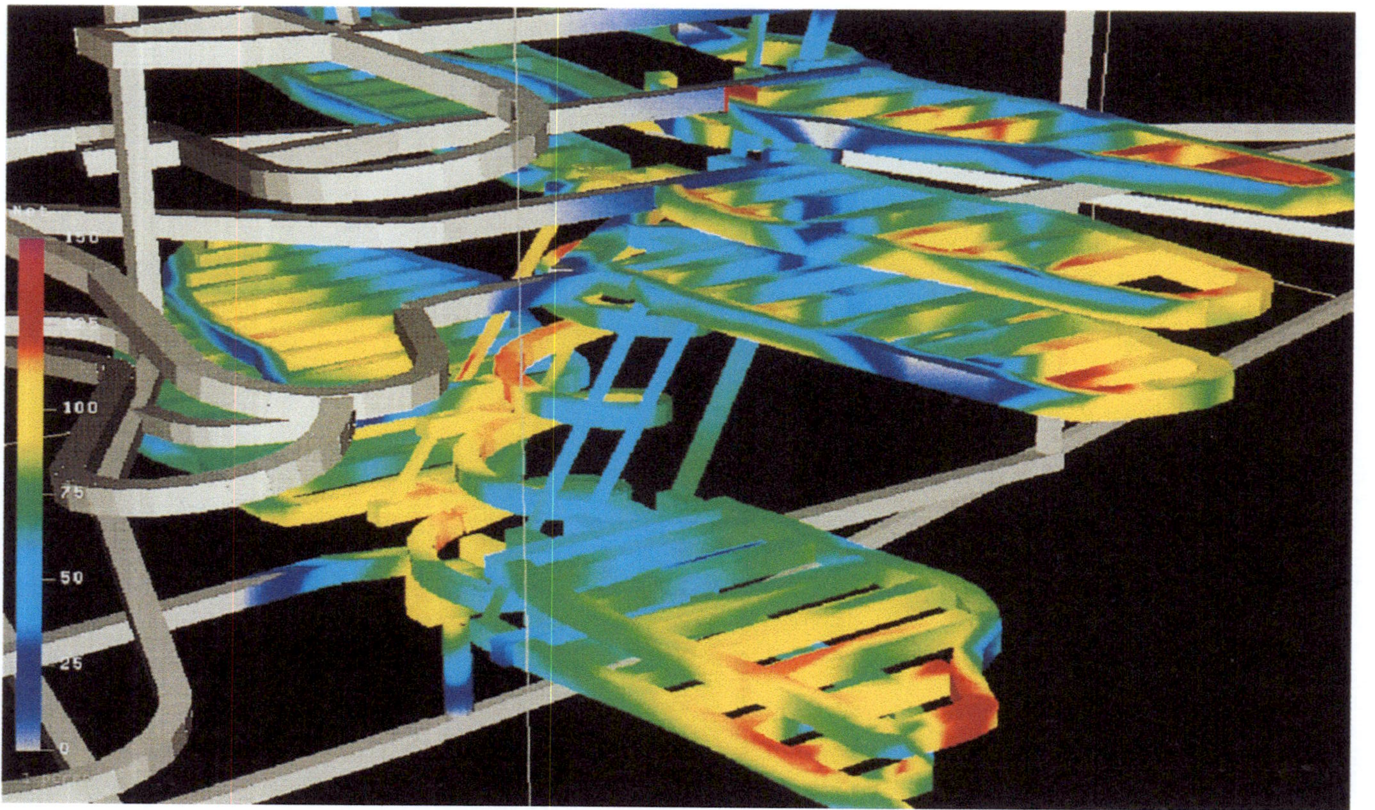

Fig. 13.7   3D view of mine excavation geometry superimposed with a color-mapping of the spatial variation of the NSR variable; the *gray colored* excavations are outside of the deposit and do not intersect the NSR variation

13.10 Mining Reserve Volumetrics

The final step involves a determination of *mining reserves*, the volume, tonnage and grades of all ore grade material intersected by the proposed ore and development excavations. Stoping volumes and stoping subunit (drift) volumes are obtained in terms of ore types, tonnages and average mineral grades, and waste rock volumes and tonnages. This involves a precise volumetric intersection of excavation geometry, mineral grade variations and geological/mineralogical volumes (see Figs. 13.7 and 13.8). Similar intersections are performed for the mine development volumes, some of which are located within the ore deposit (see table 13.1). These results, together with an established mining schedule are the basis for estimating the revenue generating capacity of the mine.

Development and operating costs are based on excavation and development volumes and their intersections with different rock types, obtained in similar fashion to the results shown in Fig. 13.8. The latter also provide a measure of the quantities and locations of rock support required. These results collectively establish the economic viability of the deposit and the selected mine plan and schedule.

```
PROJECT  : A MINE IN TURKEY
ANALYSIS : MINE FEASIBILITY STUDY

GRID MODEL              : OREGRADES
PRIMARY VARIABLE        : NSR
CUTOFF                  :   48.000
NUMERIC SCALING FACTOR  :    1.0
PRIMARY VOLUME MODEL SELECTION    : M,RS,LEVEL,980,1
SECONDARY VOLUME MODEL SELECTION  : G,03,*,*,*
VOLUME INTEGRATION INCREMENT      :    2.00
DENSITY OF DEFINED MATERIAL       :    3.200
DENSITY OF UNDEFINED MATERIAL     :    2.700
VOLUME/TONS SCALING FACTOR        : 1000.0
EXCAVATION RATIO        :    1.00
```

		<---TOTAL-->		<GREATER THAN CUTOFF>			<-LESS THAN CUTOFF->		<-UNDEF-->		<--SECONDARY->	
UNIT	COMPONENT CODE	VOL	TON	VOL	TON	PRIMARY NSR	VOL	TON	PRIMARY NSR	(%)	TON	GRADE Zn	GRADE Cu
LEVEL 980 STOPE (Primary) Intersects With –													
CBO ORE	7			76.	242.	94.66	5.	16.	30.00			12.66	4.36
MBO ORE	2			128.	409.	78.67	13.	41.	39.92			11.54	2.74
MYO ORE	6			10.	34.	86.99	1.	2.	30.83			1.41	8.53
PYO ORE	3			68.	217.	60.94	31.	99.	33.74			2.06	5.33
TOTAL		344.	1095.	282.	901.	79.01	50.	158.	34.90	4.	36.	9.18	4.01
Total Unit Tonnage = 1095253.		Average NSR Value =	70.054										

Table 13.1   Mining reserves for the 980-level stope, reported by ore type for an NSR cutoff value of 48; the 4% undefined material reported is caused by the presence of barren intrusive dikes (see Fig. 13.8); volumes and tonnages are scaled by a factor of 1/1000

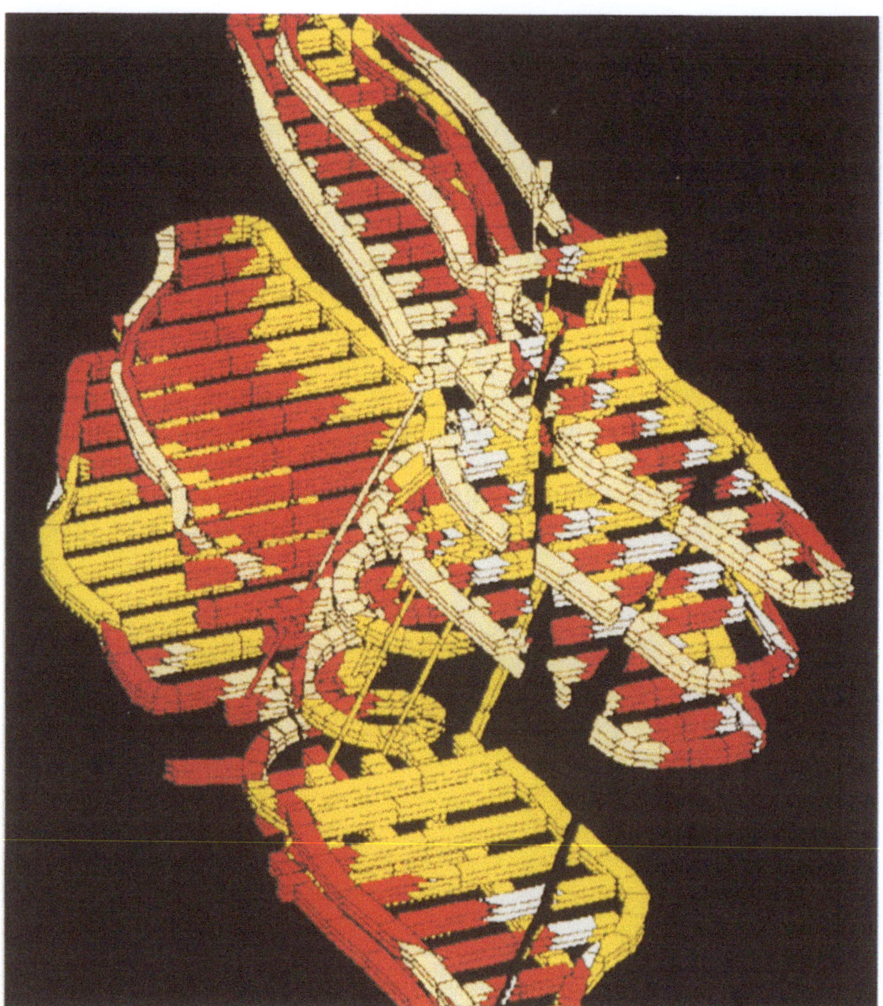

Fig. 13.8   3D visualization of the volumetric intersection of mine development and stoping excavations with the volume model representation of mineralogical ore types within the deposit; ore type CBO is *brown*, MBO is *red*, MYO is *gray* and PYO is *yellow*; gaps in the ore deposit intersection are caused by the barren intrusive dikes

13.11 Project Summary

The completion of this study within a 3 week period is a measure of the efficiencies that can be achieved with the integrated computerized approach to ore deposit characterization and mine planning. The study also emphasizes the enhanced interpretational capabilities that are provided by interacting with complex geological information in a true 3D context.

The integration of vector-based volume modeling and raster-based 3D grid modeling allows a precise characterization of complex geological, structural and mineralogical conditions. Volume modeling also provides interactive tools for rapid detailed design of underground excavation and development geometry, and equally precise volumetric analysis of ore and rock intersections. These integrated capabilities provide the necessary platform and confidence level for efficient mine planning.

14 Characterization and Development Planning for a Small Oil Reservoir

14.1 Project Overview

The majority of reservoir characterizations are performed with surface-based approaches, typically employing deformed grid data structures for 3D representation. This approach works reasonably well for relatively simple stratified conditions, but has the limitations discussed in Chapter 1 with respect to geological complexity.

The project discussed here involves the application of 3D characterization techniques to interpretation, modeling and evaluation of a small western Canadian oil reservoir. The structure is not particularly complex, however, the project demonstrates the capabilities of the integrated approach with respect to detailed interpretation of structure and provides an example of the use of these techniques for a more comprehensive reservoir study. Studies of this type are becoming increasingly important in the petroleum sector as the emphasis on optimizing recovery from existing reservoirs increases. A similar approach can also be used for regional-scale oil field studies.

The project covers property-scale evaluation of a portion of a producing reservoir. The applications include detailed reservoir modeling, oil reserve estimation, risk assessment and development planning. The small size of this reservoir and the rapid variation in reservoir quality require detailed 3D modeling of both the reservoir structure and the spatial variation of reservoir properties.

14.2 Reservoir Geology and Exploration Records

The subject of our evaluation is a portion of the Dina Pool, which belongs to the Gooseberry Lake Field in Alberta, Canada.

The reservoir occurs within the porous Dina Sandstone of the Ellerslie Member of the lower Cretaceous Mannville Group. The Dina Sandstone is overlain by the relatively impermeable limestone of the Ostracod Zone Member which forms the reservoir cap. The Dina Sandstone originated as a fluvial sand, deposited

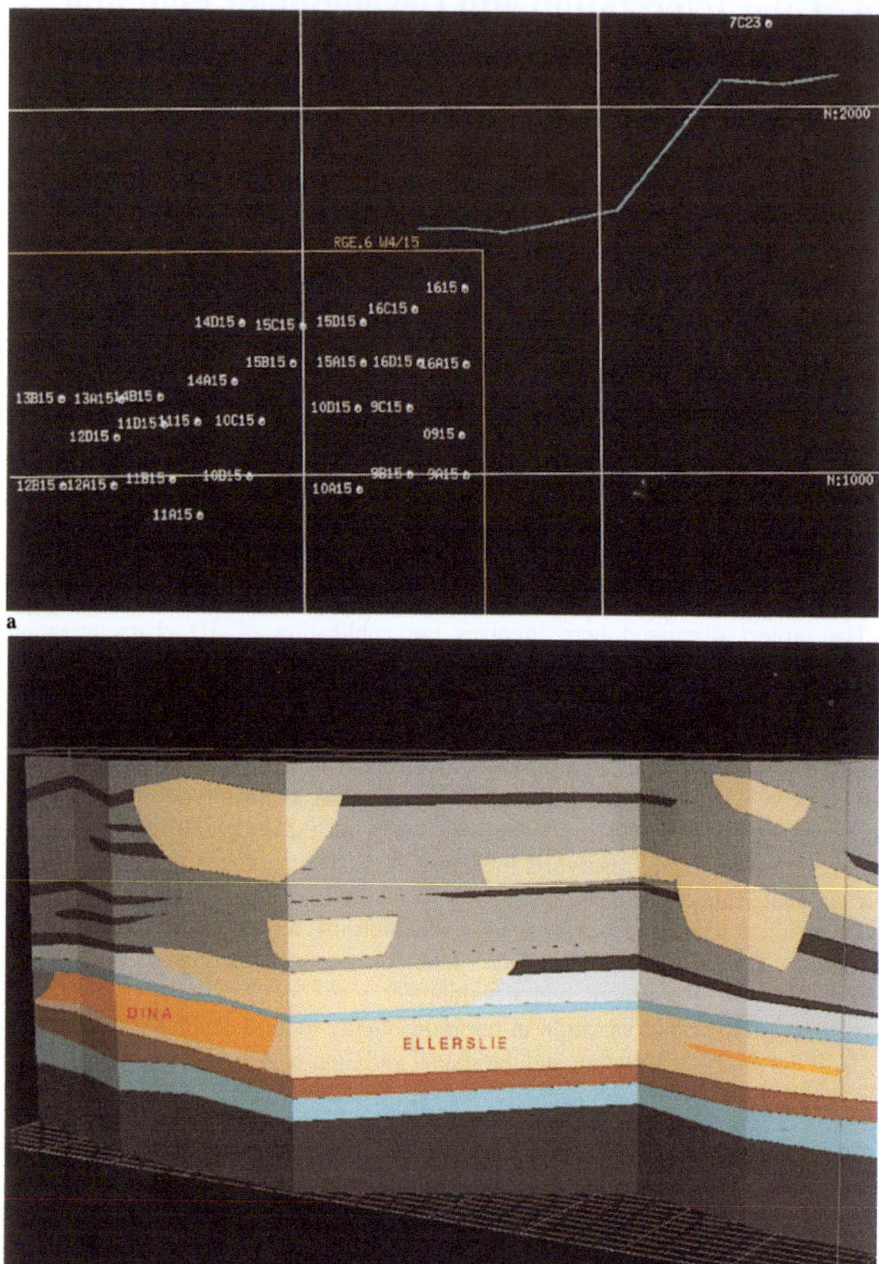

Fig. 14.1   **a** Site plan of borehole locations, northeast fence section location and property boundary;   **b** perspective of fence section interpretation on adjacent northeast properties; Dina channel is at *lower left*

unconformably on Paleozoic carbonates. It consists generally of a medium to very fine- grained, poorly consolidated, very porous and permeable quartz sandstone with minor proportions of chert, feldspar, pyrite and clay shale. The Ellerslie Member is cut by fluvial channels that contain deposits of coarse sand and gravel. Our characterization is focused on a portion of the Upper Dina Channel, which trends southwest/northeast through the property at elevations between 180 m to 200 m subsea.

The high porosity of the channel sands, together with the overlying limestone cap-rock and an impermeable *shale plug* along the northwest side, has created ideal entrapment conditions for oil. Similar conditions have been identified in earlier studies of other properties in the region. The resulting reservoir is referred to as the Dina Pool.

The channel exhibits distinctive fining upwards, becoming siltier at the top; in places the porosity has been locally occluded by calcite cement. Oil accumulates in the structurally higher portions of the sandstone and invariably sits on a very active and areally extensive aquifer within the Lower Dina Sands. Under these conditions water cuts due to coning around extraction wells generally rise rapidly, causing production-related problems.

Development of the reservoir requires a detailed knowledge of both the reservoir structure and the spatial variation of reservoir properties. This knowledge provides a basis for optimizing the configuration and extraction rates of production wells.

Extensive 3D seismic coverage provides initial indications of a *local high* in the top surface of the Ellerslie Member in the vicinity of the channel. This is most likely due to the reduced compressibility of the relatively coarse channel sands and gravels in the channel. The seismic coverage also indicates the presence of a shale plug along the northwest edge of the channel. Similar conditions have been previously identified by detailed investigation and interpretation of adjacent properties to the northeast (Fig. 14.1a). Some of this information is available to us in the form of a detailed fence section interpretation (Fig. 14.1b).

Drilling and investigation of 27 boreholes (Fig. 14.1a) on a 4-ha spacing provide us with comprehensive geophysical logging records. A number of dry holes to the north and south also provide useful information for the interpretation process. Geophysical borehole logging includes gamma ray, sonic, neutron density and electric logs. Gamma ray logs provide us with primary indications of lithology and member sequence. Sonic and neutron density logs provide measurements of porosity and observations of lithology. The electric logs provide indirect measurements of oil saturation.

The characteristics of this project include member sequence, identified by observations of *geophysical signature*, and lithology, sourced from *lithology interval picks* off the geophysical logs.

The variables of our characterization include porosity Φ, sourced from sonic and neutron density log measurements, and water saturation S_w (the reverse of oil saturation S_o) which is sourced from electric log measurements.

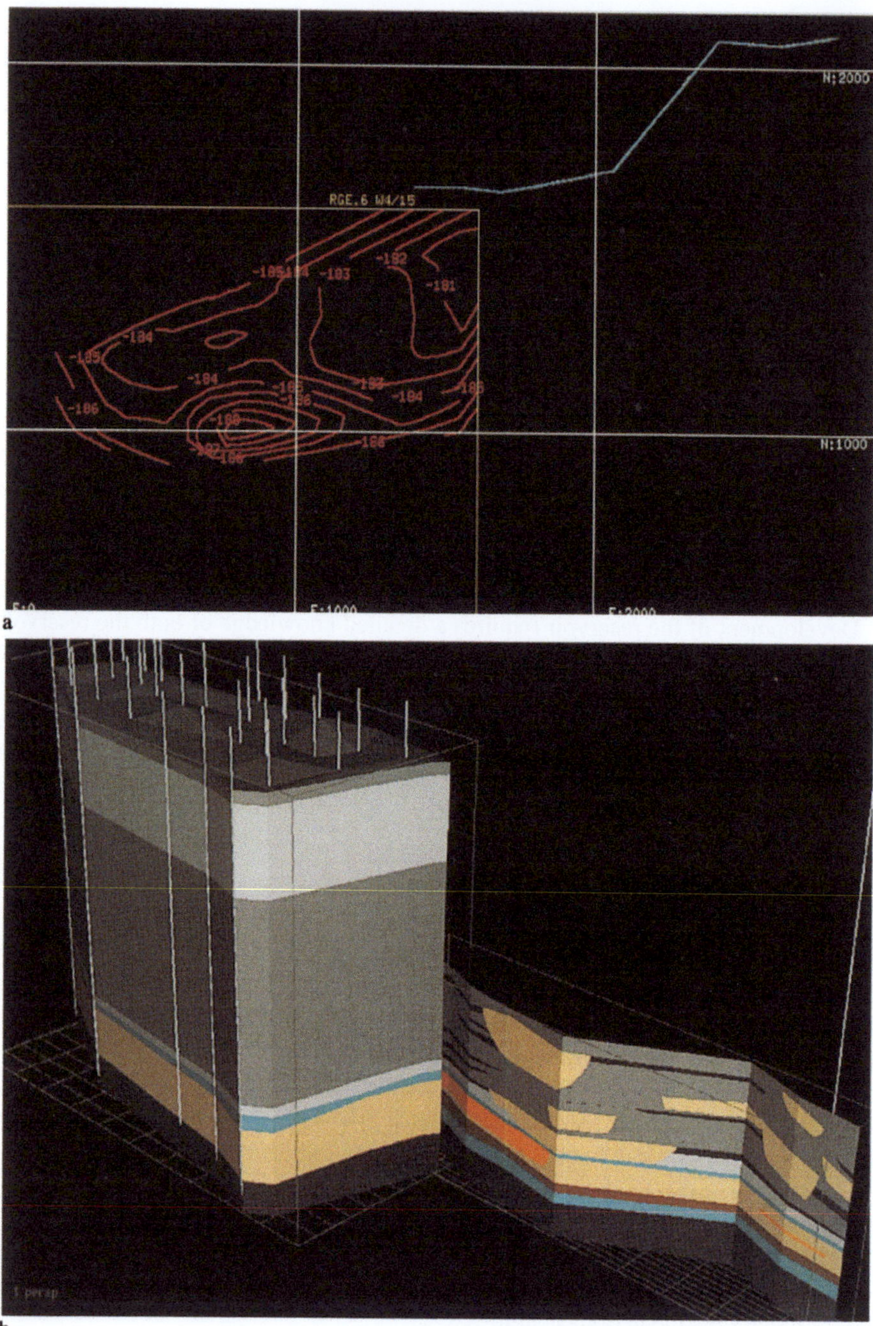

Fig. 14.2 **a** Contour analysis of the top of the Ellerslie Member in the vicinity of the channel; **b** volume model representation of the member sequence, generated by surface triangulation techniques from borehole intersections

14.3 Preliminary Modeling of Member Sequence

Our first step is a contour analysis of borehole intersections with the top of the Ellerslie Member. The results confirm the location and extent of a *local high* over the Upper Dina Channel (Fig. 14.2a) and support the concept of an oil pool.

The second step involves generation of a simple volume model representation of the member sequence. This is created by applying the surface triangulation and manipulation techniques discussed in Chapter 7 to the member intersections obtained from interpretation of borehole logs (Fig. 14.2b). The resulting member sequence model provides the necessary spatial constraints for subsequent detailed modeling of the Ellerslie Member in the channel region.

14.4 Detailed Modeling of Reservoir Structure

Interactive interpretation and the sectional approach to volume modeling (as discussed in Chap. 6) are employed to obtain a detailed lithological characterization of the Ellerslie Member in the vicinity of the channel. The initial sectional interpretations are performed on nine vertical sections with an azimuth orientation approximately transverse to the axis of the channel and a spacing of 200 m (Fig. 14.3a).

The member sequence model defines the upper and lower extents of the Ellerslie Member. Detailed geophysical log interpretations, or *lithology picks*, provide the necessary observations of lithological characteristics within the member. This information is displayed as a visual background reference for interactive interpretation of lithological boundaries on each section.

The lithology picks are used to interpret a detailed volume model representation of the Ellerslie Member which includes the Upper Dina Channel, the surrounding Dina Sandstone, the Lower Dina Sands (which create an active aquifer beneath the reservoir), and the shale plug to the northwest. The lateral extent of the interpreted channel correlates approximately with the 185 m subsea contour of the top of the Ellerslie Member. To ensure continuity with earlier studies of adjacent properties, these detailed interpretations are also correlated with the fence section interpretation of the Ellerslie Member to the northeast. Once acceptable sectional interpretations have been achieved, these are extended and interpreted in the third dimension, along the channel axis, using the interactive profile-matching techniques described in Chapter 6 (Fig. 14.4).

This detailed 3D model of lithological structure provides us with a suitable framework for a study of the spatial variation of reservoir properties. In this instance, the study is limited to the volume of oil-bearing materials within the

a

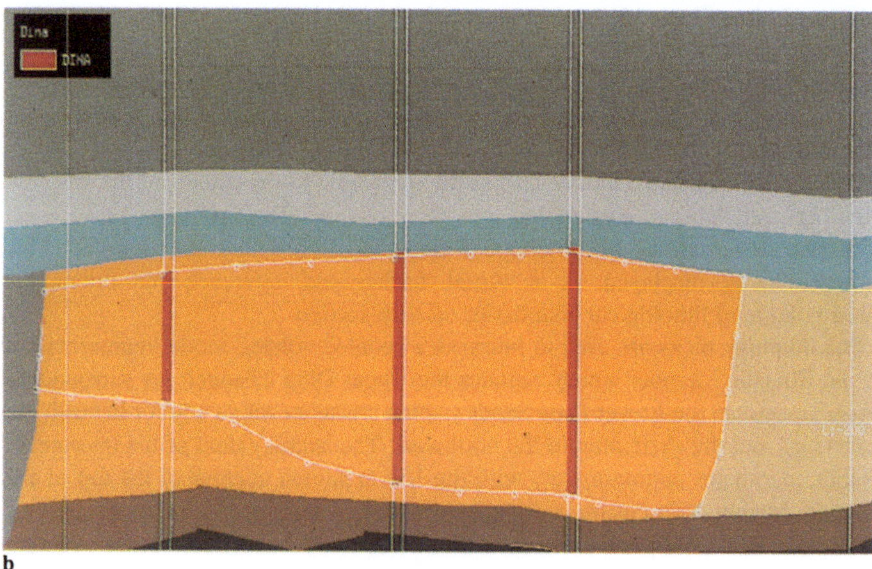

b

Fig. 14.3   a Interpretation of lithological structure within the Ellerslie Member on a vertical section across the channel axis; interpretation includes shale plug at *lower left*, limestone cap, underlying aquifer, Ellerslie sandstone at *right*, and a cross section of the Dina Channel;   **b** interpretation of the reservoir pool volume within the channel from borehole trace intersections with the pool (coded DINA)

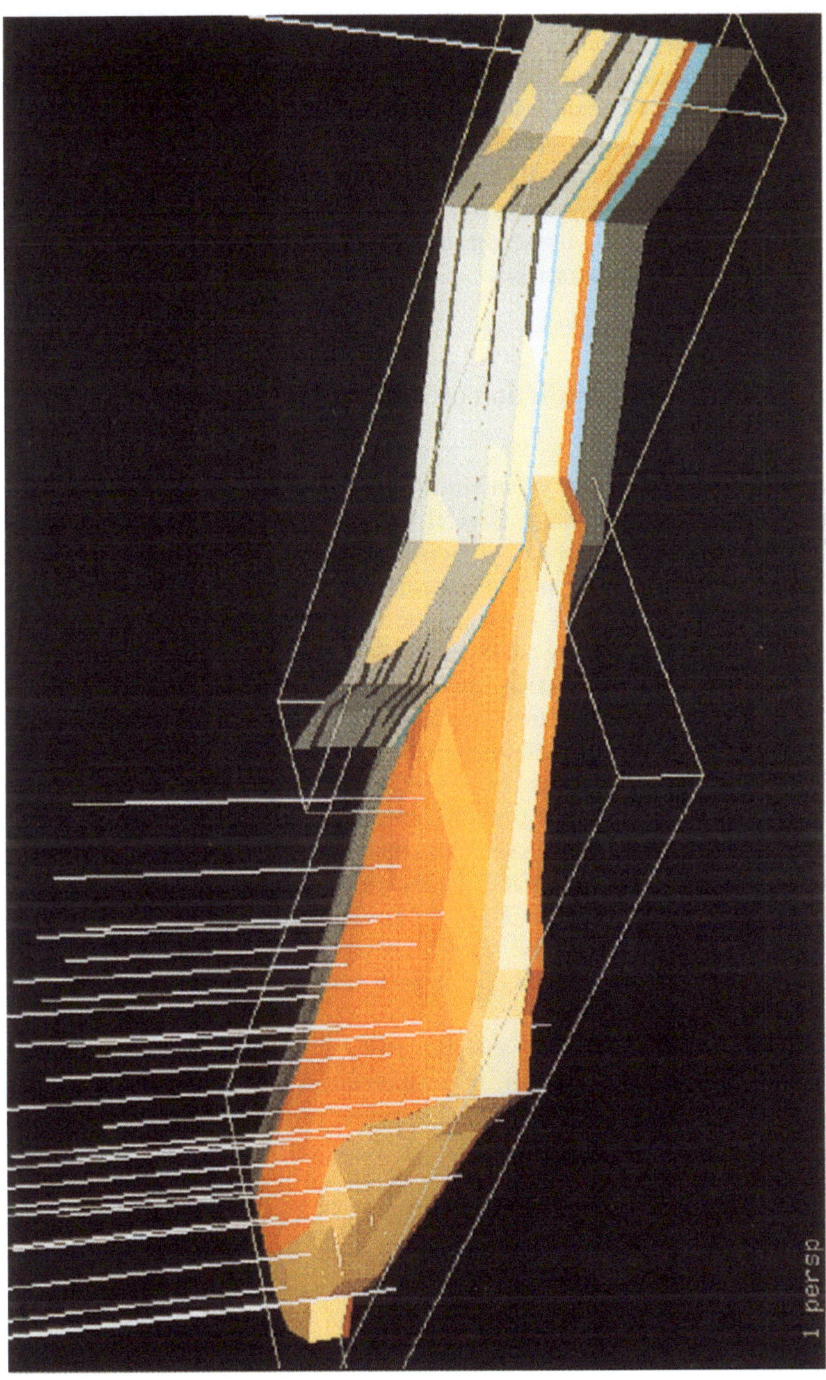

Fig. 14.4 Perspective view of the volume model representation of the lithological structure of the reservoir within the Ellerslie Member

channel. At the end of the chapter we discuss the potential advantages of extending the study beyond this volume.

Electric borehole log picks are utilized to identify occurrences of oil-bearing material. The same sectional interpretation and volume modeling procedures are then employed to define the *reservoir volume* (oil pool) within the channel (cf. Fig. 14.3b). This volume is purposely constrained by the spatial limits of the channel and clipped by the property boundary. Although there is some evidence of oil in the less porous materials to the southeast of the channel, this is ignored in the study since it is unlikely to be extractable.

14.5 Geostatistical Prediction of Reservoir Properties

Measurements of porosity Φ within the reservoir pool are obtained from the sonic and neutron density logs. Several measurements are available within the vertical thickness of the pool at each borehole location. Considerable vertical and horizontal variation of the porosity measurements is evident. Measurements of water saturation S_w, derived from electric logs, are only available as average values for a preinterpreted *pay thickness* (cf. Sect. 14.6 below) within the reservoir pool. Only one saturation value per borehole is available, which limits our prediction of water saturation to the horizontal, i.e. saturation is assumed constant over the pay thickness, but exhibits spatial variation horizontally. This also limits the potential for meaningful risk assessment, as discussed in Section 14.8 below.

Statistical analysis shows the porosity measurements to be normally distributed (Fig. 14.5a), whereas the water saturation measurements have a distribution which is slightly skewed. Scattergram analysis shows a fairly strong reverse correlation between these two properties (Fig. 14.5b). In general, low water saturation measurements (which correspond to high oil saturation) correlate well with high porosity measurements and vice versa. In each case, these results are fairly typical of reservoir properties.

Geostatistical semi-variogram analyses provide us with information on the spatial variability of the porosity Φ measurements. An initial global analysis shows a clearly discernible relationship between variability and distance over a horizontal range of approximately 800 m. However, a more detailed analysis which includes directional control detects significant horizontal anisotropy in the spatial variability of the measurements (cf. Fig. 14.6a,b), although the spatial relationships are not as well defined in these results. The continuity along the axis of the channel, with a range of influence of 1000 m, is significantly greater than that across the channel, which exhibits a range of 400 m. This anisotropy most likely results from the original fluvial deposition patterns within the channel. The vertical differentiation of the porosity measurements is insufficient, in terms of

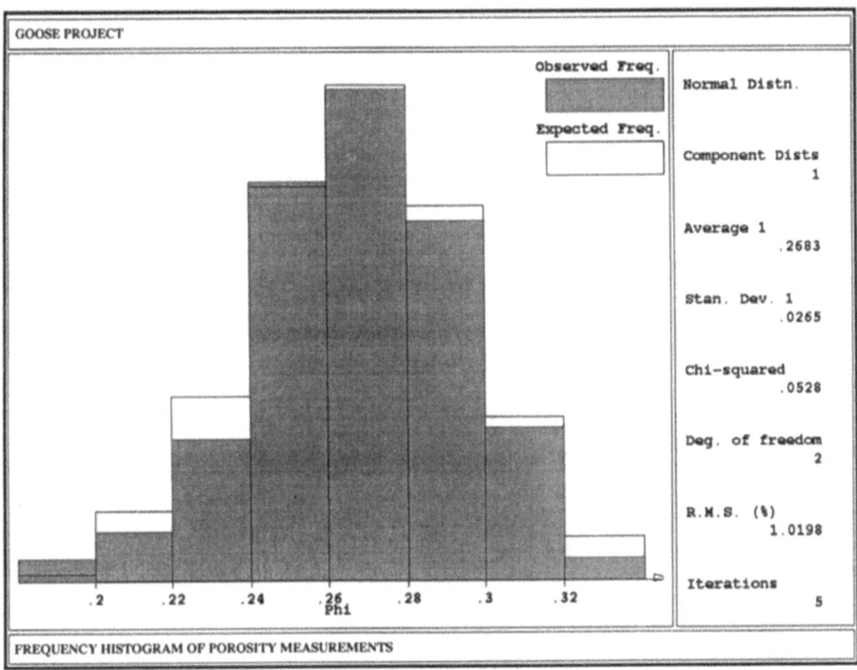

a

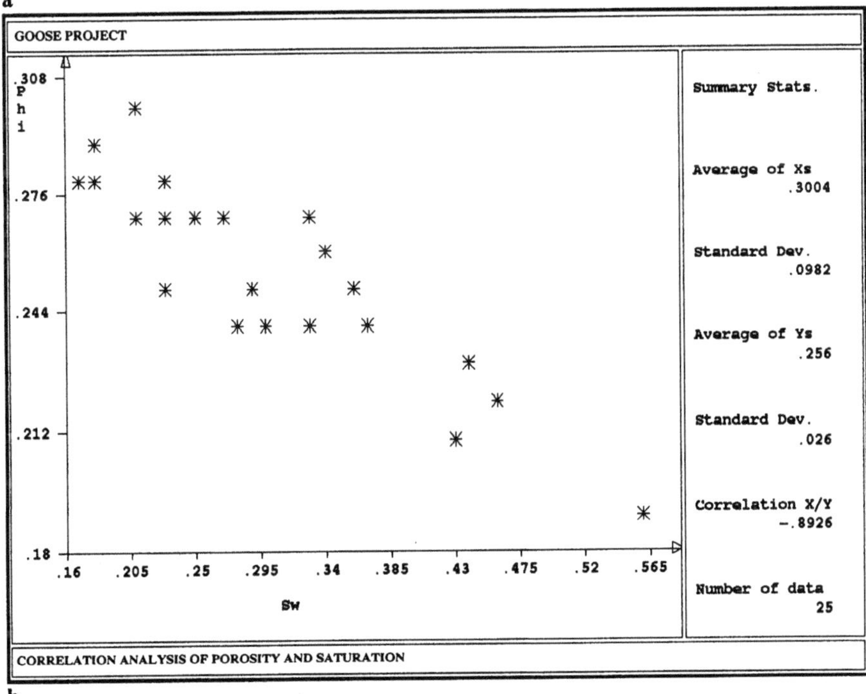

b

Fig. 14.5   **a** Frequency histogram analysis of porosity measurements; **b** scattergram analysis of the correlation of porosity and water saturation measurements

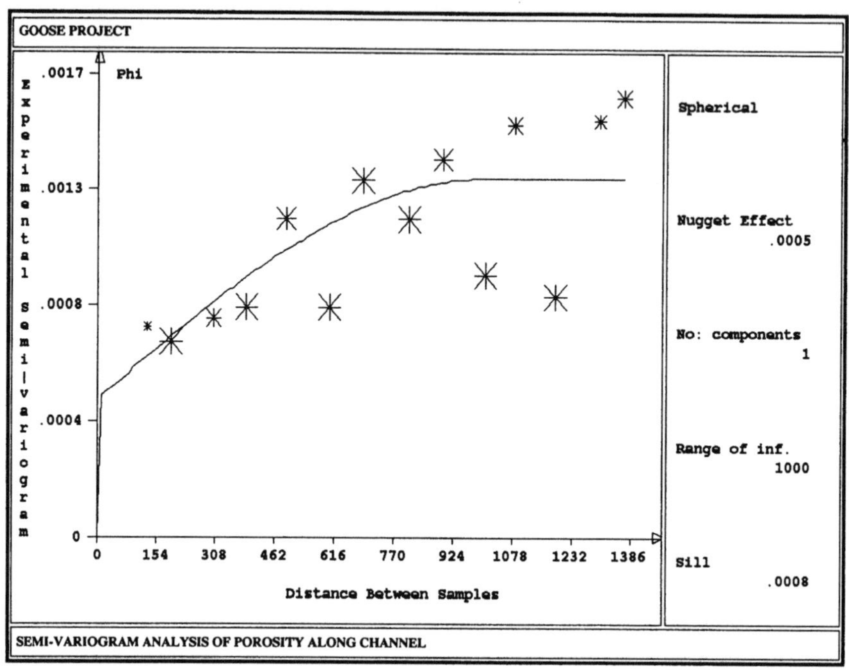

a

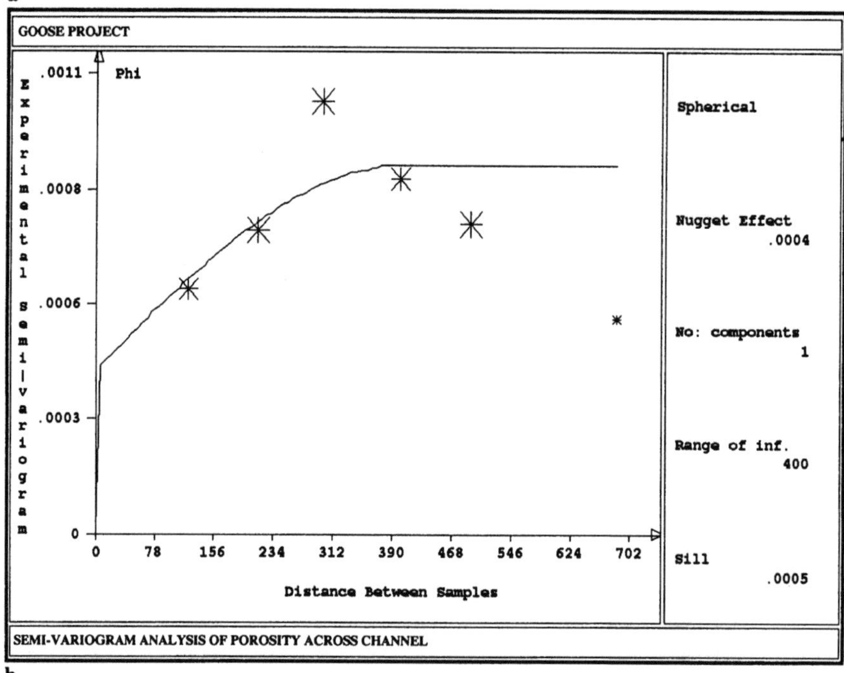

b

Fig. 14.6   **a** Semi-variogram analysis of the horizontal variability of porosity measurements along the channel axis; **b** horizontal variability of porosity across the channel axis

available measurement pairs, to obtain acceptable semi-variogram results in this direction, however a range of 12 m is indicated.

Figure 14.6a,b also shows semi-variogram prediction models, obtained by interactively superimposing spherical models onto the analytical results in the two orthogonal horizontal directions. The range of influence in the vertical direction is set at 12 m, which is adequate to encompass the average thickness of the reservoir pool. Normal 3D kriging is employed with these parameters and the porosity measurements to predict a 3D grid model representation of the spatial variation of porosity Φ throughout the reservoir pool (Fig. 14.7). The volume model representation of the reservoir pool provides the necessary spatial control for the prediction process (see Chap. 8).

The 3D grid data structure is oriented horizontally with its longitudinal Y axis approximately parallel to the channel axis. Cell dimensions of 75 m longitudinally, 35 m transversely and 2 m vertically are chosen to obtain an acceptable degree of differentiation in all directions. Figure 14.7a,b shows color-mapped porosity variations on horizontal sections near the top and bottom of the pool respectively. From these results, it is apparent that porosities are highest at the bottom of the pool and decrease rapidly towards the top. The horizontal variation indicates that porosities are also highest towards the northeast.

A similar approach is used to predict the variation of water saturation S_w within the reservoir pool. Semi-variogram analyses of the available water saturation measurements show them to have a similar spatial variability to the porosity measurements. The horizontal ranges of influence exhibit the same degree of anisotropy along the channel and in the transverse direction. This is anticipated since the properties have already been shown to exhibit a high degree of statistical correlation. The predicted results indicate that water saturation is highest towards the southwest, in other words, oil saturation, like porosity, is highest towards the northeast.

14.6 Pay Zone Interpretation

The *pay zone* is defined as the portion of the reservoir pool volume from which oil can (potentially) be profitably extracted. The lower extent of the pay zone is the *oil:water contact*, defined by the condition

$$S_w \leq 0.50.$$

The upper extent is governed by several conditions, including the cap-rock (limestone) contact, the presence of nonreservoir rock (such as shale) in the sand, and reduced porosity due to the presence of fines. Ideally, the predicted spatial variations of the various properties should be utilized as the primary reference for interpretation of the pay zone. In this instance, the pay zone limits have been

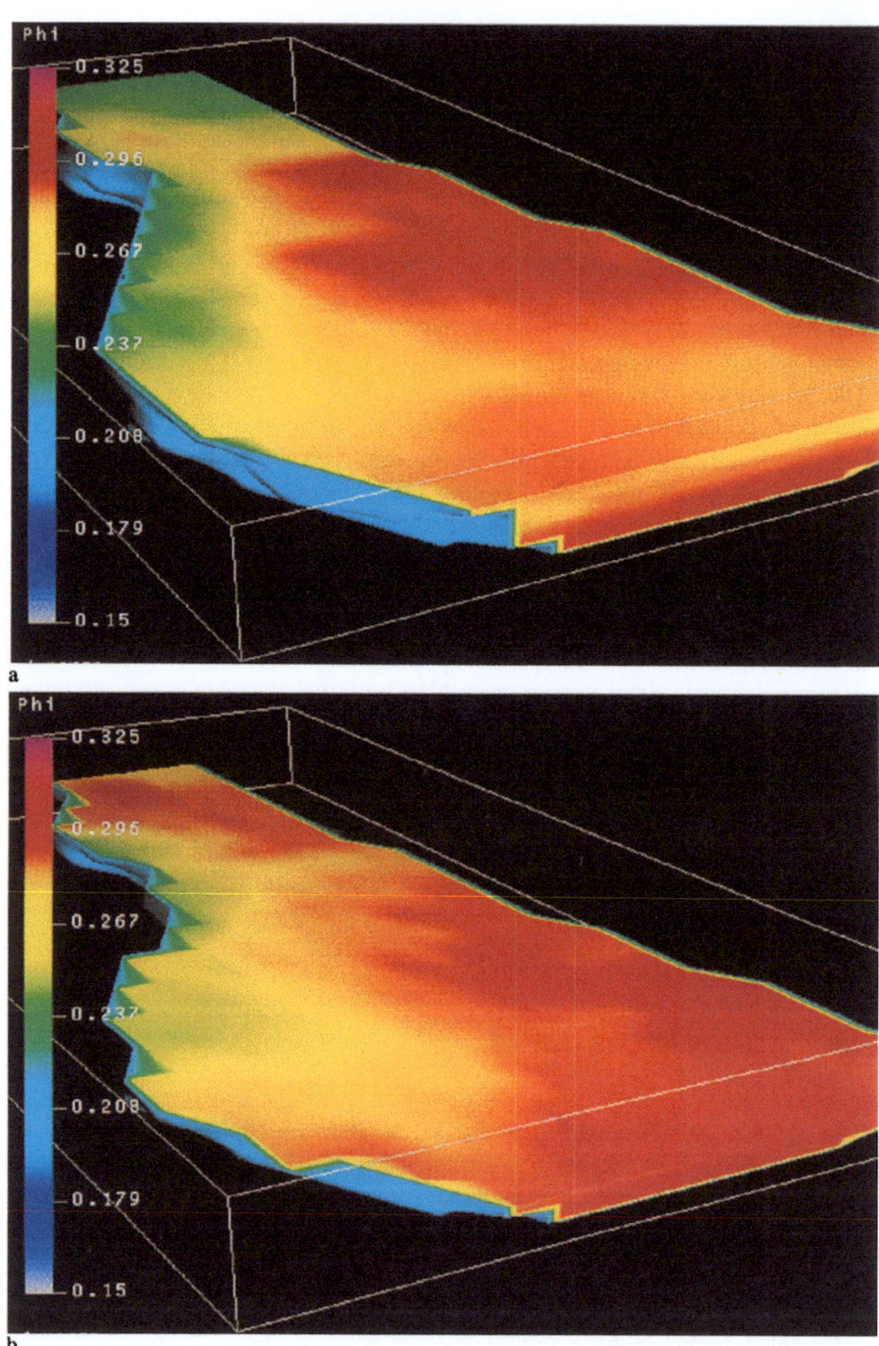

Fig. 14.7 **a** Color-mapped visualization of the predicted spatial variation of porosity on a horizontal section near the top of the reservoir pool; **b** porosity variation near the bottom of the reservoir pool

predefined, determined directly from inspection of the geophysical logs, and a lack of sufficient property measurements outside the pay zone precludes any reinterpretation on our part.

In our study, a volume model representation of the pay zone within the reservoir pool is derived by surface triangulation techniques from pay zone intervals identified in the borehole logs. By spatial integration of the predicted property variations in the 3D grid model with the volume model representation of the pay zone, we achieve a spatial characterization of the reservoir which provides a suitable platform for detailed evaluation, risk assessment and development planning (Fig. 14.8).

14.7 Reservoir Evaluation

The first step in the evaluation process is to derive the spatial variation of a *net pay* variable NP that defines the oil content per unit volume. We achieve this by applying the 3D grid model manipulation techniques discussed in Chapter 9 with the expression

$$NP = \Phi \bullet (1 - Sw).$$

The results of this manipulation provide the 3D spatial variation of the net pay variable throughout the volume of the reservoir pool. In Fig. 14.8, the net pay variation is shown color-mapped to the surface of the pay zone volume. The *net oil reserves* are obtained from a spatial analysis of the intersection of the pay zone volume with the spatial variation of the net pay variable, using the techniques discussed in Chapter 9. A shrinkage factor (0.95) and a primary recovery factor (20%) are applied to these results (Fig. 14.9a). The net oil reserves amount to 141 000 m^3 (698 677 m^3 with an average net pay value of 0.202). This result represent the *most likely* reserves, based on the available information.

14.8 Risk Assessment

In this context, we can usefully apply the *geostatistical uncertainties* attached to our prediction of reservoir properties to a determination of the *potential risk* associated with reservoir development. In the previous step, we obtained an estimate of the net oil reserves, but as yet we have no measure of how *reliable* this estimate is. Expressing this another way, we require a measure of the *potential variability* of the net oil reserves, in view of the limited amount of information available for our characterization.

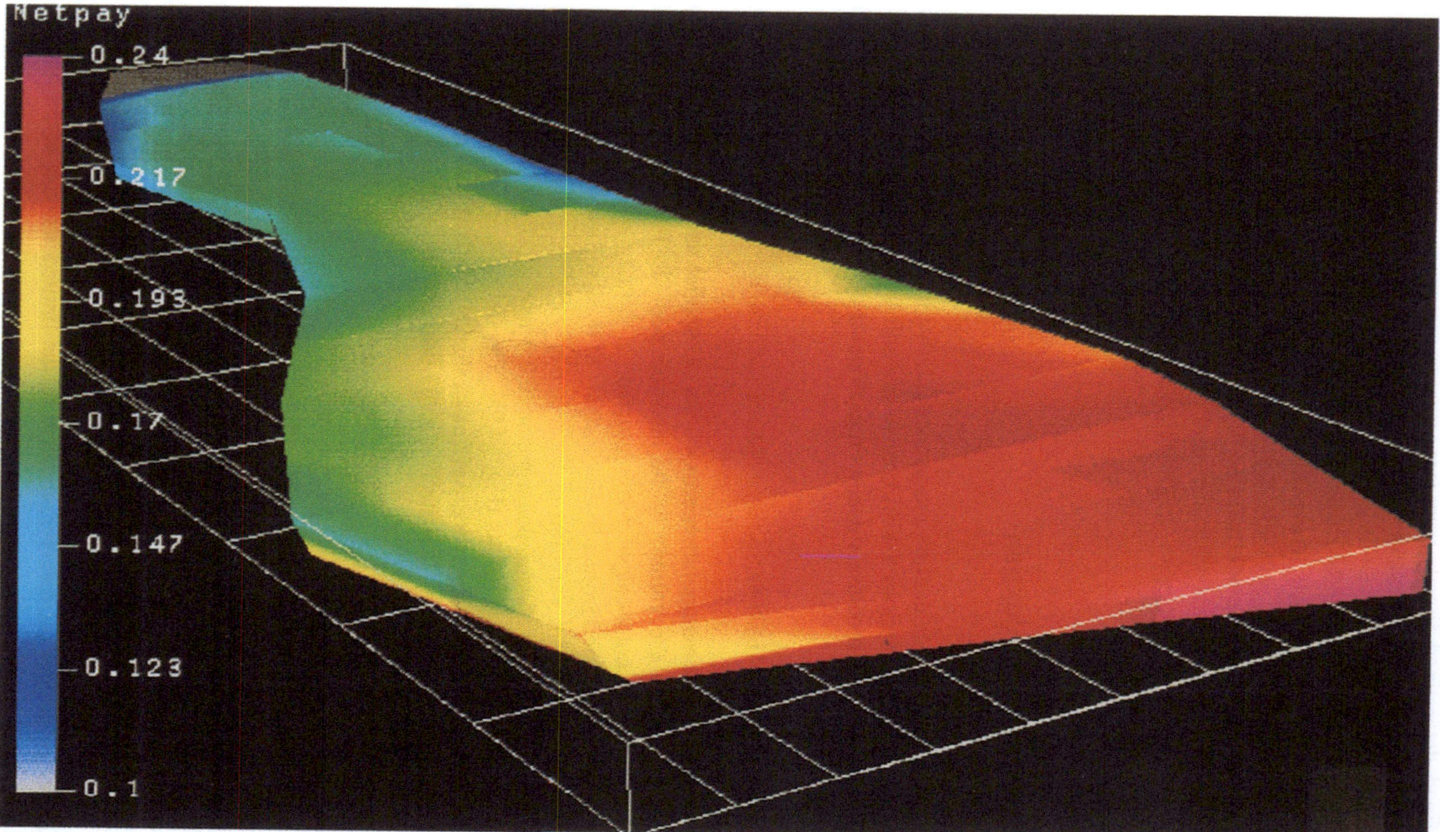

Fig. 14.8   Perspective view of the pay zone volume with the spatial variation of the net pay variable color-mapped to the surface

```
VOLUMETRICS - ANALYSIS OF MODEL INTERSECTIONS
PROJECT   : GOOSE3
ANALYSIS : NET OIL RESERVES
3D GRID MODEL              : OIL
PRIMARY ATTRIBUTE          : Netpay
THRESHOLD VALUE            :      .10
ATTRIBUTE SCALING FACTOR :    1.0
PRIMARY VOLUME MODEL SELECTION   : G,PY,*,*,*
SECONDARY VOLUME MODEL SELECTION : G,DN,*,*,*
VOLUME INTEGRATION INCREMENT     :   10.00
DENSITY                          :    1.000
VOLUME/TONS SCALING FACTOR       : 1000.0
EXTRACTION RATIO                 :      .19


<-VOLUME IDENTITY-><TOTAL><GREATER THAN THRESHOLD><LESS THAN THRESHOLD><--SECONDARY-->
UNIT            CODE  VOL    VOL    TON  PRIMARY   VOL   TON  PRIMARY ATT(1) ATT(2)
                                        Netpay                Netpay  Phi    Sw
PAY ZONE        * (Primary) Intersects With -
RESERVOIR POOL  6            3676.  3676.   .20     0.    0.    .00    .27    .26
  -------------------------------------------------------------------------------
  TOTAL            3677.  3676.  3676.   .20     0.    0.    .00    .27    .26

Total Unit Volume =  698677.    Average Unit Net Pay =     .202
```

a

b

Fig. 14.9 **a** Report of predicted net oil reserves from a volumetric analysis of the pay zone volume with the spatial variation of the net pay variable; **b** contours of prediction uncertainty (standard error) for water saturation values on a horizontal section through the reservoir interpretation

The 3D grid manipulation techniques are used to combine the *standard errors of prediction* (cf. Fig. 14.9b) for each of the reservoir properties with the predicted values (see Chap. 10). A *best case scenario* is obtained by increasing the predicted porosities by the associated standard error values and decreasing the water saturation values by their error values. The reverse procedure provides us with a *worst case scenario*. By reapplying the reservoir evaluation procedure (see Sect. 14.7 above) to these modified property variations, we obtain a measure of the potential variability of the oil reserves, to which we can attach a known confidence interval. The results of this assessment show that the net oil reserves, within the interpreted pay zone volume, have a variability of 25 000 m^3 with a confidence interval of 68%.

This variability is based solely on the prediction uncertainty for the spatial variations of porosity and water saturation. Ideally, of course, we should also reinterpret the pay zone volume for each scenario, since its volume is governed to a large extent by the property variations. In this instance, a lack of sufficient property measurements precludes this; the available water saturation measurements are only applicable to the predetermined pay zone volume. Despite the omission of this refinement, the benefits of this type of stochastic approach to evaluation of oil reserves are obvious. It provides a logical basis for production planning and assessing the risk attached to reservoir development. It also provides a basis for exploration decisions. Maximum benefit from further exploration of the reservoir is likely to be obtained by drilling where (1) the uncertainty is greatest (cf. Fig. 14.9b) and (2) the net pay variable is marginal. The new information logically reduces the variability of the reserves as it is incorporated into a revised characterization. The resulting benefit can be measured in terms of its effect on both the net oil reserves and their variability.

14.9 Development Planning

The 3D models of reservoir structure and the spatial variations of properties can be viewed on sections at any spatial orientation, or as 3D perspective displays with selective cut-aways or transparencies. The resulting enhanced appreciation of reservoir complexity is invaluable in the planning of development drilling and enhanced recovery activities. The 3D grid data structures of reservoir properties can also be exported to provide primary data for production simulation studies.

Because the models are precisely located in 3D coordinate space, we can interactively plot a proposed drilling profile (or trace) on an appropriate section or plan, with the reservoir structure and property variations displayed for background reference. The resulting profile is defined by a string of NEL coordinates that can readily be converted to downhole distance, azimuth and dip (inclination) readings for directional drilling purposes (Fig. 14.10). This capability is particularly

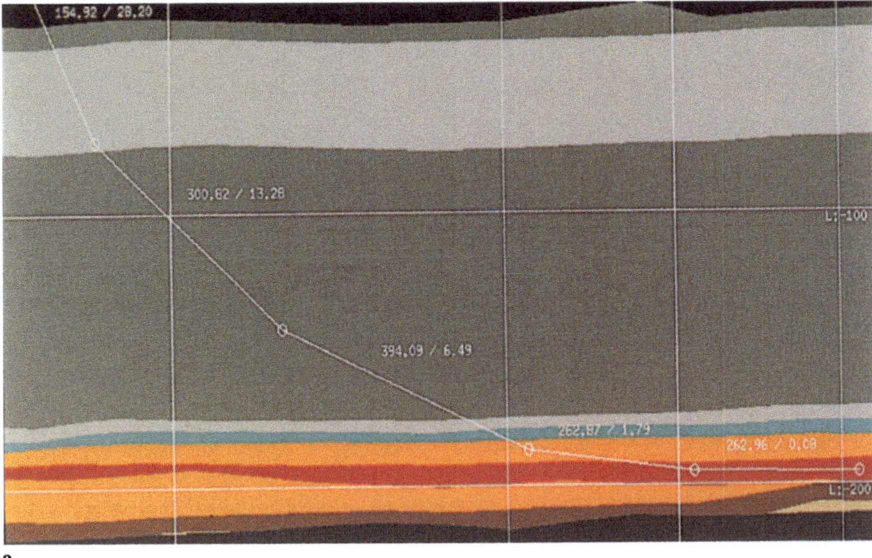

a

```
LYNX GEOSYSTEMS INC.                           Tue Nov 30 10:45:36 1993
PROJECT:GOOSE3                                              USER:simon
SUBSET: #1                                      directional drilling design


DRILLHOLE REPORT FOR SUBSET #1, HOLE:BHX, REGION:  DN, CATEGORY:DD

HOLE #        NORTH        EAST      ELVN      LGTH       Sf1        Sf2 RG'N CG
--------   ----------   ----------   --------  --------  --------  -------- ---- --
BHX           1143.44      259.35      0.00   1376.00                       DN DD

#         DIST      AZIM        DIP       #         DIST      AZIM        DIP
----   ----------  ----------  ----------   ----  ----------  ----------  ----------
0001       0.00      70.30      28.20     0002     153.92      70.30      28.20
0003     155.92      70.30      13.28     0004     454.74      70.30      13.28
0005     456.74      70.30       6.49     0006     848.83      70.30       6.49
0007     850.83      70.30       1.79     0008    1111.70      70.30       1.79
0009    1113.70      70.30       0.08     0010    1376.00      70.30       0.08
```

b

Fig. 14.10   **a** Profile design for directional drilling to intersect the reservoir pay zone, on a vertical section along the channel axis;   **b** report of reduced borehole distance, azimuth and dip readings for directional drilling

useful for the design of horizontal wells, and for determining drilling corrections during installation.

14.10 Project Summary

The project demonstrates the application of 3D computer-assisted characterization techniques to detailed reservoir modeling, evaluation and development planning. It illustrates the capabilities for modeling the spatial variations of reservoir structure and properties, and the ability to integrate these for evaluation and planning purposes. The example is relatively simple in terms of its structural complexity. In more complex situations, for example where the inter-connectivity of reservoir pools is critical or where we are dealing with faulted conditions, the ability to visualize and analyze conditions on viewplane sections at appropriate orientations is of great value.

The results provide a platform for implementing more efficient development techniques and enhanced recovery activities to maximize production and increase the ultimate recoverable reserves. The characterization presented here is constrained by limiting the measurement of reservoir properties to the volumes of immediate interest. Additional measurement and modeling of porosities and water saturation values in the marginal regions of the reservoir would provide an enhanced appreciation of potential production performance. The ability to perform meaningful risk assessment would also be enhanced.

15 Geotechnical Characterization for an Underground Powerhouse Excavation

15.1 Project Overview and Information Sources

Application of computerized characterization techniques within the geotechnical and rock mechanics sectors has been limited to date, despite the apparent advantages. The application summarized here concerns an investigation for an underground powerhouse that was initially reported by Wittke (1977) using a more conventional approach. The reported results of this earlier investigation form the basis of the current application.

Although relatively limited in comparison to the projects reported in previous chapters, this application effectively demonstrates the capability of the computerized characterization process to deal with information sources relevant to geotechnical and rock mechanics investigation. In particular, it illustrates the handling of rock joint and faulting information, and their geometrical interaction with underground excavations. The process also provides a basis for generation of analytical sections for numerical analysis, and a 3D platform for subsequent rock reinforcement design.

The proposed powerhouse excavation has the cross section shown in Fig. 15.1a and a length of 219 m. It is provisionally located at a depth of 350 m along an axis (azimuth) of N33.5E. Investigation of the site has confirmed the presence of several lithological types, considerable primary and secondary jointing of the rock mass, and one major fault. Our initial application objective is to provide a vehicle for visual appreciation of the complex geometrical interaction of the observed rock joints and faulting with the proposed underground structure. The next objective is to perform a characterization of the region by lithology, rock mass classification and geomechanical properties. Export of viewplane sections through the resulting models provides a convenient source of geometrical information for finite element analysis of the excavation effects. Finally, the characterization provides us with a vehicle for 3D design of rock reinforcement.

Lithological observations and mapping of primary and secondary rock joints and faulting are available from an exploration adit driven along the roof of the proposed powerhouse cavern. Reduced orientations of the primary rock joints and fault observations are presented in Fig. 15.1b as stereographic projections.

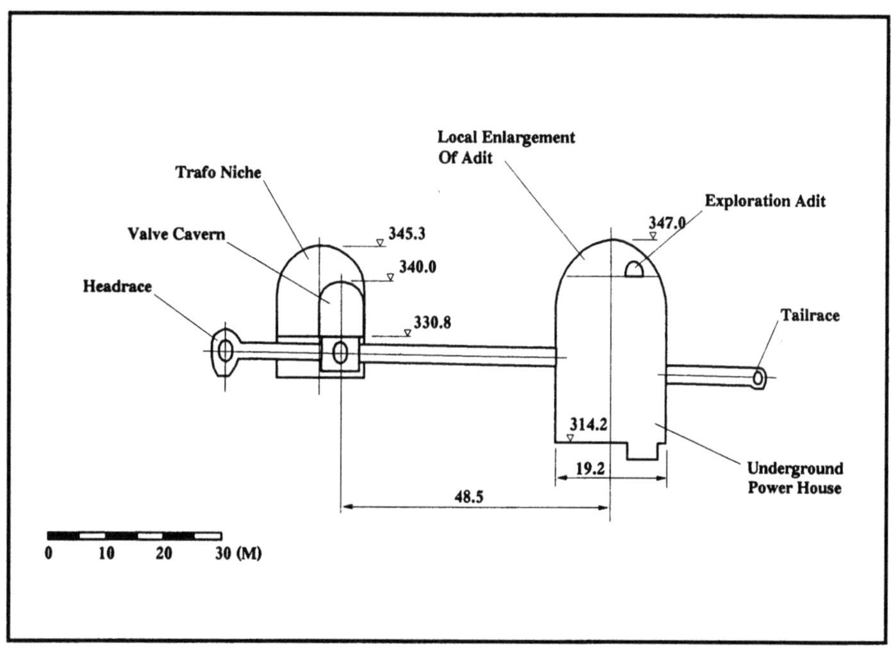

a

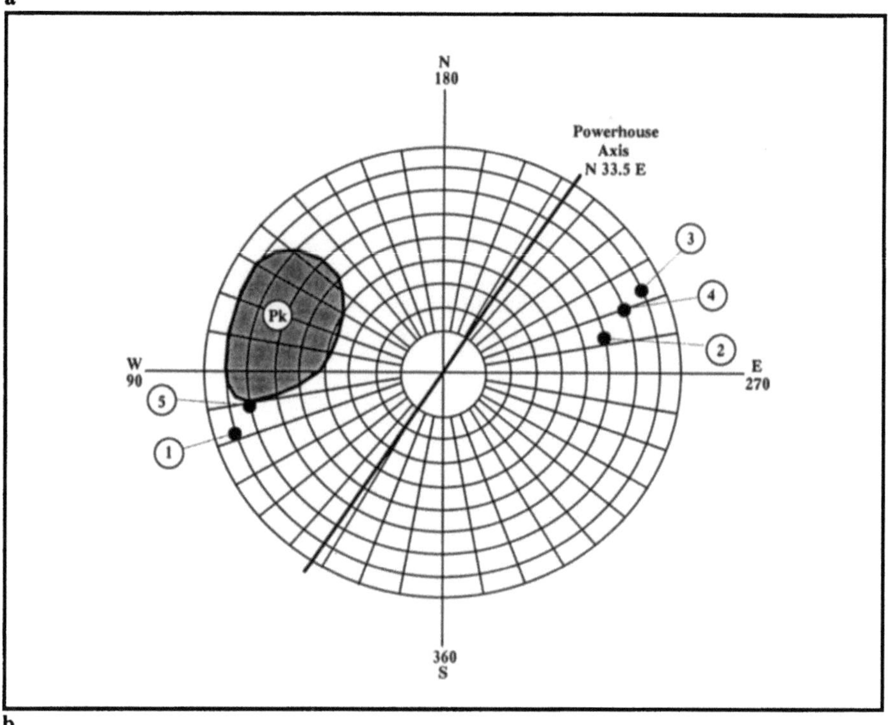

b

Fig. 15.1   **a** Cross-sectional configuration of the proposed underground powerhouse; **b** lower hemisphere stereographic projections of the major joint and fault orientations

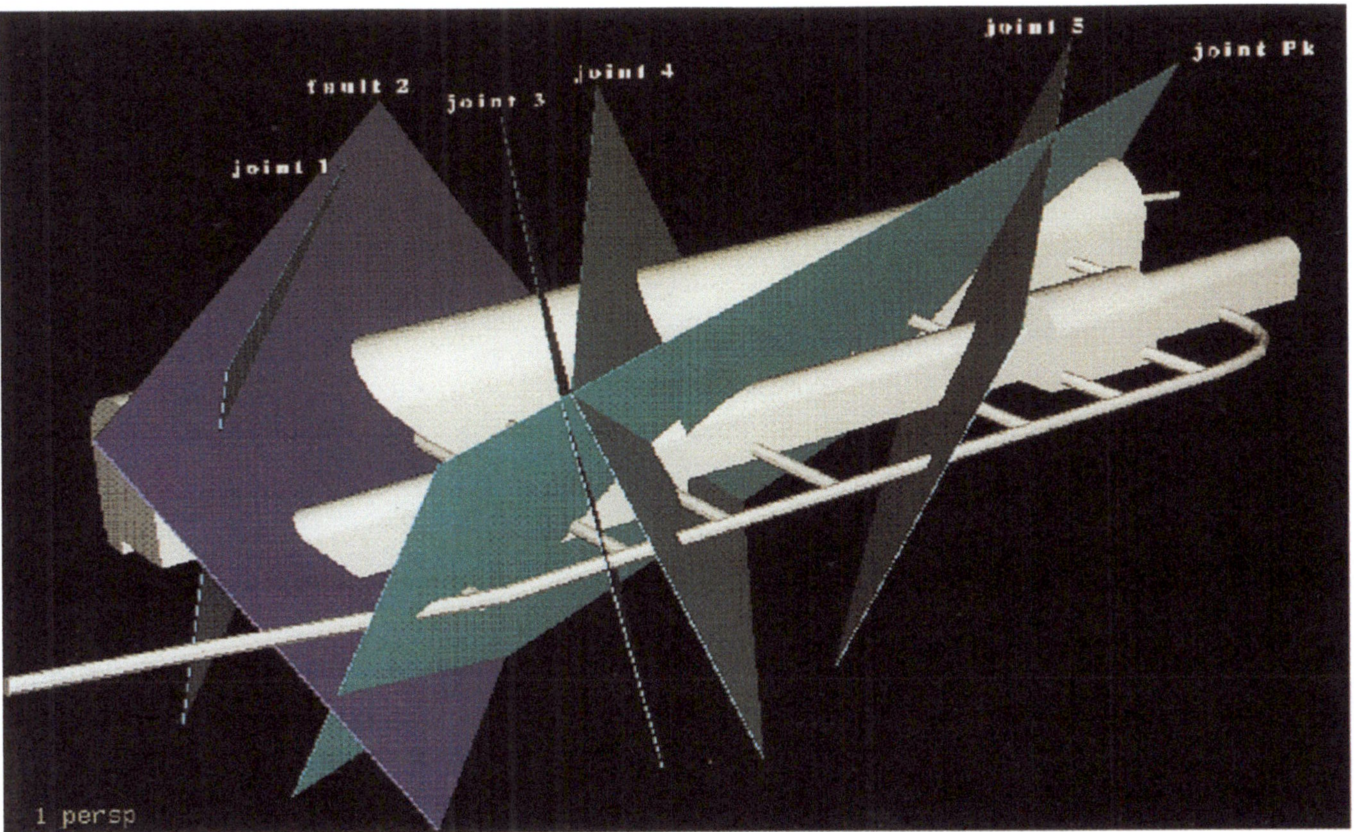

Fig. 15.2 Volume model representation of the powerhouse excavation with primary joint and fault plane orientations superimposed

These orientations and the mapped locations of joint and fault intersections with the exploration adit form the basis of our structural characterization.

Rock strength and deformability properties are available from laboratory test results and in-situ tests performed within the exploration adit. Previous analysis shows a strong correlation between these geomechanical properties and the rock mass variation throughout the region.

15.2 Structural Rock Joints and Faulting

This application illustrates the simple, yet effective, use of viewplane orientation capabilities to obtain a 3D structural model. Because observations of rock jointing and faulting are limited to the exploration adit, we have insufficient information to achieve a precise (triangulated) representation of these irregular surfaces throughout the region. We are reduced to representing them as planar surfaces at appropriate locations and orientations, a common practice in rock mechanics investigation.

The first step is to create a volume model representation of the geometry of the exploration adit, at the correct depth and orientation, using the excavation design tools discussed in Chapter 9. This provides a spatial reference for locating joint and fault observations in 3D space. To define the plane of a particular joint we simply locate the viewplane at the appropriate point along the axis of the adit, rotate it to the azimuth (strike) and inclination of the joint, and define a rectangular volume model component of nominal thickness. The azimuth and inclination of the joint are obtained from its stereographic projection (which uses a different convention for rotations (see Fig. 15.1b) where

azimuth (strike) = dip direction + 90°
inclination = 90° - dip.

The rectangular extent of the joint is defined so as to project throughout the region of interest. In the case of joint Pk, the mean orientation is used initially, although observations of the joint exhibit significant variability in orientation (cf. Fig. 15.1b). The final structural model consists of a set of planar volume model components that represent the joint and fault orientations.

By superimposing this structural model on a volume model representation of the proposed underground excavation, we obtain an enhanced appreciation of the structural interrelationships (Fig. 15.2). It is immediately apparent that joints 1 to 5 and the fault do not present significant rock stability problems in the main excavation since their orientations are approximately transverse to the primary axis. On the other hand, joint Pk intersects the southern portion of the main excavation at an acute angle, creating a potentially unstable wedge configuration with the roof and west wall of the excavation which opens upwards. Figure 15.3 is

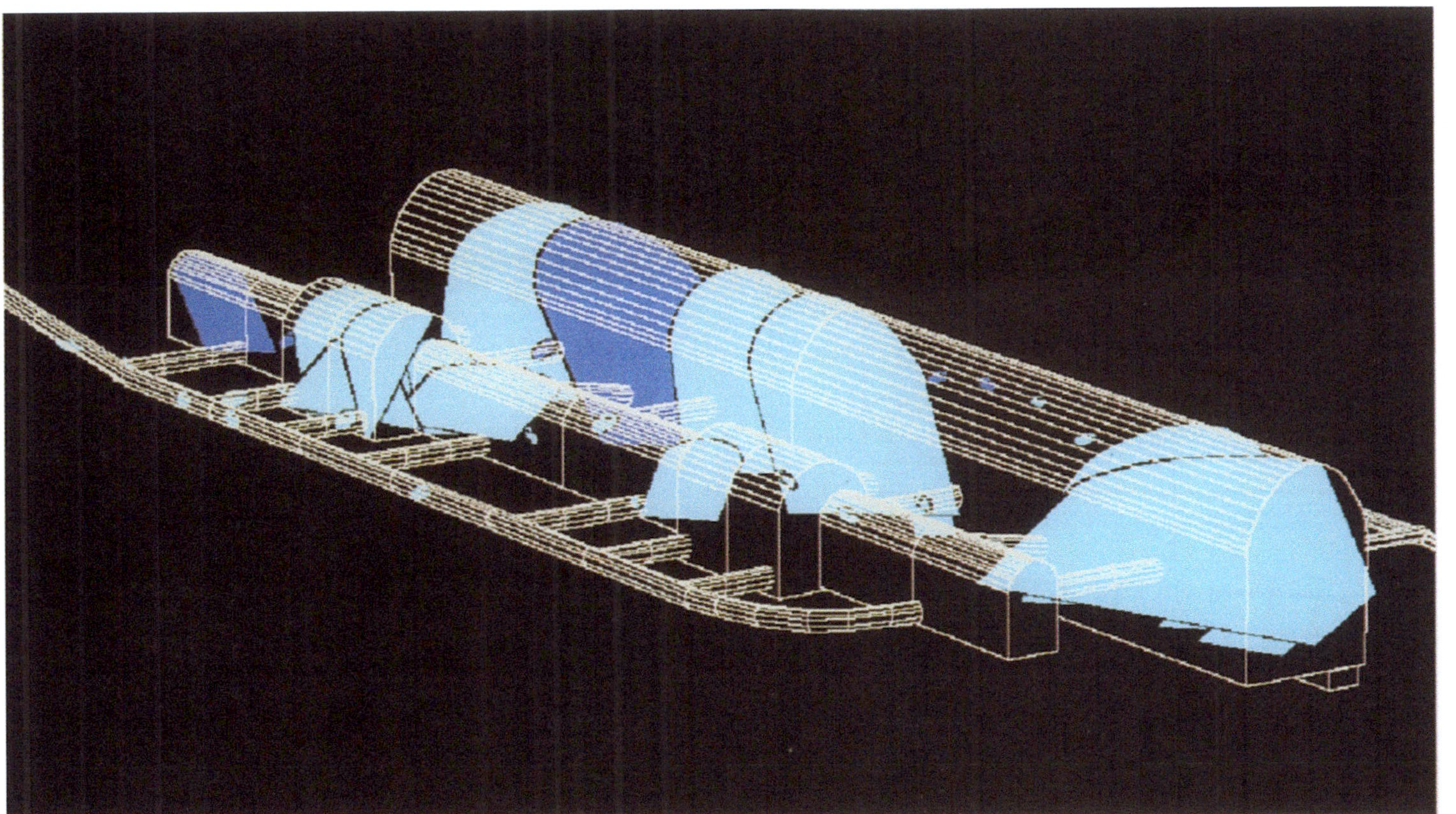

Fig. 15.3 3D visualization of the volumetric intersection of the powerhouse excavavtion with the primary joint and fault planes

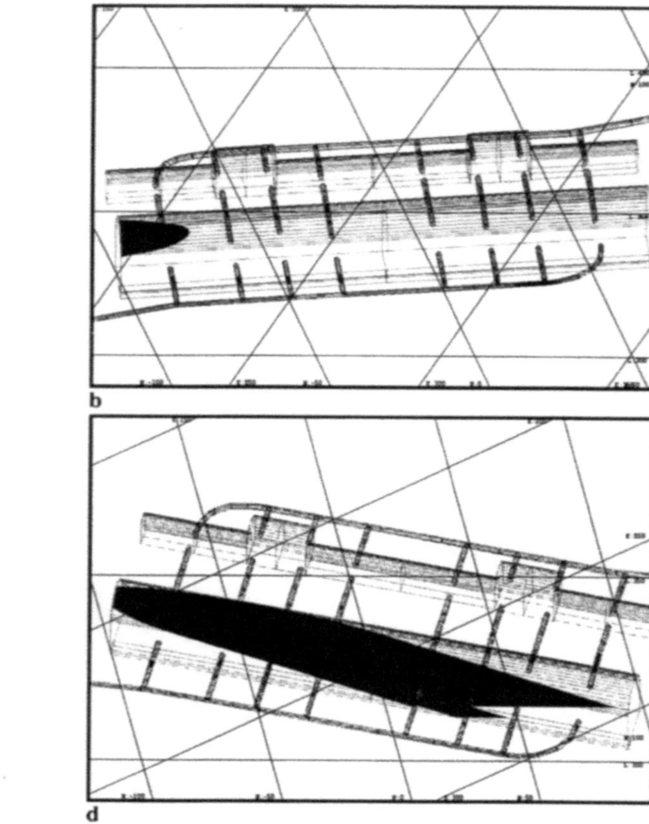

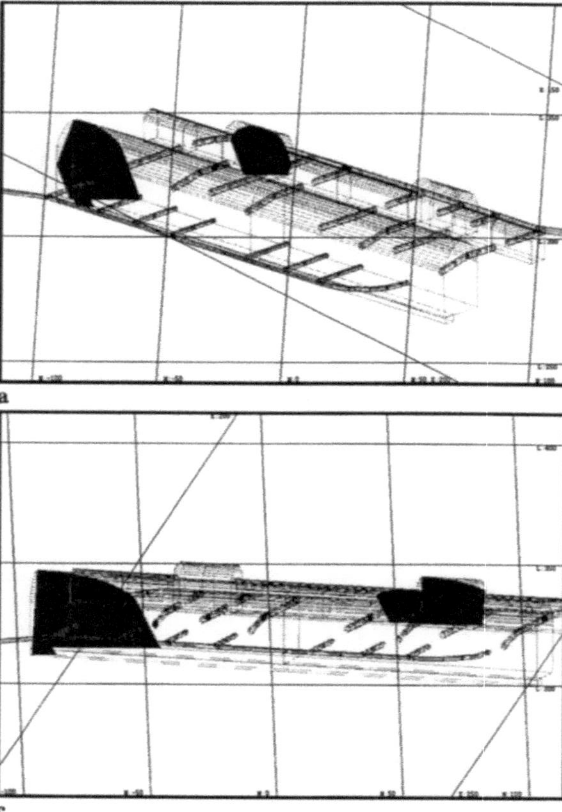

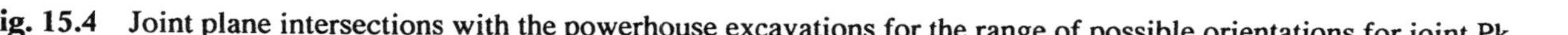

Fig. 15.4   Joint plane intersections with the powerhouse excavations for the range of possible orientations for joint Pk

a visualization of a volumetric intersection between the structural model and the excavation model, which further emphasizes this problem. It is also apparent from this second visualization that there is a convergence and intersection of several joint planes in the vicinity of the northern valve chamber excavation. This is a further indication of potentially unstable wedge conditions, although of less significance due to the reduced dimensions of the excavation in this area.

Figure 15.4 illustrates the observed variability in orientation of joint Pk, shown as a series of viewplane intersections of the range of possible joint orientations with the excavation profile. In the worst case scenario, a considerable portion of the west wall of the excavation requires reinforcement. In the absence of more detailed information, the approved construction procedure is to monitor the location of joint Pk during staged excavation of the powerhouse and to reinforce appropriately as the excavation progresses.

15.3 Lithology and Rock Mass Characteristics

The observed lithological variation of the proposed site is relatively simple. It consists of gneiss with a granite-like texture in the north (rock type AG), and gneiss with a distinctive parallel texture in the south (rock type LG). The contact between these rock types is a near-vertical surface that strikes at N160E and intersects the adit 140 m from the northern end of the excavation. The contact exhibits no evidence of structural discontinuity.

The spatial variations of geomechanical properties appear to be controlled more by the rock mass characteristics of the secondary jointing than by lithological classification. Joint mapping from the exploration adit investigation has provided observations of 537 secondary joints. Statistical and stereographic projection analysis of these results indicates the existence of five joint groups that comprise two principal rock mass characterizations, identified as class I and class II. The spatial variation of rock mass characteristics is simplified to a vertically banded formation striking approximately parallel to the average orientation of the major joints and the lithological contact. A simple volume model representation of this rock mass characterization is defined from vertical viewplanes along the primary axis of the excavation. Figure 15.5a is a visualization of a volumetric intersection between the rock mass characterization model and the excavation model that clearly illustrates the distribution of the principal rock classes throughout the excavation. In many rock mechanics investigations, the ability to quantify and visualize spatial intersections in this way can be of considerable advantage in the design and distribution of rock reinforcement. The volumes of different materials intersected can also provide a direct measure of the quantities of reinforcement required.

Analysis of the results of laboratory and in-situ tests for geomechanical properties indicates that their variation correlates approximately with the rock

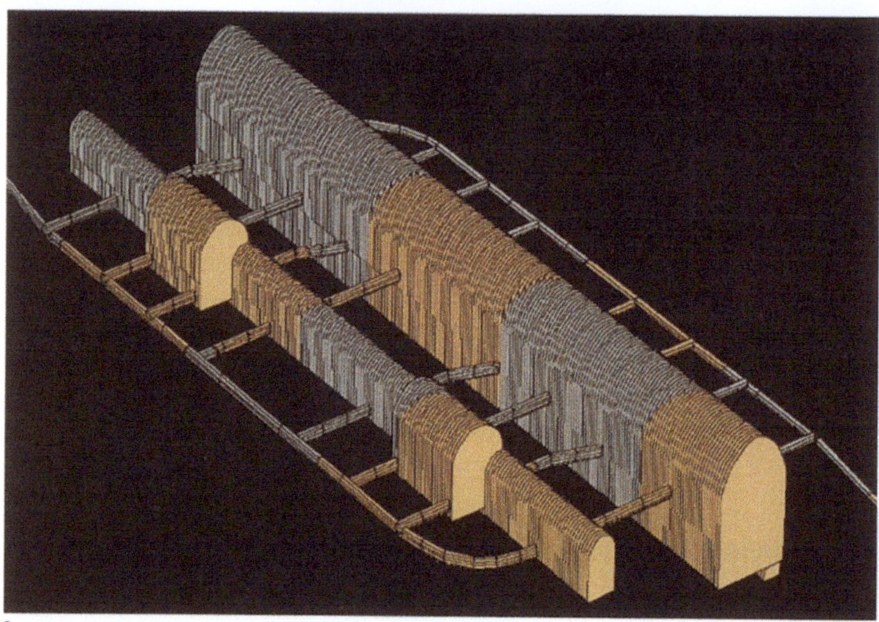

Fig. 15.5   **a** Visualization of a volumetric intersection of the powerhouse excavation geometry with the rock mass characterization model; rock mass class I is *gray*, class II *brown*; **b** transverse vertical section through the rock mass characterization, joint structures (blue) and excavation profile for export to numerical analysis

mass classification. In this case, the properties are assumed constant throughout each of the class I and class II materials due to the small number of samples available. It is worth noting that recent research has shown that geostatistical prediction can be used effectively for determining the spatial variation of rock mass discontinuities, and might have been employed to advantage here, if the original mapping observations were available.

15.4 Production of Analytical Sections

A significant advantage of 3D characterization in this context is that the information is immediately available for export to numerical analysis technologies. In this case, the objective is finite element analysis, on 2D vertical sections, of the destabilizing effect of the excavation on joint Pk. Figure 15.5b is a vertical viewplane section through the rock mass characterization model, structural model and excavation model at a point where joint Pk intersects the floor of the excavation.

The geometrical information contained in this viewplane visualization is readily transferable, in terms of real coordinates, to a map data structure which is in turn exportable in an ASCII format file. It then becomes available for finite element mesh generation in preparation for numerical analysis. We could equally well export a series of sections at appropriate intervals that would provide the geometrical basis for 3D numerical analysis. Once the results of numerical analysis are available, they can be imported for visualization and analysis in the context of the characterization models.

15.5 Rock Reinforcement Design

The characterization models also provide a vehicle for rock reinforcement design. By working on a series of appropriately scaled viewplane sections such as that shown in Fig. 15.5b, and using interactive design tools, we can readily determine the optimum locations, orientations and lengths of rock reinforcement between the excavation profile and joint Pk.

The structural characterization of joints and faults developed here is obviously an idealized simplification of the real conditions. By incorporating additional joint and fault observations obtained during progressive excavation, a much more detailed and representative structural characterization would be obtained. With sufficient observations the joint and fault surfaces would be represented by irregular, triangulated surfaces as discussed in Chapter 7, a much closer approximation of the real conditions.

15.6 Project Summary

As noted in the introduction, this project represents a relatively simple and straightforward application of 3D characterization techniques. Nonetheless, it effectively demonstrates the capabilities in a rock mechanics/geotechnical context. It also emphasizes the advantages of 3D visualization for obtaining an enhanced appreciation of the complex geometrical interrelationships between joints, faults and excavation profiles. Used appropriately, the 3D characterization techniques also provide a vehicle for generating the information required for stress and deformation analysis using finite element techniques, for visualization of the analytical results in an appropriate context, and for rock reinforcement design.

16 New Directions in Spatial Prediction, Modeling and Database Management

16.1 New Directions in Geoscience Modeling

Part I presented a generic, integrated, computerized approach to geological characterization called geoscience modeling that addresses many of the limitations and approximations of the past. Part II presents the results of applying this approach to a variety of characterization problems in the geosciences that confirm its viability and usefulness. However, development does not stop at this point; anyone who has used the approach discussed here, or any other computerized approach, is already aware of the next level of inefficiencies and limitations to be addressed.

The principal objective of this chapter is to introduce several new or emerging computer-based techniques that have the potential for significant enhancement of the geological characterization process. The techniques covered are not intended to comprise an exclusive list; there are undoubtedly others that could have equal or greater impact. The focus here is on techniques that either improve our ability to represent complex conditions with greater precision, or reduce the effort on our part required to achieve an acceptable characterization. Two of the techniques covered concern new approaches to spatial prediction, another concerns enhanced interpretation and representation of geological volumes. At the end of the chapter we consider some of the requirements for transforming what is essentially still a computer modeling process into a full 3D geoscience information system.

16.2 Direct Geostatistical Prediction for Irregular Volumes

In Chapters 5 and 8 we discussed regular geostatistical prediction techniques and their application to the spatial variation of a variable, represented as predicted values and uncertainties at points located at regular intervals within a 3D grid data structure. We established that geostatistical prediction has significant advantages over other spatial prediction techniques, a principal advantage being that it provides a measure of the uncertainty associated with any predicted value. In Chapter 8, we also established that both the predicted value and its uncertainty are dependent to some degree on the *volume* that the predicted value represents. Thus, geostatistically speaking, predicting an average value of a variable for a grid-cell volume is preferable to (more precise than) predicting a value for the grid-cell centroid (a point). This is particularly true if the objective of our characterization is a volumetric analysis of the predicted spatial variation. In general terms, the larger the volume is, the more representative the predicted value and the lower the associated uncertainty.

In many instances, it is a distinct advantage to be able to predict the average value of a variable *directly .... for any irregular volume .... no matter what its shape and size.* This approach bypasses the requirement for predicting values at grid-cell centroids, as well as any subsequent spatial analysis of volumetric intersections. The average value for any volume is predicted directly from the samples in its vicinity, using the selected prediction model. This approach has several advantages over the grid-based approach discussed in Chapter 8.

- It is computationally and procedurally more efficient.
- The resulting predicted value is more representative of the volume than the result obtained by averaging the grid-cell values within the volume.
- The associated measure of uncertainty (standard error) applies directly to the volume; the necessity for manipulating spatial variations of uncertainty to obtain best case/worst case scenarios for risk assessment is eliminated.
- The approach lends itself to *spatial query applications*; we can point to any volume located anywhere in space and interrogate its contents, move the volume to a new location and re-interrogate, and so on.

The approach has significant advantages in the context of designing excavations to maximize their contents, essentially a mine planning function. However, it also has more generic application, first, because of its enhanced prediction and uncertainty capabilities and, second, as a generic spatial query tool. We return to a discussion of the latter in the final section.

The starting point for a discussion of the underlying concepts of this approach is the set of kriging equations for determining sample weighting coefficients for prediction, thus

$$\sum w_i \bar{\gamma}(s_i, s_j) + \lambda = \bar{\gamma}(s_i, v) \quad \text{(for each sample)}$$
$$\sum w_i = 1,$$

and the kriging variance equation (from which we determine the standard error)

$$\sigma^2 = \sum w_i \bar{\gamma}(s_i, v) + \lambda - \bar{\gamma}(v, v)$$

where $\mathbf{w}_i$ is the sample weighting coefficient;

$\mathbf{s}_i$ is the sample value;

$\bar{\gamma}(\_,\_)$ is the semi-variogram function;

$\mathbf{v}$ is the volume of interest;

σ is the standard error;

λ is a Lagrangian multiplier term, required to balance the equations.

The coefficients for these equations include semi-variogram functions between a number of points, including the following:

- between each *pair of samples*, represented by the term $\bar{\gamma}(\mathbf{s}_i,\mathbf{s}_j)$;
- between each *sample* and each *point* within the volume, represented by the term $\bar{\gamma}(\mathbf{s}_i,\mathbf{v})$;
- between each *pair of points* within the volume, represented by the term $\bar{\gamma}(\mathbf{v},\mathbf{v})$.

It is obvious from these relationships that a predicted value (based on the weighting coefficients) and its uncertainty (based on the standard error term) are both dependent on the size (extent) and shape (geometry) of the volume. Thus, as we discussed in Chapter 8, for point kriging, the "volume" is represented by a single point. For regular volume kriging, the grid-cell volume is represented by an array of regularly spaced points, essentially a *point discretization* of the cell volume. By extension, volume kriging can be applied directly to any irregular volume provided we can achieve a representative point discretization of it.

Point discretization of the volume of a grid cell is easy; it has a simple rectangular shape. Derivation of an efficient discretization algorithm that is generally applicable to any irregular volume is not quite so easy. For the approach to be useful as a spatial query tool, the algorithm must be both efficient (fast) and precise. To achieve this we make use of the volumetric integration capabilities of the volume data structure that we discussed in Chapter 6. We define a series of parallel, equally spaced intersection planes through the volume of interest. On each of these planes, the boundary of intersection with the volume is determined using the procedures discussed in Chapter 6. We overlay a 2D grid of regularly spaced points on the plane and then clip it to the intersection boundary, discarding grid points outside of the boundary (Fig. 16.1a). These steps are repeated for each intersection plane; in each case, the remaining grid points are accumulated in an array that finally represents a point discretization of the volume (Fig. 16.1b).

In a similar fashion to the regular volume kriging approach discussed in Chapter 8, this array of points is then fed to the kriging equations together with the sample values and their locations. The resulting weighting coefficients are applied to the sample values to predict the *most likely average value* of the variable of interest for

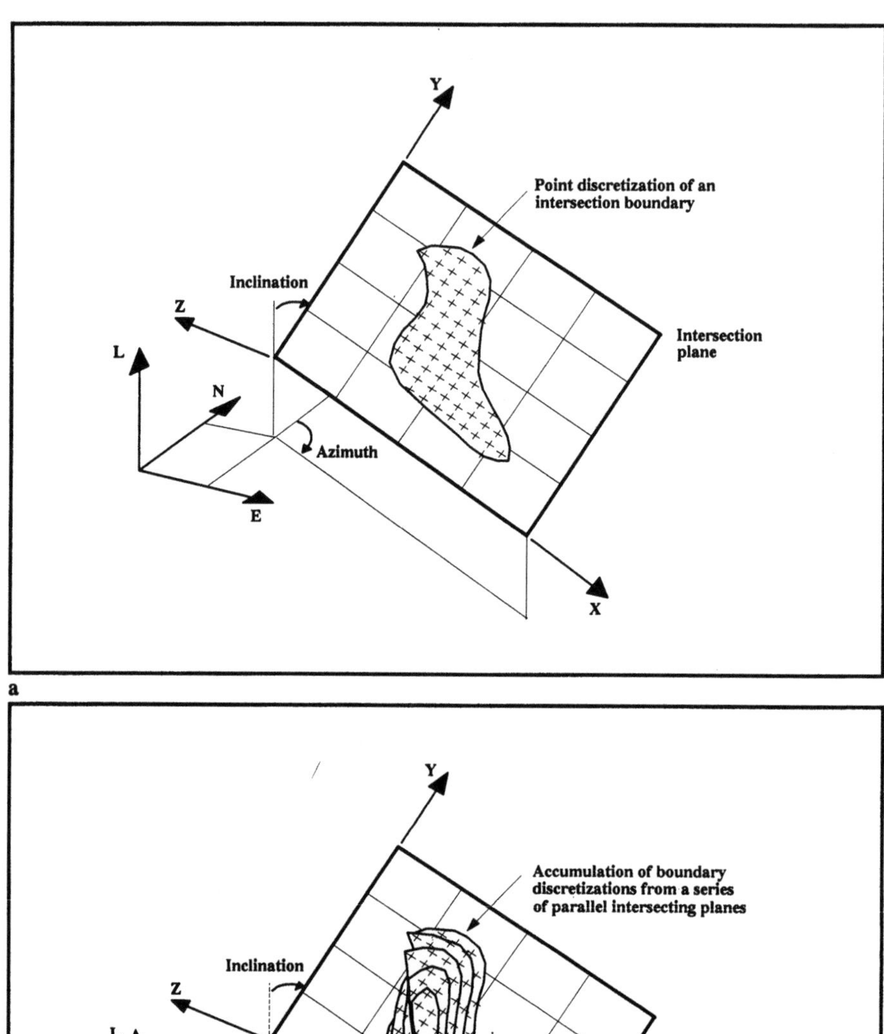

Fig. 16.1   **a** Point discretization of the boundary of intersection between an irregular volume and an intersecttion plane; **b** accumulation of a series of boundary discretizations to achieve discretization of an irregular volume

the irregular volume. The *measure of uncertainty*, or standard error, is determined directly from the kriging variance equation.

These results expand our horizons for risk assessment significantly. As discussed in Chapter 10, the predicted average for the volume has a normal distribution of values, and the shape of the probability density function is defined by the magnitude of the standard error. Thus, we can say with 68% confidence that the *real average* lies within one standard error on either side of the predicted average. This is a considerably more useful and direct result than that obtained by manipulating the spatial variation of uncertainty to obtain best case/worst case scenarios (see Chap. 10). It is also a more precise result since the geometry and extent of the volume of interest have been accounted for in the prediction process. And, finally, it has a reduced uncertainty associated with it; the potential variability of the result is significantly less than that of a point or a grid-cell volume.

Direct geostatistical prediction for irregular volumes is still an emerging technique and has not yet been widely accepted in practice. It holds considerable promise in certain application areas. In a mine planning context, the application of this approach provides a precise volume, average mineral grade and measure of uncertainty for every excavation unit in a mine plan. This allows us to derive an excavation schedule that minimizes production risk and highlights those units in a mine plan that have high uncertainty, requiring further investigation/exploration.

In a generic spatial query context, the approach provides a useful tool for direct determination of the average value and uncertainty of a variable within a predefined volume of any shape. The approach does not eliminate the need for the geostatistical prediction methods discussed in Chapters 8 through 10, or the need for the 3D grid model data structure. Without these we cannot visualize the spatial variation of a variable or determine the isosurfaces that are a necessary part of many characterizations. However, having determined an isosurface we can then apply the new approach to obtain a more precise and representative estimate of its contents.

16.3 Splining Techniques for Volume Representation

Most of the computer techniques currently employed to represent geological volumes are essentially based on definition of one or more surfaces that enclose a volume. The gridded surface techniques discussed in Chapter 1 and the more flexible triangulated surface technique discussed in Chapter 7 all assume that a volume is defined by an upper and a lower surface. The interactive vector-based volume modeling approach discussed in Chapter 6 requires a volume to be defined as a series of irregular subunits (components), each comprised of closed surfaces defined by three parallel boundaries and links between them.

We have established that a principal advantage of the volume modeling approach is its ability to deal with geological discontinuities. This is primarily due to the fact that it works with discrete subunits (an irregular volume can be truncated abruptly at a discontinuity) and voids in an interpretation can be filled in by appropriately shaped and oriented components (see Chap. 6). However, a disadvantage arises due to the fact that the boundaries defining a component consist of discrete points joined by line segments, and the boundaries in turn are similarly joined by line segments. This implies a linear interpolation of geology between points and boundaries. In practical terms, it means that a precise representation of a highly irregular but continuous volume requires definition of a large number of boundaries with many points in each.

The same degree of precision is achieved with fewer components, and thus fewer defining boundaries and points, if linear interpolation is replaced by a higher order interpolating function. This approach has the potential for significantly improving the efficiency of any interactive interpretation of geology.

Several authors have proposed the use of mathematical functions to define geological surfaces. The acknowledged disadvantage of all such approaches is their inability to deal successfully with discontinuities. However, this disadvantage is eliminated if we combine the concept of higher order interpolation functions with the concept of volume subunits, as employed by the vector-based volume modeling approach. With this approach we retain the advantages of both concepts, i.e. definition of an irregular volume requires fewer components and hence less work on our part, and we can still accommodate complex structures and discontinuities. The question remains as to what type of interpolation function to use.

A singular disadvantage of polynomial functions is their unpredictability, i.e. changing the location of a single point may result in unforeseen changes in the surface representation at a considerable distance from the point. This effectively eliminates polynomials from further consideration. A more promising result is obtained with spline functions, or NURBS (non-uniform rational B-splines). The spline is classically defined as a piecewise parametric polynomial of degree k with continuous derivatives of order k-1 at the common joints between segments. Figure 16.2a is a simplified 2D representation of a spline function. It overcomes the problem of unpredictable oscillation found in higher order polynomial functions, since the effects of changing the location of a point are restricted to the vicinity of the point. At the same time, it is capable of providing an acceptably continuous representation of a highly irregular surface.

Unfortunately, Fig. 16.2a also emphasizes a basic problem inherent to all higher order surface-fitting (interpolation) techniques. The resulting surface representation is always an approximation of the true surface defined by the points. Instead of containing the points, the surface representation is invariably a smoothed approximation that passes between them. It has been argued that this is a minor deficiency since any representation of a surface is always associated with a degree of uncertainty anyway. Although this is true to some extent, ideally the

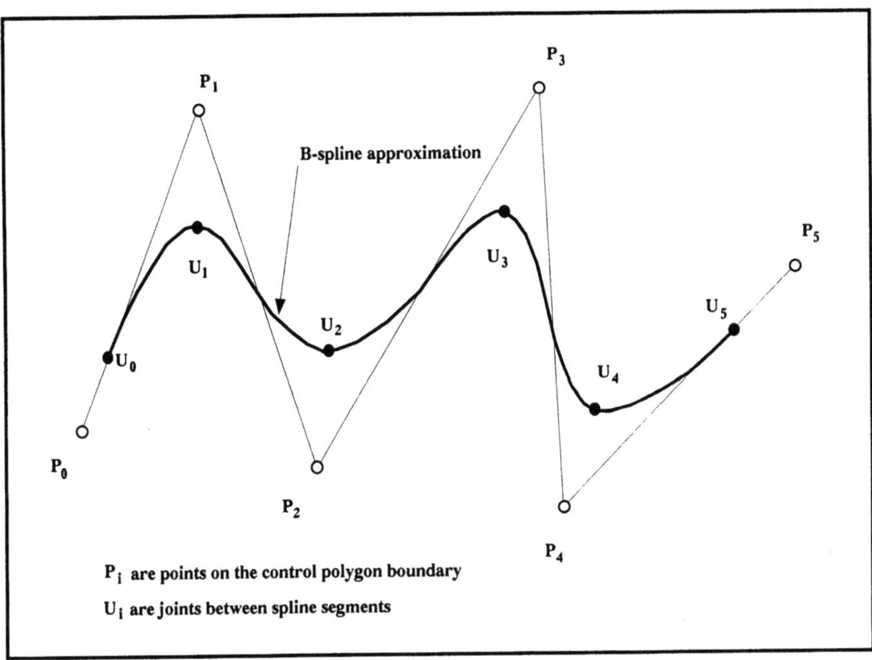

a

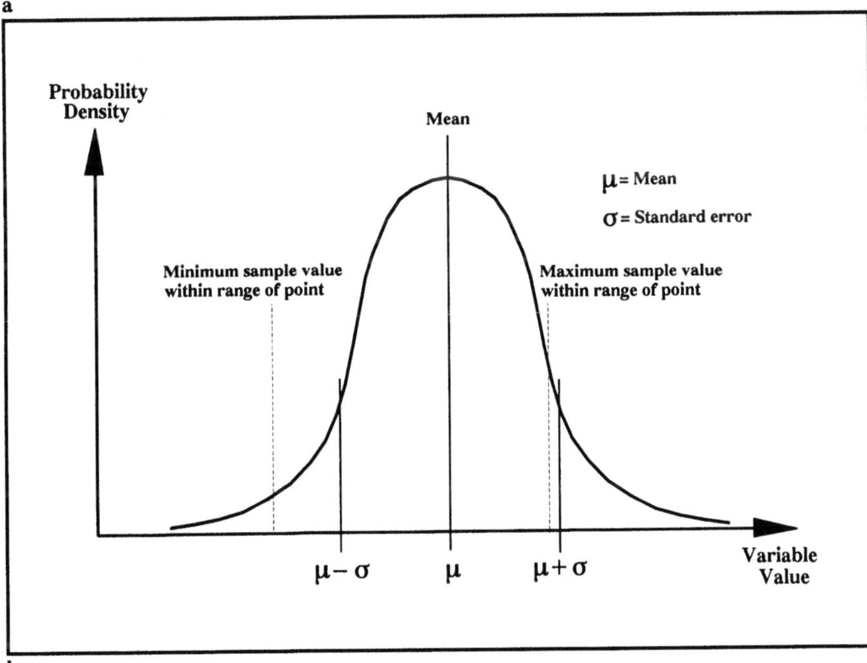

b

Fig. 16.2 **a** Smoothed B-spline representation of an irregular polygonal boundary; **b** in geostatistical prediction, the predicted value at a point (mean) is constrained between the minimum and maximum sample values within range of the point; this constraint is eliminated in the conditional simulation approach to prediction

representation of a surface should reflect its *most likely location*, based on our interpretation and/or prediction. Representation of the uncertainty involved is a parallel but separate requirement that should not be reflected by approximation of the surface representation itself. Surface approximation can also produce disconcerting side effects during interactive interpretation of a geological boundary; having defined a point on a boundary, we intuitively expect the resulting boundary to pass through it, instead of close to it. In practice, we require more of a *what you see is what you get (wysiwig)* approach.

In principle, the concept of combining the advantages of higher order interpolation functions with the volume modeling approach of representing geological volumes by a set of discrete subunits has much to offer. In practice, there are inherent deficiencies that have yet to be resolved.

16.4 Conditional Simulation and Stochastic Approaches to Spatial Prediction

Geostatistical prediction as discussed throughout this text is concerned with predicting (estimating) spatial variations in terms of an average, or most likely, value at every point considered, together with a measure of its potential variability (the standard error term). The average values implicitly involve some degree of *smoothing* of the true spatial variation, although geostatistical prediction is less prone to this than most other techniques. By integrating the prediction process with the spatial control provided by a volume model representation of geological characteristics, we eliminate the tendency to smooth the variation of a variable across observed discontinuities. However, there may be other discontinuities within the defined geological units that we are not aware of and have failed to define beforehand. In these cases, geostatistical prediction is subject to an inability to portray short-term fluctuation in a variable and has a tendency to underrepresent extreme values.

Conditional simulation is a stochastic approach to prediction that attempts to *simulate* the real conditions rather than estimating them in terms of averages and variabilities. The approach produces a spatial variation that exhibits the same degree of variability as the measured sample values, generally resulting in greater fluctuations between adjacent values than can be successfully represented by geostatistical prediction. It is particularly appropriate in cases of extreme spatial variability and results in a more precise representation of the true conditions.

The initial steps in the application of the approach are similar to geostatistical prediction, involving semi-variogram analysis of spatial variability and derivation of a suitable prediction model. From there on, the approaches diverge. Conditional simulation uses the prediction model in a stochastic context to independently derive a number of values at each point in turn. If the number of values derived is statistically   representative,   then   they   have   the   same   frequency distribution of value as the measured variability. The initial results comprise an equivalent number of spatial variations of the variable of interest. Each spatial variation is different from the next, each honors the measured sample values and their locations, and each is a statistically possible solution, i.e. it falls within the extent of the normal distribution probability curve (Fig. 16.2b). If we produce a sufficient number of these and count the number of times a particular value occurs (or is exceeded or diminished) at a point, then we have a statistical measure of its *probability of occurrence* at the point. This is a similar result to that obtained by kriging with indicator transforms, as discussed in Chapter 8. In contrast to geostatistical prediction, however, the highest and lowest values within a spatial variation produced by conditional simulation are not necessarily constrained by the highest and lowest measured sample values. They are constrained instead by statistical probability, based on the measured variability. This difference is illustrated in Fig. 16.2b, the geostatistically predicted value (the mean of the normal distribution) always falls somewhere between the minimum and maximum sample values within range of the point. The only way to derive values beyond these constraints is to add or subtract multiples of the standard error, as described in Chapter 10. The stochastic approach employed by conditional simulation allows prediction of any value within the extent of the normal distribution probability curve, however, the frequency of occurrence of the value is governed by the shape of the curve. This is a more realistic approach than assuming that we are always fortunate enough to sample the exact locations of the extreme values of a variable!

In summary, conditional simulation is a new technique with growing acceptance in the geosciences, although it is not yet widely available. It is more efficient than geostatistical prediction at representing conditions of extreme spatial variability, and to many of its users is a more statistically acceptable technique. It has the same requirements in terms of data structures, geological control, spatial analysis and visualization as the techniques presented throughout this text.

16.5 Evolution of a 3D Geoscience Information System

Several authors have already drawn parallels between the current capabilities of geoscience modeling and those of geographic information system (GIS) technology. It has been suggested that the term *geoscience information system (or GSIS)* would be more appropriate terminology for the former. Both technologies are designed to consider similar types of spatial information and to provide (essentially) similar analytical, modeling and visualization functionality, GIS in a two-dimensional context and geoscience modeling in three. The similarities end there. A GIS is at least part way to being a true spatial information system by virtue of its relational database links. It also has superior capabilities with respect to scaling of information detail for visualization purposes.

The advantages of relational database technology are well established. In the case of GIS technology, a partial integration has been achieved only as a result of a messy, hybrid solution to the problem of trying to manage spatially related information in a relational database context. This typically involves managing the spatial component of information in a proprietary database with appropriate links to samples, observations and other information stored in a relational database. This is not only inefficient, it also jeopardizes data integrity and creates a computer housekeeping nightmare. Efficiency and data integrity have been sacrificed for the advantages in functionality offered by relational database technology.

This hybrid solution has been necessitated by the current inability of relational databases to manage spatial information, particularly from a spatial query point of view. A relational database simply grinds to a halt when required to consider information associated with X and Y coordinates, each with a (theoretically) infinite range in value. As noted in Chapter 4, this problem is exponentially worse in three dimensions, and we have yet to consider the time dimension.

Recent developments in relational database technology suggest that the problem may soon disappear. The proposed solution is to simplify the problem by reducing two, three, or even four, dimensions to one by using compound *helical coordinates*. If this works, then 2D GIS technology will come of age, and the 3D GSIS concept may yet become a reality. The spatial interrogation advantages will be dramatic, for example, display/analysis of all volumes where

> **Contaminant A > X**
> **Contaminant B > Y**
> **Lithology = Sand**
> **Moisture Content = Saturated**
> **..... and so on.**

Of course, there are other requirements to be met before GSIS is a reality, but an effective spatial relational database technology is arguably the top priority. The ability to consider both irregular geological volumes and spatial variations of variables as data objects is one requirement. Being able to perform appropriate spatial queries on these objects is another. Incorporation of current GIS capabilities for relating the level of detail displayed to the scale of visualization is yet another.

The technique for direct geostatistical prediction of irregular volumes discussed at the beginning of the chapter is a potentially useful tool in a spatial query context. It would almost eliminate the need for 3D grid data structures from the characterization process. Transient 2D grids would be adequate for visualization purposes, and 3D grids would be required only for determining an isosurface of a variable.

Finally, we noted in Chapter 4 the current lack of formatting standards for transfer of spatial information. The recent growth in GIS applications has provided impetus for the development of badly needed standards. We currently waste enormous amounts of time re-entering information that already exists in electronic form because of incompatible data formats. Implementation of appropriate data transfer standards will do much to expand the application of geoscience modeling and in turn drive the development of a true GSIS technology.

In closing, geoscience modeling has evolved dramatically in recent years in terms of both technology and application. It obviously has a lot further to go yet.

Bibliography

Introduction

Philip J R (1991) Soils, natural science, and models. J Soil Sci Jan 151(1):91-98

1   The Geological Characterization Process

Bak P R, Cram D L, Prissang R (1992) Interactive evaluation of linear octree encoded deposit models. Proc XXIII Symp Application computers and operations research in mineral industry, Tucson pp 691

Bawiec W J, Grundy W D (1992) Computer generated surfaces and grids of the geometry, rock type and chemistry of a bedded mineral deposit. In: Hamilton D E, Jones T A (eds) Computer modeling of geologic surfaces and volumes. AAPG Computer Applications in Geology No 1 pp 37

Clarke D (1992) The gridded fault surface. In: Hamilton D E, Jones T A (eds) Computer modeling of geologic surfaces and volumes. AAPG Computer Applications in Geology No 1 pp 141

Cairns J L, Feldkamp L D (1993) 3D visualization for improved reservoir characterization. SPE Computer Applications June 1993

Houlding S W (1986) Benefits and misconceptions of mining computing. Int Mining J Oct:14-24

Houlding S W (1988) 3D computer modeling. Eng and Mining J Aug:45-47

Houlding S W (1991) Computer modeling limitations and new directions - part 1. CIM Bull 84(952):75-80

Turner A K (1992) Applications of three-dimensional geoscientific mapping and modeling systems to hydrogeological studies. In: Turner A K (ed) Three-dimensional modeling with geoscientific information systems. NATO ASI Series C vol 354, Kluwer, Dordrecht pp 327

US Environmental Protection Agency (1991) GEO-EAS user's guide. Environmental Monitoring Systems Laboratory, US EPA, Las Vegas pp 1.1

2   Complicating Factors for Computerization

Anderson M P (1994) Basic science required for decision making: geological setting. Subsurface restoration (in press). Lewis Publishing, Chelsea MI

Kelk B (1992) 3D modeling with geoscientific information systems: the problem. In: Turner A K (ed) Three-dimensional modeling with geoscientific information systems. NATO ASI Series C vol 354, Kluwer, Dordrecht pp 29

Matheron G F (1970) La théorie des variables régionalisées et ses applications. Tech rep fascicule 5, Les Cahiers du Centre de Morphologie Mathématique de Fontainebleu, École Supérieure des Mines de Paris

Sides E J (1992) Reconciliation studies and reserve estimation. In: Annels A E (ed) Case histories and methods in mineral resource evaluation. Geological Society Spec Publ No 63, UK pp 197

3   Features of an Integrated 3D Computer Approach

Burns K L (1992) Three-dimensional modeling and geothermal process simulation. In: Pflug R, Harbaugh J W (eds) Computer Graphics in Geology. Lecture Notes in Earth Sciences 41, Springer, Berlin Heidelberg New York pp 271

Fisher T R (1992) Use of 3D geographic information systems in hazardous waste site investigations (in press). In: Goodchild M F, Parks B, Steyaert L T (eds) Environmental modeling with GIS. Oxford University Press, New York

Houlding S W (1992) The application of new 3D computer modeling techniques to mining. In: Turner A K (ed) Three-dimensional modeling with geoscientific information systems. NATO ASI Series C vol 354, Kluwer, Dordrecht pp 303

Houlding S W (1991) Computer modeling limitations and new directions - part 2. CIM Bull 84(953):46-49

Houlding S W, Rychkun E A (1989) Technical computing: a corporate solution for the 1990's. Mining Mag May:401-411

Lasseter T J (1992) An interactive 3D modeling system for integrated interpretation in hydrocarbon reservoir exploration and production. In: Pflug R, Harbaugh J W (eds) Computer Graphics in Geology. Lecture Notes in Earth Sciences 41, Springer, Berlin Heidelberg New York pp 189

Lynx Geosystems Inc (1993) The Lynx GMS: geoscience modeling system: user documentation. Lynx Geosystems Inc, Vancouver pp D1.3

Raper J F (1992) Key 3D modeling concepts for geoscientific analysis. In: Turner A K (ed) Three-dimensional modeling with geoscientific information systems. NATO ASI Series C vol 354, Kluwer, Dordrecht pp 215

Turner A K, Kolm K E (1992) Potential applications of three-dimensional geoscientific mapping and modeling systems to regional hydrogeological assessments at Yucca Mountain, Nevada. In: Pflug R, Harbaugh J W (eds) Computer Graphics in Geology. Lecture Notes in Earth Sciences 41, Springer, Berlin Heidelberg New York pp 257

Wavefront Technologies Inc (1993) The data visualizer: user's guide. Wavefront Technologies Inc, Santa Barbara pp 7.1

Wolff R S, Yaeger L (1993) Visualization of natural phenomena. Springer, Berlin Heidelberg New York pp 57, 87, 121

4   Spatial Data Types and Structures

Lynx Geosystems Inc (1993) The Lynx GMS: geoscience modeling system: user documentation. Lynx Geosystems Inc, Vancouver pp B2.1, B3.1, D1.3, E1.1

Raper J F (1992) An atlas of three dimensional functions. In: Pflug R, Harbaugh J W (eds) Computer Graphics in Geology. Lecture Notes in Earth Sciences 41, Springer, Berlin Heidelberg New York pp 41

5   Analysis of Spatial Variability

Clark I (1979) Practical geostatistics. Applied Science Publishers, London pp 11

Englund E J (1990) A variance of geostatisticians. J Math Geol 22(4):417-455

Isaaks E H, Srivastava R M (1989) Applied geostatistics. Oxford University Press, New York pp 140, 369

Matheron G F (1970) La théorie des variables régionalisées et ses applications. Tech rep fascicule 5, Les Cahiers du Centre de Morphologie Mathématique de Fontainebleu, École Supérieure des Mines de Paris

US Environmental Protection Agency (1991) GEO-EAS user's guide. Environmental Monitoring Systems Laboratory, US EPA, Las Vegas pp 4.1

6   Geological Interpretation and Modeling

Houlding S W (1987) 3D computer modeling of geology and mine geometry. Mining Mag March:226-231

Houlding S W (1991) Computer modeling limitations and new directions - part 2. CIM Bull 84(953):46-49

7   Geological Modeling From Surfaces

Barchi M et al (1992) Computer modeling of surfaces: structural geology applications. In: Pflug R, Harbaugh J W (eds) Computer Graphics in Geology. Lecture Notes in Earth Sciences 41, Springer, Berlin Heidelberg New York pp 89

Houlding S W (1989) 3D component modeling of extensive thin seam deposits. Eng and Mining J July:28-31

Tipper J C (1992) Surface modeling for sedimentary basin simulation. In: Hamilton D E, Jones T A (eds) Computer modeling of geologic surfaces and volumes. AAPG Computer Applications in Geology No 1 pp 93

8   Geostatistical Prediction Techniques

Champigny N, Armstrong M (1991) Effect of high grades on the geostatistical estimation of gold deposits. Trans Soc Mining Metall Expl 288:1815-1818

Clark I (1979) Practical geostatistics. Applied Science Publishers, London pp 42, 99

Englund E J (1990) A variance of geostatisticians. J Math Geol 22(4):417-455

Isaaks E H (1984) Risk qualified mappings for hazardous waste sites: a case study in distribution free geostatistics. MSc Thesis Stanford University, Palo Alto

Isaaks E H, Srivastava R M (1989) Applied geostatistics. Oxford University Press, New York pp 278, 323, 351

Prissang R (1992) Three-dimensional predictive deposit modeling based on the linear octree data structure. In: Pflug R, Harbaugh J W (eds) Computer Graphics in Geology. Lecture Notes in Earth Sciences 41, Springer, Berlin Heidelberg New York pp 229

9   Spatial Analysis Techniques

Advanced Visual Systems Inc (1992) AVS user's guide. Advanced Visual Systems Inc, Waltham pp 1.15

Clarke D, Sharman K (1993) Modeling, planning and mining a geologically complex coal deposit. In: Bawden W F, Archibald J F (eds) Innovative mine design for the 21st century. Balkema, Rotterdam pp 237

Houlding S W, Stoakes M A (1990) Mine activity and resource scheduling with the use of 3D component modeling. Trans Inst Mining and Metall, UK, vol 99, A1-64:53-59

Houlding S W (1992) 3D geoscience modeling in the analysis of acid mine drainage. Proc 2nd Int Conf Environmental issues and management of waste in energy and mineral production, Calgary Vol 2 pp 741

Pflug R et al (1992) 3D visualization of geologic structures and processes. In: Pflug R, Harbaugh J W (eds) Computer Graphics in Geology. Lecture Notes in Earth Sciences 41, Springer, Berlin Heidelberg New York pp 29

Van Ooteghem R M, Rodenstein P K (1993) Computer modeling at the Sullivan Mine: a practical application in a complex operating environment. In: Bawden W F, Archibald J F (eds) Innovative mine design for the 21st century. Balkema, Rotterdam pp 671

Wavefront Technologies Inc (1993) The data visualizer: user's guide. Wavefront Technologies Inc, Santa Barbara pp 7.1

10   Uncertainty, Sampling Control and Risk Assessment

Blanchin R, Chiles J-P (1992) Geostatistical modeling of geological layers and optimization of survey design for the Channel Tunnel. In: Pflug R, Harbaugh J W (eds) Computer Graphics in Geology. Lecture Notes in Earth Sciences 41, Springer, Berlin Heidelberg New York pp 251

Houlding S W (1992) Real time grade control in mine planning and production. Proc XXIII Symp Application computers and operations research in mineral industry, Tucson pp 747

Isaaks E H (1984) Risk qualified mappings for hazardous waste sites: a case study in distribution free geostatistics. MSc Thesis Stanford University, Palo Alto

Rautman C A, Treadway A H (1991) Characterization uncertainty and its effects on models and performance. Proc. 2nd Int. High-level radioactive waste management Conf, Las Vegas Vol 2 pp 1491

Sides E J (1992) Reconciliation studies and reserve estimation. In: Annels A E (ed) Case histories and methods in mineral resource evaluation. Geological Society Spec Publ No 63, UK pp 197

11   Subsurface Soil Contamination Assessment

Fisher T R (1992) Use of 3D geographic information systems in hazardous waste site investigations (in press). In: Goodchild M F, Parks B, Steyaert L T (eds) Environmental modeling with GIS. Oxford University Press, New York

Houlding S W (1993) Subsurface contamination assessment by 3D geoscience modeling. Int J Environmental Issues in Minerals and Energy Industry, 7-13

12   Hazardous Waste Site Characterization and Underground Repository Design

Buesch D C, Nelson J E, Dickerson R P, Spengler R W (1993) Development of 3D litho-stratigraphic and confidence models at Yucca Mountain, Nevada. Proc 4th Int High-level radioactive waste management Conf, Las Vegas Vol 1 pp 943

Buesch D C, Dickerson R P, Drake R M, Spengler R W (1994) Integrated geology and preliminary cross-sections along the north ramp of the exploratory studies facility - Yucca Mountain (in press). Proc 5th Int High-level radioactive waste management Conf, Las Vegas Vol 2 pp 1055

Rautman C A, Treadway A H (1991) Characterization uncertainty and its effects on models and performance. Proc. 2nd Int. High-level radioactive waste management Conf, Las Vegas Vol 2 pp 1491

Sides E J (1992) Reconciliation studies and reserve estimation. In: Annels A E (ed) Case histories and methods in mineral resource evaluation. Geological Society Spec Publ No 63, UK pp 197

13   Ore Deposit Evaluation and Underground Mine Planning

Houlding S W, Stoakes M A (1990) Mine activity and resource scheduling with the use of
     3D component modeling. Trans Inst Mining and Metall, UK, vol 99, A1-64:53-59
Sides E J (1992) Modeling of geological discontinuities for reserve estimation purposes at
     Neves-Corvo, Portugal. In: Pflug R, Harbaugh J W (eds) Computer Graphics in
     Geology. Lecture Notes in Earth Sciences 41, Springer, Berlin Heidelberg New York
     pp 213
Van Ooteghem R M, Rodenstein P K (1993) Computer modeling at the Sullivan Mine: a
     practical application in a complex operating environment. In: Bawden W F,
     Archibald J F (eds) Innovative mine design for the 21st century. Balkema, Rotterdam
     pp 671

14   Characterization and Development Planning for a Small Oil Reservoir

Jones T A (1992) Extensions to three dimensions: introduction to the section on 3D
     geologic block modeling. In: Hamilton D E, Jones T A (eds) Computer modeling of
     geologic surfaces and volumes. AAPG Computer Applications in Geology No 1 pp
     175
Barrett R A, Bailey J (1992) Three-dimensional modeling techniques in the analysis of a
     mature steam drive. In: Hamilton D E, Jones T A (eds) Computer modeling of
     geologic surfaces and volumes. AAPG Computer Applications in Geology No 1 pp
     251

15   Geotechnical Characterization for an Underground Powerhouse
     Excavation

Mayoraz R, Mann C E, Parriaux A (1992) Three-dimensional modeling of complex
     geological structures: new development tools for creating 3D volumes. In: Hamilton
     D E, Jones T A (eds) Computer modeling of geologic surfaces and volumes. AAPG
     Computer Applications in Geology No 1 pp 261
Trenholme B S, Houlding S W (1989) 3D modeling of geotechnical and rock mechanics
     parameters for design and maintenance of mine excavations. Proc Canadian Inst
     Mining and Metallurgy AGM, Quebec City
Wittke W (1977) Static analysis for underground openings in jointed rock. In: Desai C S,
     Christian J T (eds) Numerical methods in geotechnical engineering. McGraw-Hill,
     New York pp 589

16   New Directions in Spatial Prediction, 3D Modeling and Database
     Management

Clark I (1977) Practical kriging in three dimensions. Computers Geosci, 3:173-180
Clark I, Houlding S W, Stoakes M A (1990) Direct geostatistical estimation of irregular
     3D volumes. Proc XXII Symp Application computers and operations research in
     mineral industry, Berlin Vol 2 pp 515
Fisher T R, Wales R Q (1992) Rational splines and multi-dimensional geologic modeling.
     In: Pflug R, Harbaugh J W (eds) Computer Graphics in Geology. Lecture Notes in
     Earth Sciences 41, Springer, Berlin Heidelberg New York pp 17
Keighan E (1992) Managing spatial data within the framework of the relational model.
     Oracle Corporation of Canada, intern corr, Hull
Leonard J (1992) GEOMOD: the European Community solution for 3D subsurface data
     exchange and data model. Geobyte June:59-60

Raper J F (1992) An atlas of three dimensional functions. In: Pflug R, Harbaugh J W
(eds) Computer Graphics in Geology. Lecture Notes in Earth Sciences 41, Springer,
Berlin Heidelberg New York pp 41

Index

Boldface page numbers indicate passages of primary importance

Springer-Verlag
and the Environment

We at Springer-Verlag firmly believe that an international science publisher has a special obligation to the environment, and our corporate policies consistently reflect this conviction.

We also expect our business partners – paper mills, printers, packaging manufacturers, etc. – to commit themselves to using environmentally friendly materials and production processes.

The paper in this book is made from low- or no-chlorine pulp and is acid free, in conformance with international standards for paper permanency.

MIX
Papier aus verantwortungsvollen Quellen
Paper from responsible sources
FSC® C105338

If you have any concerns about our products,
you can contact us on
ProductSafety@springernature.com

In case Publisher is established outside the EU,
the EU authorized representative is:
Springer Nature Customer Service Center GmbH
Europaplatz 3, 69115 Heidelberg, Germany

Printed by Libri Plureos GmbH
in Hamburg, Germany